T0271423

Endless Quests

Theory, Experiments and Applications
of Frontiers of Superconductivity

Peking University–World Scientific Advanced Physics Series

ISSN: 2382-5960

Series Editors: Enge Wang *(Peking University, China)*
Jian-Bai Xia *(Chinese Academy of Sciences, China)*

Peking University-World Scientific Advanced Physics Series

Endless Quests

Theory, Experiments and Applications
of Frontiers of Superconductivity

Editor

Jiang-Di Fan

Chongqing Academy of Science and Technology, China

World Scientific

NEW JERSEY · LONDON · SINGAPORE · BEIJING · SHANGHAI · HONG KONG · TAIPEI · CHENNAI

Published by

World Scientific Publishing Co. Pte. Ltd.
5 Toh Tuck Link, Singapore 596224
USA office: 27 Warren Street, Suite 401-402, Hackensack, NJ 07601
UK office: 57 Shelton Street, Covent Garden, London WC2H 9HE

Library of Congress Cataloging-in-Publication Data
Names: Fan, Jiang-Di, author.
Title: Endless quests : theory, experiments, and applications of frontiers of superconductivity /
    Jiang-Di Fan (Chongqing Academy of Science and Technology, China).
Description: New Jersey : World Scientific, 2018. | Series: Peking University-World Scientific
    advanced physics series ; v. 7 | Includes bibliographical references and index.
Identifiers: LCCN 2018014011| ISBN 9789813270787 (hardcover : alk. paper) |
    ISBN 9813270780 (hardcover : alk. paper)
Subjects: LCSH: Superconductivity--Research.
Classification: LCC QC611.92 .F26 2018 | DDC 537.6/23--dc23
LC record available at https://lccn.loc.gov/2018014011

British Library Cataloguing-in-Publication Data
A catalogue record for this book is available from the British Library.

For any available supplementary material, please visit
https://www.worldscientific.com/worldscibooks/10.1142/11000#t=suppl

Typeset by Stallion Press
Email: enquiries@stallionpress.com

Printed in Singapore

Preface

While the international community celebrated the centennial anniversary of superconductivity, we were working hard to write this book in its memory. To date, superconductivity has gone a long way to reach the point of applications, but it is still far from being widely used. Many scientists and engineers in the world during the past century have contributed their entire life to superconductivity without regret! We are just a few of the players in the "relay race of superconductivity". We wish to promote the development of superconductivity through the book we have just finished. This book consists of three parts: theory, experiment and application of superconductivity. The theory is basically outside of mainstream theoretical studies in the community of superconductivity, indicating a new method for theoretical investigations that looks promising in unveiling the mystery of superconductivity. The second part consists of first-hand experimental investigations and analyses of new superconductors discovered during the past years. The phenomenology derived from experimental work forms the foundation for verifying the validity of theoretical conjectures and provides new insights into advancing the science and technology of superconductivity. Lastly, the third part is one of the most successful applications of the high-temperature superconductor (HTS) $YBa_2Cu_3O_{7t\delta}$ (YBCO) in mass transportation: high-temperature superconductive maglev train (HTSMT). China made the first HTSMT prototype under the leadership of Jia-Su Wang and Su-Yu Wang fourteen years after the discovery of the first HTS $La_{2-x}Ba_xCuO_{4t\delta}$ (LBCO). It is likely that it will become the safest, fastest, and most comfortable and environment-friendly means of mass transportation in the years to come.

Firstly, I would like to take this opportunity to express my heartfelt thanks to all the co-authors, Drs. T.-P. Chen, Shi-Xue Dou, Jia-Su Wang, Judy Wu and Nai-Chang Yeh, who made great contributions to the book

with their first-hand investigation data. Also, my thanks go to Dr. Fu-Sheng
Pan, President of Chongqing Academy of Science and Technology, China,
who has provided me with the support and resources to complete the book.
My work on writing the book was as well supported by Southern University
and A&M College, USA, for my sabbatical leave, during which the book
was started, and also for most of research results that were completed with
Dr. Yuriy M. Malozovsky during the period when I served the institution
from 1989–2012. My appreciation goes to Dr. D. Bagayoko, Chairman of the
Department of Physics, Southern University and A&M College, USA, who
has encouraged me to look for the physical picture of Cooper pairing due
to Coulomb interaction during the difficult years when Dr. Malozovsky and
I were trying to convey to the superconductivity community our new concept that superconductivity originates from repulsive Coulomb interaction
between electrons. That is the work entitled "Sign Reversal of the Coulomb
Interaction between Quasiparticles in Momentum Space."[*] Now, it is much
easier for one to understand Cooper pairing without invoking electron-
phonon interaction that in fact leads to metal-insulator transition. Moreover, my appreciation goes to Dr. Han Zhang, Peking University, and Dr. Ju
Gao, University of Hong Kong, as well as Dr. Haiqing Lin, Chinese University of Hong Kong, who all have been my everlasting supporters in the series
of New^{3}SC Conferences which I initiated and chaired. As a matter of fact,
I have adopted the suggestion of Han Zhang for the title of this book. Lastly,
I wish to show my special appreciation to Dr. Malozovsky, my colleague
in Southern University and partner in research for our common interest in
superconductivity mechanism and origin. Since he joined my research group
in USA in 1992, he has cooperated with me together in research without
discontinuity. Dr. Malozovsky is a hard-working scientist with genius and
diligence. He remains his enthusiasm in superconductivity exploration till
today.

---

[*]J. D. Fan and Y. M. Malozovsky, International Journal of Modern Physics B, **27**, 15
(2013) 1362035.

Finally, it is my wish that the publication of this book will shed some light on unveiling the mystery of superconductivity.

J. D. Fan

Chongqing Academy of Science and Technology

Chongqing, China

May 2014

Contents

Part III: Applications   247

Part I

Theory of Superconductivity

1

Diagrammatic Iteration Approach to Electron Correlation Effects and Its Application to Superconductivity

J. D. Fan[1] and Y. M. Malozovsky[2]

1.1 Introduction

One century has passed since Kamerlingh Onnes discovered the first superconductor in 1911. How much does one really understand superconductivity today? Yes, one did make a big step forward in explaining and describing superconductivity on the basis of BCS (Bardeen–Cooper–Schrieffer) theory [1]. In comparison to semiconductors and lasers, however, the development of superconductivity is rather slow. This fact itself implies that something in one's investigations of superconductivity must be inadequate, which prevents one from well understanding it. In the community of superconductivity it seems that one tries to establish a new mechanism theory for high-temperature superconductivity (HTS), while preserving the BCS theory for conventional superconductors. We believe that an acceptable and complete theory of superconductivity mechanism must be able to cover both low- and high-temperature superconductivity and possibly predict something new, while explaining a series of experimental phenomena consistently and uniquely. The attempt to keep two mechanism theories for superconductivity in parallel is not desirable. In other words, conventional superconductivity should be a natural consequence of a general theory of superconductivity at some extreme physical and structural conditions.

---

[1]Chongqing Academy of Science and Technology, Chongqing, China.
[2]Southern University and A&M College, Louisiana, USA.

This is more or less like the relationship between relativity and classical mechanics, where in the condition of $V \ll c$, with V being the speed of motion and c the speed of light, kinetic energy $K = (m - m_0)c^2$, with m and m_0 being the mass in motion and the rest mass, respectively, inn relativity tends to the classical formula $K = mV^2/2$, or that between Planck's theory and Rayleigh–Jeans law for the blackbody radiation law in quantum mechanics, where at low frequency ν, Planck's formula is in agreement with the Rayleigh–Jeans law.

In this chapter, the main goal is to introduce a method termed diagrammatic iteration approach (DIA) to deal with many-particle systems and its application to superconductors. Before that, we have summarized several points on the issues existing nowadays in the superconductivity community. Our attention is focused on the issues produced from the BCS theory and those that followed from it.

1.2   A brief on the development of superconductivity theory

Before high-T_c superconductors (HTS) were discovered, one understood superconductivity based on the BCS theory in terms of the mean field approximation (MFA) applied to superconductivity, leading to a great simplification of Hamiltonian — the BCS Hamiltonian with the concept of Cooper pairing. Since J. G. Bednorz and K. A. Mueller discovered the first cuprate superconductor with $T_c \sim 35$ K in 1986 [2], the BCS theory has appeared to be helpless in understanding and explaining the phenomena experimentally observed in cuprates; despite the many efforts devoted to improving it with the basic assertion that attractive interaction is needed between two electrons, none of them have been really successful.

In 2001, a binary compound MgB_2 was discovered to be a superconductor at 39 K [3]. This success has promoted a great deal of new excitement in seeking new compounds with higher transition temperature and generated considerable interest in the mechanism of superconductivity. This excitement was enhanced in September of the same year by the increase in transition temperature T_c of fullerenes C_{60} to 117 K in terms of the field effect transistor setup [4]. The euphoria resulting from the discovery of the first cuprate high-T_c superconductor in 1986 has gradually faded in the past decade due to frustrations in reaching a consensus in understanding superconductivity [5]. The major issue still rests in consistently explaining a large number of anomalous properties observed in high-T_c cuprates and understanding both low- and high-temperature superconductivity in a unified framework. Following the discovery of MgB$_2$, a variety of experimental measurements has been performed all over the world, attempting to identify its properties for better understanding of the mechanism of superconductivity. Among them are, for instance, the measurements of thermoelectronic power and resistivity of Mg$_{1-x}$Al$_x$B$_2$ by Chu's group [6], indicating a decrease of T_c with Al doping and no phonon drag contribution to the thermoelectric power of Mg$_{1-x}$Al$_x$B$_2$, and identification of the isotope effect by Bud'ko et al. [7, 8], giving the isotope effect exponent $\alpha = 0.32$, smaller than $\alpha_{BCS} = 0.5$. Measurements of the superconducting gap were carried out with a variety of techniques such as NMR, tunneling, high-resolution photoemission spectroscopy, scanning tunneling spectroscopy, IR reflectivity, specific heat, etc. [9–19] It showed the gap Δ ranging from 2–8 meV; in particular, the measurements of the Raman spectroscopy [10] produce $2\Delta_1/k_B T_c = 1.6$ and $2\Delta_2/k_B T_c = 3.7$. Karapetrov [20], Lorenz [21] and co-workers found the negative change of T_c under high-pressure. Measurements of the penetration depth λ [22–26] and the electron-phonon coupling constant [24] were also performed, yielding $\lambda_{MgB_2} \sim 2 \gg \lambda_{BCS} \sim 0.4$. These measurements have confirmed that MgB$_2$ is a type-II superconductor with two gaps. Specific heat measurements [9, 27] support the s-wave symmetry of the gap [10], while existing tunneling data are not consistent with each other, which are in part attributed to junction interface flaws. From the clear isotope effect and s-wave symmetry of the superconducting gap, one claims MgB$_2$ as a BCS-like superconductor mediated by phonons. However, its structural similarity to a cuprate superconductor and rather high T_c casts a shadow on the claim.

Theoretically, based on the fundamental concept of Cooper pairing, Kortus et al. [28] calculated the electronic band structure of MgB$_2$, arrived at a consistent picture for the Fermi surface and suggested that pairing results from the strong electron-phonon interaction and the high-frequency phonon associated with the light boron element. Many of the experimenters also support this point based on the fact of the isotope effect and a BCS-like superconducting gap structure, while others [8] indicate that the gap in MgB$_2$ is too small to account for its high T_c. The existing data of electron-phonon coupling measurements are not strong enough to reach as high T_c as 39 K, in contrast to the measurement of Walti et al. [27] that gave rise to a coupling constant as high as 2.0. Hirsch suggested a universal mechanism of superconductivity from his assertion that electron and hole are asymmetrical due to the Coulomb interaction. He emphasized [29] that (Cooper) "pairing of electron carriers cannot drive superconductivity" and "holes in nearly full bands yield the low normal state conductivity and high-temperature superconductivity," whereas Chu's group indicated the contrary of Hirsch's prediction — a positive pressure effect on T_c from the experimental measurements [21, 22].

The pnictides superconductivity discovered in 2008 [30–37] arouse new excitements in the superconductivity community. This new type of super-conductors with rather high T_c has a similar structure to cuprates but carries metal and pnictide elements in the structures. Much of the interest in pnictides is because the new compounds are very different from the cuprates and may help lead to a non-BCS theory of superconductivity.

In 1987, P. W. Anderson proposed [38] an idea that a disordered antiferromagnet characterized by resonating singlet pairs of spins could be thought of as a superconducting state. This was then called the RVB model. Laughlin [39] and Kivelson [40] also suggested a related idea. Also, Chakravarty and Kivelson [41] proposed a mechanism of superconductivity in which attraction arises from repulsion between two like charges in a mesoscale structure. This interesting concept can be traced back to twenty years ago [42–44], is consistent with the over-screening idea we proposed in 1994 [45], and similar to, but different from the concept we worked out in 1994 [46] and 1997 [47]. They [41] also pointed out that "the strong electron-phonon interaction is always accompanied by self-trapping or bipolaron formation. The result is a large exponential Frank–Condon reduction of the energy scale for coherent motion of charge, inevitably leading to an insulating state." This point of view is in line with our conclusion for a given attractive interaction between two electrons due to phonons [48].

There are of course many experimental measurements that have not been cited here and many important but more or less conventional works in theory, such as the tight binding model with nearest neighbor approximation, Hubbard model, t-J model, spin bag model, spin-charge separation model, earlier van Hove singularity, nested Fermi liquid, and ab-initio calculations as well as the well-known phenomenological theory by Varma et al. [49], etc. Readers may consult Ruvalds' review [50]. We simply confine ourselves to the unconventional ideas and models which we have been interested in and working on.

Superconductivity, the one hundred-year long enigma, repeatedly arouses thoughts of science in its exploration. Despite countless forays into it, no consensus has been reached concerning the mechanism of superconductivity. We cannot help but ask whether we need to thoroughly examine the fundamental concepts and theory of superconductivity from the very beginning. After a hundred years since the discovery of superconductivity, it is time to unveil its mystery. However, we assert that on the basis of the conventional theory with whatever modification, if there is neither breakthrough in the fundamental concepts nor radical change in the physical and mathematical treatments, it will be impossible to reach a consensus.

It is helpful to look back on the history of physics. V. L. Ginzburg commented in 2000 [51], "... Actively working physicists usually take little interest in the past, and I myself am not an exception — I began studying the theory of superconductivity in 1943, but only in 1979 did I find time to look through the classical papers of Kamerlingh Onnes (1853–1926). And I found them fairly interesting. When a consensus cannot be reached after unprecedented efforts have been made, it becomes necessary and inevitable to review the history and examine the starting point, including those old concepts that have been widely accepted and believed to be true. It is likely that superconductivity is in such a situation. Take a look at the history of physics at the beginning of the twentieth century. One experienced the so-called 'ultraviolet catastrophe' in the blackbody radiation. Planck made a revolutionary change in the energy distribution by assuming that radiated energy $\varepsilon = nh\nu$, where n is integral number and ν is frequency of the radiation and h is the Planck constant, and correspondingly changed integration to summation, leading to a radiation law that is amazingly consistent with experimental data for all frequencies. Further, the Stefan–Boltzmann law can be deduced directly from Planck's radiation law. Of more importance is that the change in energy distribution from continuous to discrete brought about the birth of quantum mechanics. From this historical experience, it seems that without a breakthrough in the old concept(s) there would be, so to speak, no hope in making a substantial progress in the superconductivity mechanism and in engaging a consensus in the community." We believe that the situation in superconductivity is caused by some misconceptions, misunderstandings and the methodology that one has used to investigate superconductivity.

1.3   Dilemma of the BCS theory

It is beyond our ability and time frame to fairly summarize all the serious theoretical papers of the past century. Instead, we only mention those that draw our attention.

There are so many papers and books published during the past half-century, devoted to describing the BCS theory, that we should focus our attention on the original work [1] and the book written by J. R. Schrieffer [52]. In our understanding, the major contributions of the BCS theory to superconductivity may be summarized as follows.

(1) A simplified Hamiltonian — the BCS Hamiltonian.
(2) The BCS ground state.
(3) The concept of Cooper pairing.
(4) A transition temperature formula $T_c = 1.13\omega_D \exp(-1/N_F V)$.
(5) The isotope effect exponent of 0.5.
(6) A gap equation in a superconductor.

Afterwards, the phenomenological theory by V. L. Ginzburg and L. D. Landau [53] describes some of the phenomena of superconductivity without explaining the underlying microscopic mechanism and seems to serve as evidence of the validity of the BCS theory in the condition $e^* = 2e$, where e^* is the factor of charge carried in a superconductor and related to Cooper pairs, although Landau himself strongly opposed this relation [51]. Moreover, in terms of the Bogoliubov transformation [54], that is to diagonalize Hamiltonian, the solution of BCS theory in a homogeneous system can be obtained as well. This seems to serve as another piece of evidence of the validity of the BCS theory. In addition, Eliashberg's theory [55] is more sophisticated in describing interactions. Based on the diagrammatic method he reached the BCS results with the use of Migdal theorem. These supporting works therefore paved the road for BCS towards the Nobel Prize in 1972.

However, the belief in the validity of the BCS theory was greatly challenged by the discovery of cuprates and other new types of superconductors such as MgB_2, pnictides, etc. The new superconductors exhibit completely different phenomena in isotope effects, anomalous properties in the normal state, etc. Especially, the critical temperatures up to 135 K were never successfully interpreted in terms of electron-phonon interaction.

Jeffery W. Lynn indicated [56], "the discovery of the new oxide superconductors has generated tremendous excitement for two reasons. . . . Second, the conventional electron-phonon interaction appears not to be the origins of the superconductivity in these materials, leaving the fundamental physics open to investigation."

1.4 Examinations of the BCS theory

Combining with the notion of pairing introduced by Cooper in 1956 [57], Bardeen, Cooper and Schrieffer (BCS) published their famous paper in 1957 [1]. The key physical concept in the theory — pairing of two electrons due to an attraction between them mediated by phonons — has since been referred to as Cooper pairing. The basic mathematical treatment of the BCS theory is to apply the BCS Hamiltonian — a reduced Schrodinger equation in momentum space, to a conventional superconductor — an isotropic metal or alloy to get its solution. We will inspect its basic mathematical treatments, physical concepts, and major results as follows.

1.4.1 From second quantization to the BCS Hamiltonian

In a homogeneous electron gas (jellium model) with the plane wave approximation and the condition of the charge neutrality, the Hamiltonian of second quantization for fermions is given by [58]

$$H = \sum_{p,\sigma} \varepsilon_p C^{\dagger}_{p,\sigma} C_{p,\sigma} + \frac{1}{2} \sum_{\substack{k,k' \\ q \neq 0 \\ \sigma,\sigma'}} V_q C^{\dagger}_{k+q,\sigma} C^{\dagger}_{k'-q,\sigma'} C_{k',\sigma'} C_{k,\sigma}. \quad (1.1)$$

Equation (1.1) describes a system that meets the condition of charge neutrality. The interaction terms with $q \neq 0$ describe the fluctuations which occur because of electrons interacting only with themselves. This is because the ion-ion interactions of the background lattice and the electron-ion interactions have been canceled out due to the fact that the term of $q = 0$ corresponds to the two terms with the same but opposite values [58]. Under the jellium model of an electron gas there would be no crystalline structure. This model Hamiltonian is simpler than the original second quantization expression, but it is still too general to be solved exactly, and it is often even difficult to solve approximately. Thus, we must manage to simplify it more. Conventionally, the tight binding models with nearest neighbor

approximation and Hubbard model, etc. are examples of simplification of
this kind. The BCS Hamiltonian is another simplified model Hamiltonian
and can be simply and straightforwardly deduced as shown below in a
different way from the original work of BCS [1, 52].

Substituting $k + q = k''$ and $k' - q = -k''$, equivalent to the so-called
"pairing" condition of $k' + k = 0$ (also, $\sigma' + \sigma = 0$), into the second
term of Eq. (1.1), and then changing k'' to k' (meanwhile, completing
the summation over σ to get a factor 2), we directly obtain the reduced
Hamiltonian that is exactly the same as the BCS Hamiltonian given in
Refs. [1, 58].

$$H = \sum_{p,\sigma} \varepsilon_p C^\dagger_{p,\sigma} C_{p,\sigma} + \sum_{k,k'} V(k' - k) C^\dagger_{k'\uparrow} C^\dagger_{-k'\downarrow} C_{-k\downarrow} C_{k\uparrow}, \qquad (1.2)$$

where $V(k' - k)$ comes from V_q due to $q = k' - k$, as shown in Fig. 1.1.

Clearly, the interaction of $V(k' - k)$ connects k' with $-k$, where k' and
$-k$ are the moments of an electron and a hole, respectively. The former is
related to the creation operator $C^\dagger_{k'\uparrow}$ and the latter to the destruction (annihilation) operator $C_{-k\downarrow}$ (Fig. 1.1). For the ground state, a fully occupied
Fermi sphere, two electrons after scattering will have no option but to be
excited to two states above the Fermi surface, leaving two holes inside the
sphere. Therefore, from the diagram it is obvious that $V(k' - k)$ actually
represents the interaction between an electron and a hole. In other words,
$V(k', k) = V(k' - k)$ in the BCS Hamiltonian is the interaction in the
particle-hole channel and naturally attractive in the Coulomb framework
without invoking phonon or anything-on, whereas the interaction between
two electrons in the particle-particle channel is still repulsive as given by
the Coulomb law and adopted in the positive Hubbard model $(U > 0)$.
Of course, the $V(k', k)$ we are talking about here is the interaction in
the particle-hole channel without taking into account many-particle effects
at all.

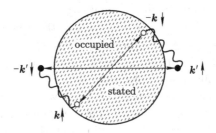

Fig. 1.1   An interaction sketch in the BCS Hamiltonian in the particle-hole channel.

The use of the condition $k' + k = 0$ (also, $\sigma' + \sigma = 0$) in the simplification of Hamiltonian is a purely mathematical treatment with the physical reality of vanishing total momentum of the particle system as described below, in which electrons are considered in pairs. In other words, the simplification with the condition of $k' + k = 0$ (also, $\sigma' + \sigma = 0$) is identical to the condition of Cooper pairing. It deserves a mention that Eq. (1.1), and hence the BCS Hamiltonian given by Eq. (1.2), are obtained from the condition under which a solid is charge-neutral, leading to positive holes, and hence the interaction of $V(k' - k)$ becomes attractive. As indicated above, the interaction terms with $q = 0$ have been neglected in Eq. (1.1), and hence in Eq. (1.2), the use of the jellium model implies that there is neither crystalline structure nor electron-phonon interactions available in the BCS Hamiltonian. Needless to say, any real superconductor does bear a lattice structure, and thus there exists electron-phonon interaction responsible for the isotope effect. However, using electron-phonon interaction in Eq. (1.2) as a mediation between two electrons to have the negative $V(k', k)$ has become groundless and is hence controversial to the prerequisite that leads to the BCS Hamiltonian. Therefore, the fact that a superconductor exhibits the isotope effect is by no means an indication of superconductivity originating from electron-phonon interaction. In this sense, we are in line with Chakravarty and Kivelson [41].

With this reinterpretation of $V(k', k)$, we can proceed to solve the BCS Hamiltonian for conventional (low-T_c) superconductors (LTS) by assuming $V(k', k) = -V_0$, as in the BCS theory. It can be seen that the rest of the work will be the same as that of the BCS theory except for the physical picture. Among the differences is that neither the original concept of Cooper pairing nor the electron-phonon interaction have been invoked, as the pair interaction $V(k', k)$ is naturally attractive in the particle-hole channel as described above, or in the quasiparticle channel in the Landau's Fermi liquid theory as described below. Moreover, this reinterpretation favors the unification of theory for both HTS and LTS. With this different physical picture, Eq. (1.2) will result in the same outcomes as the BCS theory does.

1.4.2   Grouping condition in simplifying Hamiltonian

Our investigation leads us to conclude that the condition of $k' + k = 0$ used in deriving Eq. (1.2) which is for the simplification of Eq. (1.1) and has nothing to do with Cooper pairing mediated by the so-called electron-phonon interaction. As a matter of fact, the condition is the representation of the fact that the total momentum of a particle system under investigation is zero relative to the center of mass coordinate. This can also be statistically understood and described as below for simplifying the Hamiltonian of second quantization.

For an unperturbed many-electron system the total momentum is zero relative to the center of mass coordinate. Statistically, it is always possible to find an electron with momentum k_1 if an electron has momentum $-k_1$, and also an electron with momentum $-k_2$, while another electron possesses momentum k_2, and so on. This can be summarized as $k + k' = 0$, i.e. the so-called pairing condition. It appears to be better, however, to term it the grouping condition. Moreover, the electron with k_1 may momentarily change its momentum to, for example, k_1'. Meanwhile, the electron that had the momentum $-k_1$ may not necessarily follow this change by switching to $-k_1'$. It is not difficult to imagine that it is always possible to find an electron in the system that possesses $-k_1'$ to "pair" with the first one that now changes its momentum to k_1'. Thus, the "pairing" relation is instantaneous. Grouping is neither binding nor coupling but sorting particles pair by pair in the system in order to simplify the Hamiltonian, and is possibly performed both above and below the transition temperature T_c. Therefore, if we wish to preserve the customary terminology of the 'pairing', it is necessary to distinguish the 'pairing' here from the original concept of the Cooper pairing adopted in the BCS theory.

However, physically, if there exists an attraction between two particles, it is of course possible for them to pair with each other. Nevertheless, what we are discussing here is completely a physical description of electron properties in momentum space where a "particle" should be replaced by its counterpart–quasiparticle according to Landau in his Fermi liquid theory. As stated in section 1.4.8 below, the Coulombic interaction between two quasiparticles will change its sign due to electron correlation effects. Thus, the two quasiparticles in momentum space will naturally attract each other without invoking anything, whereas the two corresponding real particles in real space ought not to pair with each other, nor may they be electrically the same as regular electrons that are in a non-superconducting state when the superconductor has fallen into the superconducting state.

1.4.3 Cooper problem

The concept of Cooper pairing is rooted in one's mind. Two paired electrons (or holes) are thought of as a boson-like particle with charge of 2e. Since 1986, one has made so many efforts in looking for a possible pairing mechanism other than phononic for HTS. Even for a Coulombic interaction-based model — the positive Hubbard model, one has brought the concept of d-wave pairing into superconductivity, attempting to understand Cooper pairing of electrons due to repulsion. It seems that except for the phenomenological theories and a few out-of-the-mainstream theories, almost all of the mechanism theories of superconductivity are looking for a non-phonon mechanism for pairing. The trend greatly resists the development of the mechanism theory.

To better understand the original definition of Cooper pairing, we have to trace it back to 1956. The Cooper problem comes from the original work of L. N. Cooper, published in 1956 [57]. A quotation of his original statement concerning this problem may be helpful. It is written as follows: "Consider a pair of electrons which interact above a quiescent Fermi sphere with an interaction of the kind that might be expected due to the phonon and the Coulomb fields. If there is a net attraction between the electrons, it turns out that they can form a bound state though their total energy is larger than zero." The combination of this kind of two electrons has since been referred to as Cooper pairing. The problem itself is often called the Cooper problem, that can be illustrated by Fig. 1.2. Comparing Fig. 1.2 with Fig. 1.1, it is clear that the Cooper problem is hypothetical and different from the BCS problem because all the electrons in the BCS problem should be within the Fermi sphere at $T = 0$, and a part of them would be excited above the Fermi surface at a finite temperature $T > 0$, staying in the vicinity of the Fermi surface as shown in Fig. 1.1.

On the other hand, it is of course possible for two electrons to be attracted to each other in a real solid as long as the frequency is close to, but smaller than, the characteristic frequency of phonon ω_q based on the electron-electron interaction potential accounting for phonons [56, 58–60]

$$V(q,\omega) = \frac{4\pi e^2}{q^2 + \lambda^2} \left[1 + \frac{\omega_q^2}{\omega^2 - \omega_q^2} \right], \qquad (1.3)$$

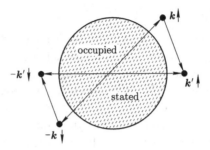

Fig. 1.2 A sketch of the Cooper problem and pairing.

where $\lambda = 4\pi e^2 N(\varepsilon_F)$, with $N(\varepsilon_F)$ being the density of states, and $1/\lambda$ is the Thomas–Fermi characteristic screening length. However, as indicated in Ref. [48], the phase transition induced under such circumstances will be insulating instead of superconducting.

In the work [57], Cooper did not set any restriction to spins of two electrons. In the BCS theory, the bound state is for those pairs of electrons with opposite moments ($k' + k = 0$) and antiparallel spins (singlet). Based on our investigation [61], however, if two electrons are in the singlet state, there must exist a repulsive bare interaction between them (in real space) to form a bound state (in momentum space). This rather surprising conclusion is in agreement with our recent theoretical result regarding the sign reversal of the Coulombic interaction [62]. Moreover, two electrons in a bound state should preserve the properties of being a fermion and cannot be treated as a boson. This assertion is supported by the experimental observation [63] of the Fermi surface of superconductors, including cuprate oxides. Similarly, two electrons with antiparallel spins and the repulsive Coulomb interaction between them in a hydrogen molecule or ^{3}He atom are bound together in the vicinity of a nucleus, but one has never treated them as a boson.

It seems to deserve a reminder that the repulsive Coulomb interaction between two electrons is a representation in real space while the bound state associated with the two electrons is a description in momentum space and formed due to an attraction, caused by electron correlation effects, between two quasielectrons — the counterparts of the two electrons.

1.4.4 The ground state

Based on the idea of deriving Eq. (1.2) from Eq. (1.1), i.e., particles are grouped pair by pair, one may discuss the BCS Hamiltonian more rigorously by suggesting a two-particle initial state denoted by $|\Psi_2\rangle = c_{k,\sigma}^{\dagger} c_{-k,\sigma'}^{\dagger} |0\rangle$, where $|0\rangle$ is the vacuum state, $|\Psi_2\rangle = |k, -k\rangle$, and σ, σ' represent spin-up and spin-down, respectively. Under the operation of the second term of Eq. (1.2), one obtains a new two-particle state given by

$$c_{k',\sigma}^{\dagger} c_{-k',\sigma'}^{\dagger} c_{-k,\sigma'} c_{k,\sigma} (c_{k,\sigma}^{\dagger} c_{-k,\sigma'}^{\dagger} |0\rangle) = c_{k',\sigma}^{\dagger} c_{-k',\sigma'}^{\dagger} |0\rangle = |k, -k\rangle = |\Psi_2'\rangle.$$
(1.4a)

Here, no hole can be created at all because there is no existence of a Fermi surface. The physical process of Eq. (1.4a) may exactly be represented by Fig. 1.1. Yet, for a many-fermion system, Pauli's exclusion principle applies and all states below the Fermi level are occupied at $T = 0$. The ground state of an N-electron system may be constructed pair by pair as

$$|\Psi_{0,N}\rangle = \prod_{|q|<k_F} c_{q,\sigma}^{\dagger} c_{-q,\sigma'}^{\dagger} |0\rangle.$$
(1.4b)

Letting Eq. (1.2) operate on Eq. (1.4b), one has

$$c_{k',\sigma}^{\dagger} c_{-k',\sigma'}^{\dagger} c_{-k,\sigma'} c_{k,\sigma} |\Psi_{0,N}\rangle = c_{k',\sigma}^{\dagger} c_{-k',\sigma'}^{\dagger} c_{-k,\sigma'} c_{k,\sigma} | \prod_{|q|<k_F} c_{q,\sigma}^{\dagger} c_{-q,\sigma'}^{\dagger} |0\rangle$$

$$= c_{k',\sigma}^{\dagger} c_{-k',\sigma'}^{\dagger} |0\rangle_{|k|>k_F} \times |k_1, -k_1; \ldots; 0, 0; k_N, -k_N\rangle_{|k|<k_F}$$

$$= |0, 0; \ldots; k', -k'; \ldots; 0, 0\rangle_{|k|>k_F} \times |\Psi_{0,N-2}\rangle_{|k|<k_F}$$

$$= |\Psi_2'\rangle_{|k|>k_F} |\Psi_{0,N-2}\rangle_{|k|<k_F}.$$
(1.4c)

For simplicity, the spin notations above have been ignored, and the restrictions of $|k'| < k_F$ and $|k'| > k_F$ are due to the requirement of Pauli's exclusion principle. Concurrent with a new two-particle state $|k', -k'\rangle$ that is produced above the Fermi level, the original two-particle state $|k, -k\rangle$ in the Fermi sphere has been removed, or a two-hole state has been created in the background of the Fermi sea. They are denoted by $|\Psi_2'\rangle_{|k|>k_F}$ and $|\Psi_{0,N-2}\rangle_{|k|<k_F}$, respectively. They are just the two particle-hole excitations displayed in Fig. 1.1. It is worthwhile to mention here that there is a summation over k and k' in the BCS Hamiltonian, so that one will obtain a coupled state with N-particle-hole excitations consisting of $N/2$ terms like Eq. (1.4c).

1.4.5   2e-factor issue

It is widely believed that Cooper pairing is evidenced by the 2e factor observed in the magnetic flux quantization in a hollow cylindrical superconductor [64, 65] and in the Josephson effects [66]. It was found that if a superconducting hollow cylinder is cooled below the transition temperature in the presence of an axial magnetic field, the magnetic flux is trapped in the cylinder after switching off the external field. However, as indicated by Tinkham [67], this is the situation when the flux is trapped in a thick-walled cylinder as in the experiments of Doll and Nabauer [64] and of Deaver and Fairbank [65]. Blatt [68] also indicated that the flux quantum is exactly $hc/2e$ only if the cylinder is thick compared to the penetration depth. If the penetration depth is comparable with the thickness of the cylinder, there are measurable deviations from this unit. One has seemingly not paid attention to this important fact while accepting the concept of Cooper pairing in describing superconductivity. The argument is whether or not two electrons were really combined together as if they were a single boson-like particle and whether or not the 2e factor would always appear in whatever experimental configuration. The deviation itself is just the evidence that the 2e factor has nothing to do with the electrons' pairing. In fact, the difference of the flux quantum between $hc/2e$ (superconductor) and hc/e (conductor) must have its own physical and geometrical origins.

Historically, under the influence of Landau's work [69], few physicists, including V. L. Ginzburg, took an interest in the theory of superconductivity in the years after World War II. A phenomenological theory of superconductivity that was since called Ginzburg–Landau (GL) theory was worked out [70, 71], in which $e^* = 2e$ was shown to be true by Gor'kov [72], but Landau did not agree to it at all. As indicated by Ginzburg [51], "Interestingly, this simple idea ($e^* = 2e$) concurred to no one, in particular, neither to me nor to Landau." Apparently, Ginzburg has changed his point to favor the concept of Cooper pairing [51]. After 15 years since it was published, BCS theory eventually won the Nobel Prize due to the 'evidences' of those experiments mentioned above and the support of a series of investigations by Gor'kov [72], Bogoliubov [73], Valatin [74], and others [75], in which virtually the same results were obtained.

To our best knowledge, the earliest challenge to the belief of experimentally observed 2e factor serving as the evidence of Cooper pairing comes from Kai-Jia Cheng and Shu-Yu Cheng who explained the origin of 2e appearing in the phase factor in terms of Lorentz force in 1992 [76, 77], and indicated that the $2e$ factor has nothing to do with Cooper pairing. Motivated by this, we have shown as well in our earlier works [78] that the $2e$ factor appearing in the fraction of $hc/2e$ observed in the captioned experiments has nothing to do with Cooper pairing. Instead, it actually originates from electrons' satisfying both gauge invariance and the London equation for a superconductor: each of them contributes a factor e/hc, summing up to $2e/hc$ and appearing as $hc/2e$ in the phase factor. In contrast, only gauge invariance should be satisfied for a conductor, thus only the factor e/hc is observed in experiments. The whole proof in Ref. [78] is simple and rigorous in terms of Berry's phase. It is easy to examine its validity. Then, we similarly showed where the $2e$ factor is from in the Josephson effect [79] by using the Berry phase and the first London equation.

Both of the two papers [78, 79] have clearly shown that in the presence of a field A the supercurrent is related to the so-called mechanical momentum $\Pi = p - eA/c$ [80], instead of canonical momentum p only. It is well-known that the mechanical momentum Π, other than the canonical momentum p, satisfies gauge invariance. In fact, the relation of supercurrent to the mechanical momentum Π is the origin of the second London equation $j(x) = -(e^2 n/mc)A(x)$ in a superconductor, inside of which the supercurrent is absent [78]. The gauge invariance applies to both metals and superconductors. However, the London equation holds for a superconductor only.

Therefore, the $2e$ factor observed experimentally with a superconductor is not evidence of Cooper pairing. Once again, the argument here is attributed to the necessary and sufficient condition in logic: if two electrons were really to be bound together as an entity — as a Cooper pair is widely thought of as, the charge of the entity would be $2e$, whereas the $2e$ factor experimentally observed may or may not be the signature of the Cooper pair. Needless to say, the Cooper pair configuration is not the final in the superconducting state as shown in [81]. According to the Pauli exclusion principle on any state k or $-k$, there can be two electrons with opposite spin orientations. While the electron with momentum k and spin-up pairs with the one with $-k$ and spin-down, the electron with k and spin-down must pair with the one with $-k$ and spin-up. Therefore, a more stable quartet configuration is formed. In that paper, we have rigorously shown that the quartet configuration possesses twice the lower ground state energy than that for a Cooper pair and thus is the stable ground state.

It is worthwhile to point out that, without invoking electron-phonon interaction and in the Coulombic framework for a conventional superconductor, interaction is weak and hence the assumption of $V(k' - k) = -V_0$ is reasonable and acceptable in the BCS Hamiltonian. The rest of the work would be completely the same as that in the BCS theory. However, for the cuprate oxides, pnictides, etc., the treatment of the BCS is no longer applicable because the strength of interaction is already beyond the validity of the BCS theory. Correspondingly, the diagrammatic method on the basis of Feynman diagrams is the unique way to deal with high-T_c superconductivity. Afterwards, low-T_c superconductivity is a natural consequence at some specific physical and structural conditions. We are going to describe it in more detail later.

1.4.6   The isotope effect

One of the most important and striking successes of the BCS theory is the isotope effect delivered from the theory. It is also the major point of those who insist on the phononic model mechanism of superconductivity. It is true that one has absolutely no reason to ignore the existence of phonons in superconductors. However, the issue here is that phonons lead to the isotope effect, but superconductivity displaying the isotope effect may not necessarily stem from phonons. Otherwise, how can one logically explain the vanishing isotope effect near the optimal T_c of cuprate superconductors in which phonons definitely exist? This issue was discussed above in Sec. 1.4.1 where the BCS Hamiltonian was derived under the jellium model with further simplification of zero total momentum of the system under investigation. The conflict between jellium model used for receiving the BCS Hamiltonian and the assumption of electron-phonon interaction serving as the mediation to form Cooper pairs is obvious.

In fact, the isotope effect is deduced in the BCS theory from the correct selection of the Debye frequency ω_D to be the cutoff of the energy spectrum, where ω_D is set to be the upper limit of the integration over energy. A Coulombic mechanism does not at all rule out such a selection of the same cutoff in an LTS in which the Fermi energy ε_F is very high. Meanwhile, in an HTS ε_F may be comparable with ω_D and there is no reason to simply select ω_D as the cutoff. Indicating this fact may be helpful in qualitatively understanding the "anomalous" isotope effects observed in cuprates. Indeed, there must be electron-phonon interaction as long as there exists a lattice for any superconductor. This fact itself implies that the use of electron-phonon interaction for the pairing mechanism is invalid, while adopting the BCS Hamiltonian for conventional superconductors. Of course, in a real superconductor, there must be both Coulomb and phonon fields. The co-existence of phonon and Coulomb fields and competition between them present such a complicated phenomenon that the isotope exponent α may be positive (BCS-like), vanishing or negative [82]. If one understands superconductivity in such a way that the mechanism of superconductivity and the origin of the isotope effect are separated, the situation will become much clearer and the unification of theories of superconductivity will be much easier. The current trend to respectively understand low- and high-T_c superconductivity along two tracks appears inappropriate. The currently prevailing criterion that observation of the isotope effect is the manifestation of the phonon mechanism is obviously unsuitable. It is apparently insufficient in terms of the mere phonon mechanism to understand a variety of isotope effects observed: the isotope exponent α is 0, equal to or less than 0.5, or even negative. The BCS theory gives rise to $\alpha = 0.5$, but only a very limited number of conventional superconductors follows this regulation. Many of them carry a value of α far from 0.5. Even α of the recently discovered MgB_2 is equal to 0.3. Clarification of this misunderstanding is very vital for a unified description of superconductivity.

1.4.7   *The D-wave mechanism of pairing*

There have been more and more experimental observations that present a d-wave-like symmetry of the superconducting gap in cuprates. One seems to be easily led to a belief in the d-wave mechanism of superconductivity, which is used to explain the pairing mechanism between two electrons when there is a strong Coulomb repulsion between them. It seems that one has unknowingly puzzled with two important concepts: symmetry of gap and mechanism of pairing. As a matter of fact, symmetry of a superconducting gap has nothing to do with the mechanism of superconductivity. A denial of the d-wave model of the superconducting pairing mechanism does not rule out the possibility of the d-wave-like symmetry of the gap that has been widely observed in experiments. This important point should be clarified. One does not need to doubt the experimental accuracy of the d-wave-like symmetry of the gap from measurements that have perhaps been repeated numerous times all over the world. The point is that the

d-wave-like symmetry cannot be used to explain the pairing mechanism. If one accepts the Coulombic origin of superconductivity, there is no need to invoke the d-wave model of the pairing mechanism and the disputes between s- and d-waves symmetry will be in vain.

Moreover, the currently overwhelming d-wave model of the mechanism of superconductivity seems to ignore the logic. According to the conventional point of view, in a superconductor electrons must be first paired at low temperature leading to a superconducting phase transition, and simultaneously, to a gap Δ in the energy spectrum. Here 'pairing' is the cause and the gap is its consequence. Whatever kind of symmetry the gap may possess is only the result of 'pairing'. However, the d-wave model of the mechanism of superconductivity attempts to use the d-wave-like symmetry of the gap, $D_{x^2-y^2}$ with a sign change to explain how two electrons can be paired while a strong Coulombic repulsion exists between them. It is simply nothing else but to put the cart before the horse. In addition, as shown in our earlier work [62], it is possible for Cooper pairing and hence quartet configuration to naturally form with the Coulomb interaction because of the sign reversal of the interaction in momentum space induced by many-particle correlation effects. If one keeps in mind that the interaction between two quasiparticles (quasielectrons) and the interaction between two particles (electrons) in a weakly coupled electron system are the same except for a sign reversal [62], it is unnecessary to look for a mechanism for pairing.

1.4.8   *Sign reversal of interaction in momentum space*

It is usually believed that the interaction between two quasiparticles in the Fermi liquid theory [58, 83–88] is the same as the renormalized one between two particles. Nevertheless, this is not completely true. In terms of an exact equation for the thermodynamic potential due to interaction between particles, it has been shown [62] that the interaction between two quasiparticles is negative to the renormalized interaction between two particles. Thus, the repulsive interaction between two particles in real space becomes an attractive one between two quasiparticles in momentum space.

The historical misunderstanding of the Coulomb interaction in momentum space lies in the belief that it is only a Fourier transformation of it when it is transformed from real to momentum space, leading to the search for a mechanism of attraction between two electrons for more than a half-century. With the new understanding of Coulomb interaction between two quasiparticles (quasielectrons) it is naturally attractive in momentum space, forming Cooper pairs and further quartet configuration at the transition temperature and hence bringing about superconducting phase transition. It is worthwhile to indicate that the superconducting phase transition we are describing here is for the reality of a superconductor in which there exists a lattice structure, and so does the weak electron-phonon interaction in comparison with the attractive Coulomb interaction between two quasielectrons in momentum space. Clearly, in the Coulomb channel there is no need of electron-phonon interaction to mediate the interaction between two quasielectrons in momentum space. This eliminates the difficulty of there being neither crystal structure nor electron-phonon interaction when using the BCS Hamiltonian. The BCS Hamiltonian from the model of a homogeneous gas of electrons (jellium model) can be used to describe conventional superconductors without invoking electron-phonon interaction. In this sense the BCS theory works for conventional superconductivity only if the concept of Cooper pairing without the mediation of electron-phonon interactions was adopted. To really describe superconductivity at both low- and high-temperature transition in a unified framework, one has to use field theory in terms of the diagrammatic method to deal with phase transition as presented below. In such a way, superconductivity originates from the Coulomb interaction and the electron-phonon interaction causes isotope effects. For conventional superconductors the electron-phonon interaction is comparable with the Coulomb interaction, and thus the isotope effect is appreciable, whereas for HTS, strong Coulomb interaction prevails, the isotope effect is too weak to be detected. In such a description, the variety of the well-known isotope effects can be qualitatively explained and understood.

1.5   Issues in electronic theory

To reveal where the problem comes from in conventional electronic theory, we have to trace back to the limitations of the validity of the theory.

1.5.1   *Interelectronic coupling*

We will present some results directly from Mahan's book [58] in the upcoming two sections. First of all, the inter-electronic coupling constant is defined as

$$\alpha = e^2/\varepsilon v_F = 1/a_B k_F, \tag{1.5}$$

where e is for electron charge, v_F is for Fermi velocity, k_F is for Fermi momentum with the convention of assuming $\hbar = h/2\pi$ to be a unit, and $a_B = \varepsilon/e^2 m$ — the effective Bohr radius with ε being the dielectric constant. It is known that

$$k_F^3 = 3\pi^2 n, \quad \text{(3D)} \tag{1.6a}$$

$$k_F^2 = 2\pi n_s, \quad \text{(2D)} \tag{1.6b}$$

where n is the density of electrons for 3D and n_s for 2D, respectively. It can be seen that the higher the density of electrons, the smaller the coupling constant. The condition of $\alpha \ll 1$ corresponds to weak coupling, as it is in a conventional metal, while the fact that $\alpha \sim 1$ or $\gg 1$ represents a strong coupling case as observed, for example, in heavy fermion systems, cuprate superconductors, etc. The conventional electronic theory based on RPA or local density approximation is, strictly speaking, valid for the case of weak coupling only. For many metals, RPA is a good approximation, and conventional electronic theory works very well because the density of electrons in the metals is generally large enough to ensure a small coupling constant α. The conventional superconductors can also be well-described by a theory based on the mechanism of weak coupling such as the BCS theory. However, the category of cuprate superconductors are formed due to doping, the density of carriers (electrons or holes) is always so low that the inter-electronic coupling constant is quite large. Therefore, it is not strange that the BCS theory works well for conventional superconductors, but fails for cuprates.

1.5.2   *Parameter r_s*

Another important parameter r_s that is thought of as a criterion of the electron density is defined as below:

$$\frac{4\pi}{3}r_s^3 = 1/(na_B^3), \quad \text{(3D)} \tag{1.7a}$$

$$\pi r_s^2 = 1/(na_B^2), \quad \text{(2D)} \tag{1.7b}$$

where $r_s = r/a_B$ is the radius of the sphere, in Bohr radius units, in which an electron charge is enclosed. It can be seen that the higher the density n or

n_s, the smaller the parameter r_s. In terms of this definition, kinetic energy E_k is calculated to be $2.21/r_s^2$ (3D) or $1/r_s^2$ (2D), whereas the interaction part in the ground state energy is divided into two parts: exchange and correlation. The exchange part has been exactly calculated for the Coulomb interaction, that is the Fourier transformation of Coulomb interaction of $e^2/\varepsilon r : V(k) = 4\pi e^2/\varepsilon k^2$ (3D) and $V(k) = 2\pi e^2/\varepsilon k$ (2D). The second part, however, was estimated [89–93] in RPA or in terms of the generalized RPA including the well-known Hubbard correction (see the so-called Hubbard local field factor) which explicitly accounts for the exchange and short-range correlations, i. e., only the single scattering between electrons is taken into account among the possible scattering patterns between electrons. The efforts were made to calculate each contribution. The result for the ground state energy E_g is given by [58]

$$E_g = \frac{2.21}{r_s^2} - \frac{0.916}{r_s} - 0.096 + 0.0622 \ln r_s + \cdots, \qquad \text{(3D)} \qquad (1.8a)$$

and

$$E_g = \frac{1}{r_s^2} - \frac{1.204}{r_s} + 0.2287 + \varepsilon_r, \qquad \text{(2D)} \qquad (1.8b)$$

where the first term is kinetic energy, the second is exchange energy, and all the rest of the terms together construct correlation energy E_c. In Eq. (1.8b) ε_r is the so-called ring energy, which is usually evaluated approximately in consideration of the ring-diagrams and the second-order exchange diagrams. Many authors have obtained the results shown in Eqs. (1.8a) and (1.8b) [58, 89–94].

It is seen that when the density of electrons is high ($r_s \ll 1$), kinetic energy prevails over the potential energy including exchange and correlation energies and the potential energy is hence negligible. Thus, the free-particle model works well and the corresponding results from the free-particle model are accurate. The conventional perturbation theory was developed based on the free-electron model. It is obvious that the free-electron model is good for such a high electron density as it is in conventional metals, but not accurate for a low density of electrons in which r_s is close to or greater than 1 and the coupling constant α is also greater than 1. It is easy to see that when $r_s > 2, |E_x| > E_k$. Therefore, as the density of electrons keeps decreasing, E_c is no longer negligible, nor is the free particle model valid.

1.5.3 Correlation effects

The different types of the elastic or inelastic, magnetic (spin-dependent) or non-magnetic processes of scattering may cause electrons to become localized or nearly localized. In particular, if the material under investigation has a low density of electrons and/or a high density of defects (or impurities), inter-electron coupling constant α is close to or greater than 1. Meanwhile, the parameter characterizing the scattering of electrons $k_F l$ is also close to 1, and $l = v_F \tau$ is the mean free path. Therefore, strictly speaking, the conventional treatment of electrons in bulk electronic materials does not apply to a dilute electron gas because the methods, based on the use of the lowest order of perturbations, i.e., RPA in interaction and the Born approximation in scattering, are no longer valid for this case where correlation effects play a major role since these methods treat electrons as if they were free. In other words, the so-called vertex correction — higher-order correction in the self-energy part and/or two-electron interaction kernel must be taken into account in the diagrammatic method. The incorporation of correlation in the interaction (scattering) allows one to expand the existing methods and to develop a new method in the electronic theory. The question is how one can go beyond RPA with the diagrammatic method.

It is also well-known [95] that correlation effects in a low dimension are enhanced due to the reduction of freedom of motion. Therefore, RPA is inadequate in a low dimension even though the coupling constant may still be small so that the problem seems to pertain to the regime of weak coupling, where the conventional electronic theory works well.

1.5.4 Anisotropy of structures

The behavior of electrons in materials such as cuprate superconductors and nanoscale materials are characteristic of the essential spatial constraint [96]. These materials have a layered 2D (L2D) or a quasi-2D structure. For instance, cuprates, different heterostructures as well as quantum wires, boxes, wells, etc. have layered or multilayered structures. In these materials, correlation effects in inter-electronic interactions and scattering, such as scattering on impurities, lattice defects or phonons, are much stronger than that in a 3D case. As an example, let us consider the most important and fundamental Coulomb interaction in different dimensions.

In momentum space, the bare Coulomb interaction reads as

$$V(\boldsymbol{k}) = \frac{2\pi e^2}{\varepsilon k}, \qquad\qquad (2D) \qquad\qquad (1.9a)$$

$$V(\boldsymbol{k}) = \frac{4\pi e^2}{\varepsilon k^2}, \qquad\qquad (3D) \qquad\qquad (1.9b)$$

$$V(\boldsymbol{k}) = \frac{2\pi e^2}{\varepsilon k_{||}} \frac{\sinh ck_{||}}{\cosh ck_{||} - \cos ck_z}, \qquad (L2D) \qquad\qquad (1.9c)$$

where c is the interlayer spacing (distance) and ε is the dielectric constant; and $k_{||}$ in Eq. (1.9c) for an L2D structure stands for a 2D wave vector in any layer and k_z is its z component perpendicular to a layer plane. The third expression is applicable to the structure of the cuprate superconductors.

It is seen that the same Coulomb interaction potential in real space exhibits different formalisms in momentum space due to the dimensionality of structure. This feature must lead to a peculiarity of the electron behavior in the materials that possess either an L2D or a quasi-2D structure. Correspondingly, this structural peculiarity should be taken into account in a theory that can provide a better description and understanding of the electron properties.

The transport property of electrons is one of the major concerns in science and technology. A lot of relevant phenomena have not been well understood although some descriptions, based on conventional theory, are given in literature [96, 97] and references listed in it. For example, linearity of resistivity with temperature in cuprates, positive or negative pressure effects on T_c in cuprate superconductors, marginal Fermi liquid behavior [49, 97, 98] in cuprates, Aharonov-Bohm effects, weak localization of electrons, as well as correlation energy of a dilute electron gas, etc. and where the existing works do not provide a good understanding that is consistent with experimental observations or there is no theory at all. Therefore, a development of existing theories and methods is a must for an essential understanding of the properties of the superconductors.

1.6   Diagrammatic iteration approach (DIA)

1.6.1   *Methodology*

In general, there are two methods to deal with an electron system: Hamiltonian and field theory. Conventionally, one in general seeks a proper Hamiltonian for a system of electrons and then finds out its solution. The

BCS theory is a good example of attempts of this kind for conventional superconductors. Since the cuprate HTSs were discovered, one has been attempting to find a new Hamiltonian for them on the basis of the concept of Cooper pairing and the BC Hamiltonian. A variety of t-J models [99], for instance, are examples of this attempt and the d-wave pairing model [100, 101] is another trial in establishing the pairing mechanism with a strong repulsive Coulomb repulsion between electrons. Noting the important role of the Coulomb interactions, P. W. Anderson proposed his idea of the possibility of HTS based on the Hubbard model [38, 102]. However, no matter what kind of Hamiltonian one can adopt for HTS, the major problem lies in interaction. One can reasonably assume a constant for interaction in the case of weak coupling such as in a conventional superconductor. This is exactly what the BCS theory does. But one cannot still assume a constant interaction for strongly coupled electrons in HTS without taking the risk of losing key information. The Hubbard model is a typical example of assuming a large constant U for the rather complicated interaction potential. Needless to say, almost no analytical solution can be found in quantum mechanics except for a few extremely simple cases where the interaction potential is either zero or constant within a limited region. The BCS Hamiltonian for an attractive weak interaction is one such solvable case. Apparently, it is impossible to get an analytic solution if the interaction $V(\mathbf{k}', \mathbf{k})$ is a function of $\mathbf{k}$ and $\mathbf{k}'$. In addition, it is widely believed that the Hubbard model with a positive interaction U (constant) is a good model to describe a strongly correlated electron system and thus is adequate for high-temperature superconductivity. This model with the positive U is clearly and definitely Coulombic without taking into account the many-particle effect and is also apparently a description of the electron system under investigation in the particle-particle channel. The constant U is an oversimplified interaction potential that can lead to a qualitative description only. The Hubbard models with a large constant interaction potential U look simple enough, but to date a solution to it has been obtained only in one dimension. Therefore, even if the assumption of a constant interaction potential is valid in some special cases, we are still left with the difficulty of obtaining its solution. The major issue here is how much one can do with a Hamiltonian and how much information one is able to extract from it.

Nevertheless, the field theory appears to be better in dealing with a many-particle system in the case of strong coupling. By means of the diagrammatic method, if one manages to construct the interaction diagrams and then convert it to an integral equation, one may solve it to look for a pole condition that corresponds to a phase transition. By examining susceptibility that is related to conductivity one is able to determine whether the transition is superconducting or isolating. In this method, the single-particle self-energy, interaction kernel and vertex correction are convoluted together. The relation of the final interaction potential after multiple scattering among the particles to the bare interaction is given by the Dyson equation that is also a convoluted relation. Moreover, Ward's identity connects the self-energy with the interaction kernel through a functional derivative with respect to Green's function of electron. In the case of weak coupling, as is well-known, the perturbation approach works excellently. For example, based on this method, Eliashberg [103] received great success for conventional superconductors in obtaining the consistent results with the BCS theory. However, in strong coupling, the Eliashberg theory needs a development to account for the vertex correction that was neglected in the case of weak coupling. The situation the field theory faces resembles what the BCS meets. The major challenge here comes from the fact that in such a strong coupling case the conventional perturbation approach does not apply. This is just similar to the case that the prerequisite of $\lambda = N_F V_0 < 1$ for the BCS theory is not met in cuprate superconductors or other HTS.

The difficulty that one meets in trying to go beyond RPA, however, rests in the lack of an acceptable approach to take into account higher-order vertex correction. The partial summation that has been used in this approach often gives rise to diverse and sometimes controversial outcomes and conclusions because some of the higher-order terms may be canceled out and a partial summation may happen to pick up only the terms that lead to divergence. The work [47] by the authors has suggested a diagrammatic iteration approach (DIA) to make a full summation all over the possible diagrams. An application of this approach to a model system of cuprates, L2D, presents an alluring prospect that both low- and high-temperature superconductivity can be well understood in the same framework without invoking different models. Therefore, we are making a brief introduction to it. Interested readers may consult Ref. [47].

A survey of literature on superconductivity theories during the past couple of decades clearly shows a severe shortage of works based on the field theory. The importance of the field theory in condensed matter physics is unquestionable since it is the only way to approach the detailed information of interactions in a many-particle system.

1.6.2 *Introduction to DIA*

In weak coupling, it is well-known that the Hamiltonian method is in good agreement with the diagrammatic approach. In other words, the BCS and Eliashberg theories [103] give rise to consistent results. In strong coupling, however, the BCS theory fails, and so does the Eliashberg theory. In order to obtain a result beyond RPA, a pre-development of the conventional perturbation approach is necessary. Namely, a technique that goes beyond RPA must firstly be developed. The DIA established in the previous work [47] is what one needs. However, it seems that the DIA method is not yet well-known to date although it was published more than one decade ago. We will briefly introduce it to readers here for the completeness of this work.

Now, let us brief the reader on the methodology of DIA. The well-known Migdal's expression [58, 104–107] for the single-particle self-energy incorporating the vertex part can be written as

$$\sum(p) = i \int \frac{d^4k}{(2\pi)^4} G(p-k)D(k)\Gamma(p,k), \qquad (1.10)$$

where p and k are the four-dimensional vectors, $G(k)$ and $D(k)$ are the dressed electron and boson Green's functions, respectively. The boson Green's function satisfies the so-called Dyson equation $D(k) = D_0(k) + D_0(k)\tilde{\chi}(k)D(k)$, with $D_0(k)$ being the zeroth-order boson Green's function and $\tilde{\chi}(k)$ being the irreducible polarizability that has the standard form as given below [105, 107]. For example, in the case of the Coulomb interaction $D_0(k) = V(\boldsymbol{k})$, where $V(\boldsymbol{k})$ is the matrix element of the bare Coulomb interaction. In the case of the electron-phonon interaction $D_0(k) = D_{0ph}(\boldsymbol{k},\omega) = 2g_0^2(\boldsymbol{k})/[\omega^2 - \omega_0^2(\boldsymbol{k}) + i\delta]$, where $g_0(\boldsymbol{k})$ and $\omega_0(\boldsymbol{k})$ are the un-renormalized (original) matrix element of the electron-phonon interaction and the phonon characteristic frequency, respectively. The irreducible polarizability is given by

$$\tilde{\chi}(k) = -2i \int \frac{d^4p'}{(2\pi)^4} G(p')G(p'-k)\Gamma(p',k), \qquad (1.11)$$

where $\Gamma(p,k)$ is the three-point vertex function and, in the ladder approximation, is defined as [58, 104–109]

$$\Gamma(p,k) = 1 + i \int \frac{d^4p'}{(2\pi)^4} K(p,p')G(p')G(p'-k)\Gamma(p',k). \qquad (1.12)$$

As an example, it is easy to derive the well-known dielectric response function $\varepsilon(k) = 1 - V(\mathbf{k})\tilde{\chi}(k)$ in terms of the Dyson equation by noting that $D_0(k)/D(k) = \varepsilon(k)$.

The diagram corresponding to Eq. (1.10) is represented by the left diagram of Fig. 1.3, while the right one stands for its RPA approximation, where Γ is taken to be 1. The black corner is termed the vertex correction, which is negligible in the weak coupling for electron-phonon interaction as proven by Migdal [104]. The dashed line representing a boson field is understood as a dressed line under the RPA assumption. The well-known Eliashberg theory is based on this diagram (the right one on Fig. 1.3) without taking into account higher-order diagrams that are not shown in Fig. 1.3.

However, when coupling is strong, from the viewpoint of the diagrammatic method, the high-order corrections are no longer negligible and the problem now becomes how to find out the diagrammatic structure of the vertex function. In principle, there are infinite Feynman diagrams that represent all of the possible interaction patterns. Of more importance is now to have a rule or regulation of constructing them. An intuitive way is to follow the order of the boson line. A preliminary evaluation shows that the convoluted relationship between the self-energy Σ, kernel of interaction K and the vertex function Γ leads to a variety of developments of the diagrams, as shown below, and the weight of each of the diagrams, even if they are in the same order, is quite different, and that the contribution from a diagram in a higher order may be larger than the contribution from a low-order diagram. A variety of partial summations may lead to completely different or conflicting results. Therefore, only a complete summation over all of the possible diagrams is reliable and acceptable. To this end we have

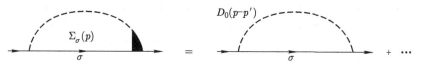

Fig. 1.3   Diagram of the spin-dependent single particle self-energy, $\Sigma_\sigma(p)$ (left) and its RPA representation (right).

developed a method to reach this goal, going out of RPA step by step. This is the so-called diagrammatic iteration approach (DIA) briefly described as follows.

The core of the DIA is iteration. Namely, starting from the known relations — the single-particle self-energy in RPA, wherein Γ is taken to be 1, one obtains the two-particle kernel of interaction K in terms of Ward's identity, $K^{\sigma\sigma'}(p,p') = -i\delta\Sigma_\sigma(p)/\delta G_{\sigma'}(p')$, where G is the Green's function of electron, p — four-dimensional momentum and σ — spin. Then, with the newly obtained K and the RPA $\Gamma = 1$, one can calculate the vertex part Γ according to Eq. (1.11) at the first step beyond RPA. The vertex part obtained here can be substituted into Eq. (1.10) for Γ to re-calculate the self-energy at the first step beyond RPA. By now, all the quantities of Σ, K and Γ at the first step beyond RPA have been calculated. Next, in terms of the self-energy Σ and vertex Γ obtained at the first step beyond RPA, one is able to re-calculate K and Γ at the second step beyond RPA, then Σ again. In doing so can one sequentially obtain all the desired information step by step to any step beyond RPA. Of course, from Eq. (1.11) the irreducible polarizability can also be calculated at each step in terms of the corresponding vertex part Γ. However, mathematical complexity makes it unrealistic to do so analytically to higher orders. Of more difficulty is that one is unable to get the summation of the complicated kennels of all the orders even if one could get several single analytical expressions. Thus, a diagrammatic technique — DIA — is recommended.

The DIA was developed in our earlier works [47, 109]. The first thing needed is to find the diagrammatic functional derivative operation. Clearly, the functional derivative identity $\delta G_\sigma(p)/\delta G_{\sigma'}(p') = \delta(p - p')\delta_{\sigma\sigma}$ is true. It is understood that the functional derivative with respect to the electron Green's function is represented by a cut diagrammatically on the solid line that stands for the electron Green's function. If a diagram containing a solid line represents the self-energy Σ, the resulting new diagram after the cut and being starched out clearly represents the corresponding kennel of interaction. This is consistent with $K^{\sigma\sigma'}(p,p') = -i\delta\Sigma_\sigma(p)/\delta G_{\sigma'}(p')$, (see the first diagram on the right side of the equal sign on Fig. 1.4).

It is worthwhile to indicate that one may have two different models in dealing with the interaction potential: a given or renormalized interaction. The first model means a simple boson propagator represented by a dashed line as shown on the right hand side of Fig. 1.3. In the BCS Hamiltonian, the interaction is simply replaced by a constant $-V$. That is a typical example

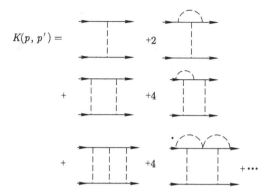

$$K(p, p') =$$

Fig. 1.4   Diagram of the kernel of interaction for the model of a given interaction.

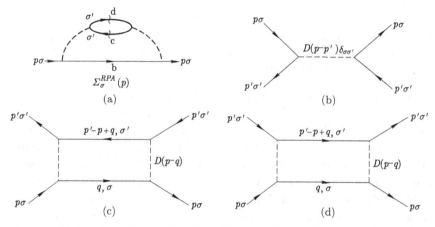

Fig. 1.5   (a) The RPA diagram for the self-energy of a particle-hole excitation (the solid loop): the diagrammatic functional derivative of the self-energy $\Sigma_\sigma(p)$ with respect to the electron Green's function is represented by the wavy lines at b, c and d. The resulting diagram due to the cutting at any electron line and stretching corresponds to the functional derivative of the self-energy, $\delta\Sigma_\sigma(p)/\delta G_{\sigma'}(p')$ i.e. the kernel of interaction. (b) The screened exchange interaction of the first order; (c) the direct interaction of the second order and (d) the 'Cooper' diagram. The broken lines correspond to the boson Green's function in the RPA.

of a given interaction. The second model takes into account contributions from particle-hole excitations, one of which is represented by a solid loop on the dashed line as shown on the top of Fig. 1.5.

For the first model — a given interaction, by means of the diagrammatic functional derivative technique, a cut on the electron line of the RPA

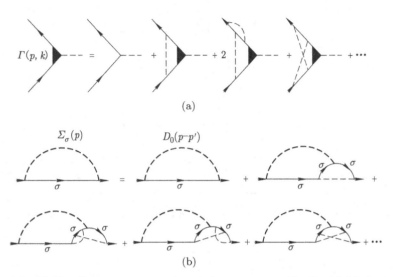

Fig. 1.6   (a) The diagram representation of the vertex part $\Gamma(p, k)$ and (b) the spin-dependent single-particle self-energy $\Sigma_\sigma(p)$ for the model of a given interaction. The dashed lines are the boson's Green function.

self-energy diagram on the right side of Fig. 1.3 obviously results in the first-order exchange kernel of interaction as shown by the first diagram on the top right of Fig. 1.4, in terms of which one can construct the first-order vertex part as displayed by the second diagram on the top right of Fig. 1.6(a). Correspondingly, we can construct the single-particle self-energy diagram in the first order (at the first step beyond RPA as well) as shown by the second diagram on the top right of Fig. 1.6(b). Now, we turn to use the diagrammatic functional derivative for this first-order self-energy diagram on which there are now three segments of the solid line to be cut, producing hence three sub-diagrams of the interaction kernel represented by two identical diagrams given by the rightmost one on top and the left diagram in the second row in Fig. 1.4. In terms of those three newly obtained sub-diagrams of the interaction kernel we can construct the next three sub-diagrams of the vertex correction on Fig. 1.6(a) and self-energy on Fig. 1.6(b). As a showcase, Fig. 1.4 exhibits the first thirteen sub-diagrams of the kernel of interaction obtained by the diagrammatic cut technique.

However, for the second model — a renormalized interaction, a cut of any electron line on the RPA diagram of the single-particle self-energy $\Sigma_\sigma^{RPA}(p)$ is shown on Fig. 1.5 by a wavy line (b, c, d). Each of the cuts corresponds to a functional derivative $\delta \Sigma_\sigma^{RPA}(p)/\delta G_{\sigma'}(p') = iK_{\sigma\sigma'}(p,p')$, giving a sub-diagram of the kernel of interaction. Diagrams (b), (c), (d) from cutting at b, c, d in $\Sigma_\sigma^{RPA}(p)$ are shown in the same figure, Fig. 1.5. The diagrams are: (b) the screened exchange interaction of the first order; (c) the screened direct interaction of the second order; (d) the Cooper-like diagram. The last one has a logarithmic singularity similar to, but different from the Cooper diagram because its initial and final moments in scattering are the same, whereas they are different in the Cooper diagram.

As an analog to the procedures performed for the first model, one receives sequential sub-diagrams of the kernel of interaction for the second model, as exhibited on Fig. 1.7(a), where the second and third sub-diagrams on the right side of the equal sign originate from the particle-hole excitation loop on Fig. 1.5 and are not present on Fig. 1.4 for the first model — a given interaction. Moreover, it is seen that while going beyond RPA, the appearance of diagrams follows "steps", instead of the usual "orders" of the boson line. The diagrams in different orders may appear at the same step. This can be more clearly seen at higher steps and implies that a lower-order diagram may not be as weighty as a higher one. In terms of the obtained kennels of interaction we can construct the diagrams of the vertex part, Fig. 1.7(b), and hence the corresponding self-energy diagrams as shown on Fig. 1.7(c).

As seen above, the core of DIA is to do iterations. Starting from the lowest order — RPA, we are able to get all the higher-order diagrams step by step. In fact, a limit number of iteration kernels (usually, up to the third step) is sufficient to find out the rule of constructing other higher-order diagrams of interaction kernels based on those we have already obtained. Therefore, by means of the DIA, all the manipulations can be done diagrammatically. We believe that this is a useful, powerful and a reasonable approach at present to a strongly-coupled particle system. We will apply it to an electron system of superconducting materials: both conventional and high-temperature superconductors to illustrate its effectiveness and validity.

1.6.3   *Applications of DIA to an electron gas*

1.6.3.1   *For a given interaction in an isotropic 3D gas*

We are now applying the DIA to an isotropic 3D electron gas for a given interaction — a model system of conventional superconductors. Here we

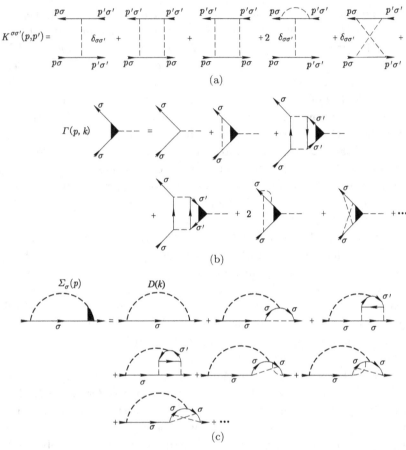

Fig. 1.7   (a) Diagram representation of the kernel of interaction $K_{\sigma\sigma'}(p,p')$, (b) the vertex corrections $\Gamma(p,k)$, and (c) the single-particle self-energy $\Sigma(p)$ for the model of a renormalized interaction with the particle-hole excitations.

present some of its major results from the application to showcase its validity.

Since the kernel of interaction is now spin-dependent, a distinction between the spin-symmetric $K_c(p,p')$ and spin-asymmetric interactions $K_s(p,p')$ is needed. The symmetric (in charge channel) and asymmetric (in spin channel) kernels of interaction are defined below:

$$K_c(p,p') = K^{\uparrow\uparrow}(p,p') + K^{\uparrow\downarrow}(p,p'), \qquad (1.13a)$$

and

$$K_s(p, p') = K^{\uparrow\uparrow}(p, p') - K^{\uparrow\downarrow}(p, p'), \qquad (1.13b)$$

where $K_c(p, p')$ and $K_s(p, p')$ are kernels of interaction in charge and spin channels, respectively. On account of the property $K^{\sigma\sigma'}(p, p') = K^{-\sigma-\sigma}(p, p')$, which is valid in the paramagnetic phase, i.e., $G_\downarrow(p) = G_\uparrow(p) = G(p)$, as can be shown, in the case of a given interaction (first model), $K_c(p, p') = K_s(p, p')$ [47].

To solve Eq. (1.12), it is necessary to carry out the summation over all of the essential contributions under the integral in Eq. (1.12). For convenience, let us replace in Eq. (1.12) the kernel $K_c(p, p')$ by the effective two-particle interaction $I(p, p', k) = I(p, p'; p + k, p' - k)$, i.e. $K_c(p, p') \to I(p, p', k)$ that is equivalent to the notation of $I(p_1, p_2; p_3, p_4)|_{p_3 = p_1 + k, p_4 = p_2 - k}$ in the Fermi liquid theory [57, 85, 86].

To evaluate $I(p, p', k)$, we rearrange the diagrams of the interaction kennel on Fig. 1.4 and put them into Fig. 1.8(a) which is a series of ladder diagrams. We sort the diagrams of interaction kernels in Fig. 1.8(a) and group all of the possible interaction (scattering) patterns (diagrams) and use an equivalent diagram to represent the grouped diagrams as shown on Fig. 1.8(b), where the first diagram stands for the sum of all the scattering diagrams whose graphs cannot be divided by a vertical line into two parts which are joined by two electron lines directed to one side [87, 111] and denoted by $\tilde{I}(p, p')$. The rest of the diagrams in Fig. 1.8(b) are given by the combinations as shown. If we now adopt the effective two-particle interaction $I(p, p', k)$ to replace $K_c(p, p')$ on Fig. 1.8(c), we hence get the diagrammatic equation of $I(p, p', k)$, which is in a convoluted format like the Dyson equation and represents the finally resulting two-particle interaction. According to Fig. 1.8(c), the equation for it can be written as [52, 87, 111]

$$I(p, p', k) = \tilde{I}(p, p') + i \int \frac{d^4q}{(2\pi)^4}$$

$$\times \tilde{I}(p, q) G(q) G(p' + p - q) I(q, p', k). \qquad (1.14)$$

For a given interaction, taking into account the spin index in Eq. (1.10), assuming $D(k) = D_0(k)$ and $\Gamma(p, k) = 1$ in it, and using the identity $\delta G_\sigma(p)/\delta G_{\sigma'}(p') = \delta(p - p')\delta_{\sigma\sigma''}$ for Ward's relation $K^{\sigma\sigma'}(p, p') = -i\delta\Sigma_\sigma(p)/\delta G_{\sigma'}(p')$, we derive the expression for the kernel at the first step

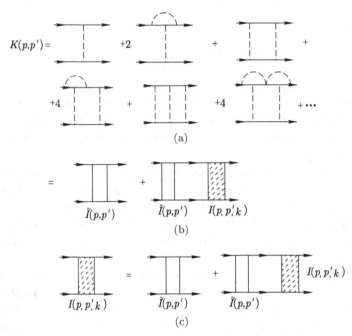

Fig. 1.8   The diagram representation in the model of a given interaction: (a) The interaction kernel $K(p, p')$ in the particle–particle channel up to the second order, (b) a series
of the ladder diagrams corresponding to $K(p, p')$ and (c) the diagrammatic equation
for $I(p, p', k)$, the effective two-particle interaction. The broken lines stand for a given
interaction. The two vertical lines represent $\tilde{I}(p, p')$, the irreducible interaction kernel,
and the hatched area stands for $I(p, p', k)$, the effective two-particle interaction or the
effective two-particle vertex function.

(the first order as well, in this case) beyond RPA as

$$K^{\sigma\sigma'}(p, p') = K_1^{\sigma\sigma'}(p, p') = D_0(p - p')\delta_{\sigma\sigma'}. \tag{1.15}$$

Substituting Eq. (1.12) with the spin indices and the kernel of Eq. (1.15)
into Eq. (1.10), we obtain the self-energy incorporating the vertex correction
at the first step (order) beyond RPA. Using Ward's identity one more time
for the single-particle self-energy that incorporates the vertex correction
with the kernel of Eq. (1.15), we express the interaction kernel up to the
second step (order) as

$$K^{\sigma\sigma'}(p, p') = K_1^{\sigma\sigma'}(p, p') + K_2^{\sigma\sigma'}(p, p') + K_3^{\sigma\sigma'}(p, p'), \tag{1.16}$$

where $K_1^{\sigma\sigma'}(p, p')$ is given by Eq. (1.15),

$$K_2^{\sigma\sigma'}(p, p') = 2D_0(p - p')\delta_{\sigma\sigma'}\left\{i \int \frac{d^4q}{(2\pi)^4} D_0(p - q)G_\sigma(q)G_\sigma(p' - p + q)\right\},$$
(1.17a)

and

$$K_3^{\sigma\sigma'}(p, p') = i\delta_{\sigma\sigma'} \int \frac{d^4q}{(2\pi)^4} D_0(p - q)D_0(q - p')G_\sigma(q)G_\sigma(p' + p - q).$$
(1.17b)

Instead of doing so further, we use diagrams to represent the above equations. The diagrams for these interactions are given in Fig. 1.4. Also, it is seen from Fig. 1.4 that the first diagram on the top after the equal sign corresponds to the first-order exchange interaction, whereas two others (the second) and the third, correspond to the second order. As a matter of fact, the second diagram, $K_2(p, p')$, corresponds to the first-order exchange interaction incorporating the vertex correction of the first-order. The third diagram, $K_3^{\sigma\sigma'}(p, p')$, in Eq. (1.17b), the so-called "crossed" diagram, is very important because it represents the interaction in the Cooper channel for a superconductor and leads to the well-known logarithmic singularity of the two-particle vertex function [52, 104, 110] in the case of an attractive interaction $D_0(k) < 0$. The diagrams of the vertex part and the self-energy with the interaction kernel, described by Eqs. (1.16) and (1.17), are given in Fig. 1.6.

As an example, we can evaluate kernels $K_2(p, p')$ and $K_3(p, p')$ in the case of the non-retarded interaction, i.e., when $D_0(k) = D_0(\mathbf{k})$, where $D_0(\mathbf{k})$ is a given frequency-independent interaction. We assume that the electron Green's function is given by $G^{-1}(\mathbf{k}, i\omega) = i\omega - \xi_k$, where $\xi_k = \varepsilon_k - \mu$ is the single-particle energy measured relative to the chemical potential. This implies that we use Green's function of a perfect Fermi gas. After the frequency summation, Eqs. (1.17a) and (1.17b) become, respectively, forms

$$K_2(\mathbf{p}, i\omega; \mathbf{p}', i\omega') = -2D_0(\mathbf{p} - \mathbf{p}')\sum_{\mathbf{q}} D_0(\mathbf{p} - \mathbf{q})\frac{n_F(\xi_\mathbf{q}) - n_F(\xi_{\mathbf{q}-\mathbf{p}+\mathbf{p}'})}{\xi_\mathbf{q} - \xi_{\mathbf{q}-\mathbf{p}+\mathbf{p}'} - i(\omega - \omega')},$$
(1.18a)

and

$$K_3(\boldsymbol{p}, i\omega; \boldsymbol{p}', i\omega') = \sum_q D_0(\boldsymbol{p} - \boldsymbol{q})D_0(\boldsymbol{q} - \boldsymbol{p}')\frac{1 - n_F(\xi_q) - n_F(\xi_{\boldsymbol{p}'+\boldsymbol{p}-\boldsymbol{q}'})}{i(\omega + \omega') - \xi_q - \xi_{\boldsymbol{p}'} + \boldsymbol{p} - \boldsymbol{q}},$$

(1.18b)

where $\delta_{\sigma\sigma'}$ is omitted. Equation (1.18b) indeed has a logarithmic singularity $\sim \ln[\omega_c/\max\{\omega, T\}]$ for $\boldsymbol{p}' = -\boldsymbol{p} \approx \boldsymbol{p}_F$, where ω_c is the characteristic cutoff energy of the given interaction.

We consider the case when the given interaction is similar to the electron-phonon interaction, i.e. $D_0(\boldsymbol{k}, \omega) = 2g_{0\boldsymbol{k}}^2\omega_{0\boldsymbol{k}}/[\omega^2 - \omega_{0\boldsymbol{k}}^2 + i\delta]$. Now, we can evaluate $\widetilde{I}(p, p')$, the irreducible kernel of interaction, and consider $K_2(p, p')$ which, from Eq. (1.17a), can be written as

$$K_2(\boldsymbol{p}, i\omega; \boldsymbol{p}', i\omega') = -2D_0(\boldsymbol{p} - \boldsymbol{p}', i(\omega - \omega'))xT$$
$$\cdot \sum_{\omega_{qn'}q} D_0(\boldsymbol{p} - \boldsymbol{q}, i(\omega - \omega_{qn}))G(\boldsymbol{q}, i\omega_{qn})G(\boldsymbol{q} - \boldsymbol{p} + \boldsymbol{p}', i(\omega_{qn} - \omega + \omega')).$$

(1.19)

We first consider the case of $\omega_c \ll \varepsilon_F$, where the characteristic boson cutoff frequency is much less than the Fermi energy. Carrying out both the frequency summation in Eq. (1.19) in the region $|\omega - \omega_{qn}| < \omega_c$ and the integration with respect to $q|\boldsymbol{p} - \boldsymbol{q}| < q_m$, where $q_m \sim p_F$ is the characteristic cutoff momentum of the boson spectrum, we can approximate Eq. (1.19) for $K_2(p, p')$ as $K_2(p, p') \approx -2D_0(p - p')\lambda N_F\omega_c/\varepsilon_F \approx 2\lambda^2 N_F\omega_c/\varepsilon_F \ll \lambda$, where $\lambda = -\langle D_0(\boldsymbol{p} - \boldsymbol{p}', 0)\rangle$ is the electron-boson interaction constant. λ is related to ζ_b, the dimensionless electron-boson interaction constant, as $\lambda = \zeta_b/N_F$ with

$$\zeta_b = \frac{2}{(2\pi)^3} \int \frac{d^2p}{v_F} \int \frac{d^2p'}{v_F'} \frac{g_0^2(p - p')}{\omega_0(p - p')} \Big/ \int \frac{d^2p}{v_F}$$

and the density of states per spin N_F (for example, in an isotropic 3D or 2D electron spectrum $N_{F3D} = m^*p_F/2\pi^2$ or $N_{F2D} = m^*/2\pi$, respectively, where m^* is the effective mass and p_F is the Fermi momentum). $K_2(p, p')$ given by Eq. (1.19) is small for $\omega_c \ll \varepsilon_F$. Thus, we can estimate $\widetilde{I}(p, p')$ as $\widetilde{I}(p, p') \approx -\lambda(1 - 2\lambda N_F\omega_c/\varepsilon_F)$, and confine ourselves to the simple first-order vertex, i.e., $\widetilde{I}(p, p') \approx K_1(p, p') = D_0(p - p') \approx -\lambda$, when $\omega_c \ll \varepsilon_F$. This is in agreement with the Migdal theorem [58, 87, 104]. It is seen that due to the fact that $\omega_c \ll \varepsilon_F$, this conclusion is valid even without assuming that λ (or ζ_b) is small. In the case that ω_c is not small (i.e., $\omega_c \approx \varepsilon_F$), the irreducible kernel can be approximated as $\widetilde{I}(p, p') \approx D_0(p-p')(1-2\lambda N_F) = \widetilde{D}_0(p-p') \approx -\widetilde{\lambda}$, where $\widetilde{\lambda} = \lambda(1-2\lambda N_F) < \lambda$ is the effective electron-boson interaction constant. This implies that the vertex correction does affect the electron-boson interaction constant when $\omega_c \approx \varepsilon_F$.

Correspondingly, substituting $\widetilde{I}(p, p')$ into Eq. (1.14), we see that in Eq. (1.14), $I(p, p', k) = I(Q)$ is a function of $Q = p' + p$ only, where $Q = p + p'$ is the total momentum of the system of two particles. Further, using this fact and carrying out the frequency summation in Eq. (1.14), we obtain the solution to Eq. (1.14) as

$$I(Q, \omega_Q) \approx -\frac{\widetilde{\lambda}}{1 + \widetilde{\lambda} \sum\limits_{q, |\xi_q| < \omega_c} \dfrac{1 - n_F(\xi_q) - n_F(\xi Q - q)}{\omega_Q - \xi_q - \xi_{Q-q} + i\delta}}, \tag{1.20}$$

where the analytic continuation $(i\omega_Q \rightarrow \omega_Q + i\delta)$ was made and $\omega_Q = \omega + \omega'$. Equation (1.20) leads to the well-known result concerning the instability of the normal state in the case of a given attractive interaction $(\widetilde{\lambda} > 0)$. As an example, we can examine Eq. (1.20) at $Q = 0$ and $T = 0$. If we assume that ω_Q in Eq. (1.20) is real and positive, we have [58, 87]

$$I(\omega_Q) \approx -\frac{1}{\widetilde{\lambda}} - \widetilde{\lambda} N_F \left[\ln \frac{2\omega_c}{\omega_Q} + i\frac{\pi}{2} \right]. \tag{1.21}$$

It can be seen that the quantity $I(\omega_Q)$ has a pole (in the upper half of the complex plane) only in the case of an attractive interaction at the point $\omega_Q = i\Omega$, where $\Omega = 2\omega_c \exp(-1/\widetilde{\lambda} N_F)$. The pole of $I(Q, \omega_Q)$ as a function of $|Q|$, when Q is non-zero $(Q \neq 0$, and $v_F |Q| \ll |\omega_Q|)$, exists for $\omega_Q = i\Omega(1 - v_F^2 |Q|^2/6\Omega^2)$ [58, 87]. The absolute value of ω_Q decreases as $|Q|$ increases. For the case of $T \neq 0$, Eq. (1.20) allows us to evaluate T_c, the temperature at which a pole first appears in $I(Q, \omega_Q)$ at $|Q| = \omega_Q = 0$. It follows from Eq. (1.20) that $I^{-1}(0, 0) = 0$ at

$$T = T_c = 1.13\omega_c \exp\left(-\frac{1}{\widetilde{\lambda}} N_F \right), \tag{1.22}$$

which is exactly the same transition temperature obtained in the BCS theory [1]. In addition, we can see that in the case of $\omega_c \approx \varepsilon_F$, the maximum value of $\widetilde{\lambda} N_F \approx 1/8$. Therefore, the maximum value of $T_c \approx \varepsilon_F \exp(-8)$. This means that the vertex correction to the electron-boson coupling constant in the case of $\omega_c \approx \varepsilon_F$ suppresses the value of T_c essentially as indicated by Grabowski and Sham [111].

Now we can estimate $\Gamma(p, k)$, the vertex correction given by Eq. (1.12). Using Eq. (1.20) and taking $\Gamma(p', k) = 1$ in the integral in Eq. (1.12), we find the vertex correction of the first order as

$$\Gamma^{(1)}(\boldsymbol{p}, i\omega; \boldsymbol{k}, i\omega') = 1 + \Delta\Gamma^{(1)}(\boldsymbol{p}, i\omega; \boldsymbol{k}, i\omega'), \tag{1.23}$$

where

$$\Delta\Gamma^{(1)}(\boldsymbol{p}, i\omega; \boldsymbol{k}, i\omega') = -T \sum_{\omega_{p'n}, \boldsymbol{p}'} I(\boldsymbol{p}, i\omega; \boldsymbol{p}', i\omega_{p,n})$$
$$\cdot G(\boldsymbol{p}', i\omega_{p'n}) G(\boldsymbol{p}' - \boldsymbol{k}, i(\omega_{p,n} - \omega')), \tag{1.24}$$

with $I(\boldsymbol{p}, i\omega; \boldsymbol{p}', i\omega')$ given by Eq. (1.20). Now, performing the frequency summation, one estimates Eq. (1.24) as

$$\Delta\Gamma^{(1)}(\boldsymbol{p}, i\omega; \boldsymbol{k}, i\omega') \approx -\sum_{\boldsymbol{p}'} \chi^0_{\boldsymbol{p'k}}(\omega') I(\boldsymbol{p}, i\omega; \boldsymbol{p}', \xi_{\boldsymbol{p'}-\boldsymbol{k}} + i\omega'), \tag{1.25}$$

where

$$\chi^0_{\boldsymbol{pk}}(\omega) = [n_F(\xi_{\boldsymbol{p}}) - n_F(\xi_{\boldsymbol{p-k}})]/(\xi_{\boldsymbol{p}} - \xi_{\boldsymbol{p-k}} - i\omega) \tag{1.26}$$

and $n_F(\omega) = [\exp(\omega/T)+1]^{-1}$ is the Fermi distribution function. Firstly, we consider the case of $\omega_c \ll \varepsilon_F$ to evaluate the vertex correction. It is understood that the region of $v_F|\boldsymbol{k}| \gg \omega$ is the only one that is important. It is obvious from Eq. (1.25) (using Eq. (1.20) in it) that in the denominator of $I(\boldsymbol{p}, i\omega; \boldsymbol{p}', \xi_{\boldsymbol{p'}-\boldsymbol{k}} + i\omega')$ appears the quantity $\sim \tilde{\lambda} N_F \ln(\omega_c/\max\{v_F|\boldsymbol{k}|, \omega'\})$ when $\omega_c > \max\{v_F|\boldsymbol{k}|, \omega'\}$. The denominator of $I(\boldsymbol{p}, i\omega; \boldsymbol{p}', \xi_{\boldsymbol{p'}-\boldsymbol{k}} + i\omega')$ in Eq. (1.25) in the region of $v_F|\boldsymbol{k}| > \omega_c \gg \omega'$ is equal to ≈ 1, whereas in the more important region $\omega_c > v_F|\boldsymbol{k}| \gg \omega'$ the denominator of $I(\boldsymbol{p}, i\omega; \boldsymbol{p}', \xi_{\boldsymbol{p'}-\boldsymbol{k}} + i\omega')$ does not have a singularity because $\omega_c \ll \varepsilon_F$. Such a small $|\boldsymbol{k}| < \omega_c/v_F$ plays no role in the consideration of the effect of the vertex correction in the calculations of the self-energy and the irreducible response [58, 87, 104] (Migdal's theorem). Thus, the denominator of $I(\boldsymbol{p}, i\omega; \boldsymbol{p}', \xi_{\boldsymbol{p'}-\boldsymbol{k}} + i\omega')$ in Eq. (1.20) is equal to ≈ 1 in the case $\omega_c \ll \varepsilon_F$ and $|\boldsymbol{k}| \gg \omega_c/v_F$. This means that in Eq. (1.24) we can replace $I(p, p')$ by $\tilde{I}(p, p') \approx D_0(p - p') \approx -\lambda$. Therefore, using Eq. (1.24), we find an estimate of $\Gamma^{(1)}(\boldsymbol{p}, i\omega; \boldsymbol{k}, i\omega')$ given by Eq. (1.23) to be $\Gamma^{(1)}(p_F, 0; \boldsymbol{k}, \omega')|_{v_F k \gg \omega'} \approx 1 - \lambda N_F \omega_c/\varepsilon_F$. This is in agreement with the Migdal theorem [58, 87, 106, 112]. Thus, the effect of vertex corrections to the self-energy and the irreducible response is negligible in the case of $\omega_c \ll \varepsilon_F$.

Next, we consider the case that the boson frequency is comparable to the Fermi energy, $\omega_c \approx \varepsilon_F$. In this case the vertex corrections are not expected to be negligible (the Migdal theorem is violated in this case) and should be included in all the orders of perturbation theory. Thus, using Eqs. (1.23) and (1.25) in Eq. (1.12), we can find the vertex correction up to the second order given by

$$
\begin{cases}
\Gamma^{(2)}(\boldsymbol{p}, i\omega; \boldsymbol{k}, i\omega') = 1 + \Delta\Gamma^{(1)}(\boldsymbol{p}, i\omega; \boldsymbol{k}, i\omega'), \\
- \sum_{\boldsymbol{p}'} \chi^0_{\boldsymbol{p}',\boldsymbol{k}}(\omega') I(\boldsymbol{p}, i\omega; \boldsymbol{p}', \xi_{\boldsymbol{p}'-\boldsymbol{k}} + i\omega') \Delta\Gamma^{(1)}(\boldsymbol{p}', \xi_{\boldsymbol{p}'-\boldsymbol{k}} + i\omega'; \boldsymbol{k}, i\omega').
\end{cases}
$$
(1.27)

Noting that in the summation over $\boldsymbol{p}'$ in Eq. (1.27) the major contribution comes from $|\boldsymbol{p}'| \approx p_F$, we can therefore, in Eq. (1.27), replace $\Delta\Gamma^{(1)}(\boldsymbol{p}', \xi_{\boldsymbol{p}'-\boldsymbol{k}} + i\omega'; \boldsymbol{k}, i\omega')$ by its average over the Fermi surface $\langle \Delta\Gamma^{(1)}(\boldsymbol{p}', \xi_{\boldsymbol{p}'-\boldsymbol{k}} + i\omega'; \boldsymbol{k}, i\omega')\rangle_{|\boldsymbol{p}'|=p_F}$. Then Eq. (1.27) can be rewritten as

$$
\Gamma^{(2)}(\boldsymbol{p}, i\omega; \boldsymbol{k}, i\omega')
$$
$$
= 1 + \Delta\Gamma^{(1)}(\boldsymbol{p}, i\omega; \boldsymbol{k}, i\omega')\langle \Gamma^{(1)}(\boldsymbol{p}', \xi_{\boldsymbol{p}'-\boldsymbol{k}} + i\omega'; \boldsymbol{k}, i\omega')\rangle_{|\boldsymbol{p}'|=p_F}. \quad (1.28)
$$

Similarly, we obtain the vertex correction up to the third order $\Gamma^{(3)}(p, k)$, and so on to the n-th order $\Gamma^{(n)}(p, k)$, and finally express them in terms of $\Delta\Gamma^{(1)}(p, k)$. It is obvious that the expression of $\Gamma^{(n)}(p, k)$ is a series of $\Delta\Gamma^{(1)}(p, k)$ and that $\Gamma(p, k) = \lim_{n\to\infty} \Gamma^{(n)}(p, k)$. Making the summation of the series (ladder summation), we evaluate the vertex part as

$$
\Gamma(\boldsymbol{p}, i\omega; \boldsymbol{k}, i\omega') \approx 1 + \frac{\Delta\Gamma^{(1)}(\boldsymbol{p}, i\omega; \boldsymbol{k}, i\omega')}{1 - \langle \Delta\Gamma^{(1)}(\boldsymbol{p}, \xi_{\boldsymbol{p}-\boldsymbol{k}} + i\omega'; \boldsymbol{k}, i\omega')\rangle_{|\boldsymbol{p}|=p_F}}, \quad (1.29)
$$

where $\Delta\Gamma^{(1)}(\boldsymbol{p}, i\omega; \boldsymbol{k}, i\omega')$ is given by Eq. (1.25). In the denominator of Eq. (1.29), in terms of Eqs. (1.20) and (1.25), $\langle \Delta\Gamma^{(1)}(\boldsymbol{p}, \xi_{\boldsymbol{p}-\boldsymbol{k}} + i\omega'; \boldsymbol{k}, i\omega')\rangle_{|\boldsymbol{p}|=p_F}$ can be represented by

$$
\langle \Delta\Gamma^{(1)}(\boldsymbol{p}, \xi_{\boldsymbol{p}-\boldsymbol{k}} + i\omega; \boldsymbol{k}, i\omega')\rangle_{|\boldsymbol{p}|=p_F} \approx \frac{\widetilde{\lambda}}{2} \frac{\chi_0(\boldsymbol{k}, i\omega)}{1 + \frac{\widetilde{\lambda}}{2}B(i\omega)}, \quad (1.30)
$$

where $\chi_0(\boldsymbol{k}, i\omega) = 2\sum_{\boldsymbol{p}} \chi^0_{\boldsymbol{p}\boldsymbol{k}}(\omega)$ is the free-particle polarizability with $\chi^0_{\boldsymbol{p}\boldsymbol{k}}(\omega)$ given by Eq. (1.26), and

$$
B(\omega) = \sum_{q, |\xi_q| < \varepsilon_F} \frac{1 - 2N_F(\xi_q)}{i\omega - \xi_q}. \quad (1.31)
$$

Now we can evaluate the irreducible response given by Eq. (1.11). Using Eq. (1.29) with Eqs. (1.30) and (1.31) in Eq. (1.11), after the analytical continuation to the real axes $(i\omega \rightarrow \omega + i\delta)$, the irreducible response is evaluated as

$$\tilde{\chi}(k,\omega) \approx -\frac{\chi_0(k,\omega)}{11 - (\tilde{\lambda}/2)\chi_0(k,\omega)/[1 + (\tilde{\lambda}/2)B(\omega + i\delta)]}, \qquad (1.32)$$

with $B(w)$ given by Eq. (1.31).

Equation (1.32) is valid in the region of $v_F|\mathbf{k}| \ll \omega_c \approx \varepsilon_F$ and $v_F|\mathbf{k}| \gg \omega$.

Now, we may further separate the real from the imaginary part of $\tilde{\chi}(\mathbf{k},\omega)$ in Eq. (1.32) and have

$$\tilde{\chi}(\mathbf{k},\omega)|_{\omega<v_Fk<\varepsilon_F} = \mathrm{Re}\tilde{\chi}(\mathbf{k},\omega)|_{\omega<v_Fk<\varepsilon_F} + i\mathrm{Im}\tilde{\chi}(\mathbf{k},\omega)|_{\omega<v_Fk<\varepsilon_F'}, \tag{1.33}$$

$$\mathrm{Re}\tilde{\chi}(k,\omega)|_{\omega<v_Fk<\varepsilon_F} \approx -\frac{\lambda N_F}{1 + \tilde{\lambda}N_F/A(\omega)}, \tag{1.34a}$$

and

$$\mathrm{Im}\tilde{\chi}(k,\omega)|_{\omega<v_Fk<\varepsilon_F} \approx -\frac{\pi N_F}{[1 + \tilde{\lambda}N_F/A(\omega)]^2}\left\{\frac{\omega}{v_Fk} - \tilde{\lambda}^2 N_F^2\tanh\frac{\omega}{2T}\right\}, \tag{1.34b}$$

where $A(\omega) = 1 + (\tilde{\lambda}/2)\mathrm{Re}B(\omega) = 1 - \tilde{\lambda}N_F\ln[\omega_F/\max(\omega,T)]$ and $\tilde{\lambda} = \tilde{\lambda}/A(\omega)$. Equations (1.33) and (1.34) in the model of a given interaction are valid for both the charge and spin channels. In the static ($\omega = 0$) and long wavelength limit, Eq. (1.34a) tends to the form of the result of Landau's Fermi liquid theory [86], $\tilde{\chi}(0,0) = -2N_F/(1 + F_0^{s,a})$ with $F_0^{s,a} = \tilde{\lambda}N_F/A(0) \approx \tilde{\lambda}N_F/[1 - \tilde{\lambda}N_F\ln(\varepsilon_F/T)]$ being the Landau's Fermi liquid parameters in the charge and spin channels at the zero frequency. Thus, $F_0^{s,a}(\omega) = \tilde{\lambda}N_F/A(\omega)$ are the frequency-dependent Landau's Fermi liquid parameters in the charge and spin channels. Moreover, the identity $F_0^s(\omega) = F_0^a(\omega)$ holds in the framework of the model of a given interaction.

In addition, as can be seen from Eq. (1.34a), the multiple scattering produces the strong frequency dependence of the response function. The imaginary part of the polarizability contains an anomalous Fermi liquid term $\mathrm{Im}\tilde{\chi} \sim \omega/T$ as well as a normal Fermi liquid response $\mathrm{Im}\tilde{\chi} \sim -\omega/v_Fk$

in the region $\omega \ll v_F k < \omega_c \approx \varepsilon_F$. The contribution of the anomalous Fermi liquid term $\mathrm{Im}\tilde{\chi} \sim \omega/T$ to the irreducible response strongly depends on the value of the coupling constant $\tilde{\lambda} = \tilde{\lambda}/A(\omega)$ which is also frequency and temperature dependent. This term presents an opposite sign relative to the normal Fermi liquid term $\sim -\omega/v_F k$. In the next section we will show that the particle-hole excitations lead to the similar form of the anomalous Fermi liquid response but with a sign opposite to that from the model of a given interaction. It should be noted that the anomalous Fermi liquid term does not depend on the sign of interaction because $\mathrm{Im}\tilde{\chi}_{an} \sim \tilde{\lambda}^2\omega/T$. Therefore, Eq. (1.34a) and Eq. (1.34b) hold for both the attractive $(\tilde{\lambda} > 0)$ and repulsive $(\tilde{\lambda} < 0)$ interactions. Of course, in the case of a repulsive $(\tilde{\lambda} < 0)$ given interaction no interacting instability occurs and hence there is no phase transition. This is in agreement with the fact that there would be no solution to the BCS Hamiltonian if the interaction in it were positive.

It is therefore summarized as follows. In the case of a given interaction the interacting instability happens only if the interaction is attractive $(\tilde{\lambda} > 0)$. This is just the Cooper instability where at $T \neq 0$ the calculated transition temperature, at which the instability occurs, is the same as given by BCS theory. Correspondingly, the irreducible response has the normal Fermi liquid behavior as well as a weak anomalous behavior which the Fermi liquid does not possess. In the weak interaction, $\lambda \ll 1$, the anomalous response is not appreciable and may not be experimentally observed. Moreover, the anomalous response has the opposite sign to the normal one. This anomalous response is different from that obtained for the renormalized interaction on account of the particle-hole excitations as shown below.

Obviously, the BCS theory based on the Hamiltonian method is unable to provide the information of the response described in Eq. (1.33).

1.6.3.2 *For a renormalized interaction in an isotropic 3D gas*

We now turn to study the effect of the particle-hole excitations on the pairing instability in an interacting Fermi gas. In this section we consider the second model that deals with the renormalized interaction. In this case the renormalized boson propagator satisfies Dyson's equation $D(k) = D_0(k) + D_0(k)\tilde{\chi}(k)D(k)$, where $D_0(k)$ is a given (original) interaction and $\tilde{\chi}(k)$ here is the irreducible polarizability including the vertex correction.

We start from the RPA in order to evaluate the kernel $K^{\sigma\sigma'}(p,p')$ and use the equation for the self-energy in the RPA, i.e., Eq. (1.10) with $\Gamma(p,k) = 1$ and $\widetilde{\chi}(k) = \chi_0(k)$, where

$$x_0(k) = -i \sum_\sigma \int \frac{d^4q}{2\pi^4} G_\sigma(q) G_\sigma(q-k) \tag{1.35}$$

is the "free-particle" response function with the dressed electron Green's function $G_\sigma(p)$. Using the identity $\delta G_\sigma(p)/\delta G_{\sigma'}(p') = \delta(p-p')\delta_{\sigma\sigma'}$, we have the expression for the kernel at the first step of iteration beyond the RPA:

$$K^{\sigma\sigma'}(p,p') = K_1^{\sigma\sigma'}(p,p') + K_2^{\sigma\sigma'}(p,p') + K_3^{\sigma\sigma'}(p,p'), \tag{1.36}$$

where

$$K_1^{\sigma\sigma'}(p,p') = D(p-p')\delta_{\sigma\sigma'}, \tag{1.37a}$$

$$K_2^{\sigma\sigma'}(p,p') = -i \int \frac{d^4q}{(2\pi)^4} D^2(p-q) G_\sigma(q) G_{\sigma'}(p'-p+q), \tag{1.37b}$$

and

$$K_3^{\sigma\sigma'}(p,p') = -i \int \frac{d^4q}{(2\pi)^4} D^2(p-q) G_\sigma(q) G_{\sigma'}(p'+p-q), \tag{1.37c}$$

with the boson propagator $D(k)$ in the RPA, i.e., $D(k) = D_0(k)/[1 - D_0(k)\chi_0(k)]$. The analytic derivation of Eqs. (1.37a)–(1.37c) is given in Appendix A. The diagrams for these interactions are given in Fig. 1.5, where the bisection (cutting) of any electron line stands for the procedure of the diagrammatic functional derivative $\delta G_\sigma(p)/\delta G_{\sigma'}(p') = \delta(p-p')\delta_{\sigma\sigma'}$ as before. It is seen from Fig. 1.5 that the first diagram corresponds to the first-order screened exchange interaction, whereas the two others correspond to the second-order interactions in the particle-hole channel. The second diagram represents the second-order screened direct interaction. The last one for Eq. (1.37c) in the particle-hole channel is very important because it stands for the interaction in the "Cooper" channel (this diagram in fact differs from the Cooper one because its initial and final momenta in scattering are the same) and has a logarithmic singularity, too. Indeed, as done for Eq. (1.18), using the electron Green's function given by $G^{-1}(\boldsymbol{k}, i\omega) = i\omega - \xi_k$, for the non-retarded interaction that $D(k) = D(\boldsymbol{k})$, where $D(\boldsymbol{k})$ is frequency-independent, after the frequency summation in Eqs. (1.37b) and (1.37c) we have

$$K_2(\boldsymbol{p}, i\omega; \boldsymbol{p}', i\omega') = \sum_q D^2(\boldsymbol{p} - \boldsymbol{q}) \frac{n_F(\xi_q) - n_F(\xi_{q-p+p'})}{\xi_q - \xi_{q-p+p'} - i(\omega - \omega')},$$

(1.38a)

and

$$K_3(\boldsymbol{p}, i\omega; \boldsymbol{p}', i\omega') = -\sum_q D^2(\boldsymbol{p} - \boldsymbol{q}) \frac{1 - n_F(\xi_q) - n_F(\xi_{p+p'-q})}{i(\omega + \omega') - \xi_q - \xi_{p+p'-q}},$$

(1.38b)

where $K_3(\boldsymbol{p}, i\omega; \boldsymbol{p}', i\omega')$ has the same logarithmic singularity as in Eq. (1.18b), but appears with a negative sign. In addition, in Eq. (1.38) we have omitted the summation over the spin indices.

Let us compare Eq. (1.36) with Eq. (1.16). Equation (1.36) includes three different types of interaction: the first one is the screened exchange in the first order and the two others are of the second order from the particle-hole channel. For the first model there is no contribution from the particle-hole channel. In addition, in Eq. (1.36) appears $D(k)$, the renormalized boson propagator in the RPA. In contrast, in Eq. (1.15) appears $D_0(k)$. Thus, the second model in the RPA includes more contributions than the first one. Substituting Eq. (1.12) into Eq. (1.10) with the kernel from Eq. (1.36), we obtain the self-energy beyond the RPA. Using Ward's identity one more time for the self-energy, incorporating the vertex correction with the kernel from Eq. (1.36), we get the kernel of interaction beyond the RPA up to the second step. Thus, $K^{\sigma\sigma'}(p, p')$, the kernel of interactions, including all the contributions up to the second order, can be written as

$$K^{\sigma\sigma'}(p, p') = K_1^{\sigma\sigma'}(p, p') + K_2^{\sigma\sigma'}(p, p') + K_3^{\sigma\sigma'}(p, p')$$
$$+ K_4^{\sigma\sigma'}(p, p') + K_5^{\sigma\sigma'}(p, p'),$$

(1.39)

where $K_1^{\sigma\sigma'}(p, p'), K_2^{\sigma\sigma'}(p, p')$, and $K_3^{\sigma\sigma'}(p, p')$ are given by Eqs. (1.37a), (1.37b), and (1.37c), respectively, and

$$K_4^{\sigma\sigma'}(p, p') = 2D(p - p')\delta_{\sigma\sigma'} \left\{ i \int \frac{d^4q}{(2\pi)^4} D(p - q) G_\sigma(q) G_\sigma(p' - p + q) \right\},$$

(1.40a)

and

$$K_5^{\sigma\sigma'}(p,p') = i\delta_{\sigma\sigma'} \int \frac{d^4q}{(2\pi)^4} D(p-q)D(q-p')G_\sigma(q)G_\sigma(p'+p-q),$$

(1.40b)

are two of the diagrams in the second step and are of the second order. Equations (1.40a) and (1.40b) are similar to Eqs. (1.18a) and (1.18b), respectively, except that in Eqs. (1.40a) and (1.40b), $D(k)$ is the renormalized propagator in the RPA, whereas in Eqs. (1.17a) and (1.17b), $D_0(k)$ is the given interaction. In addition, Eq. (1.39) includes very important terms $K_2^{\sigma\sigma'}(p,p')$ and $K_3^{\sigma\sigma'}(p,p')$ due to the particle-hole excitations, but Eqs. (1.17a) and (1.17b) do not. The diagrams for the vertex part and the self-energy in this model up to the second order are shown in Fig. 1.7.

As shown in Eq. (1.13a) and Eq. (1.13b), $K_c(p,p')$ and $K_s(p,p')$ are kernels of interaction in the charge and spin channels, respectively. Using Eq. (1.16) in Eq. (1.13), we can see that $K_c(p,p') = K_s(p,p')$ for the case of a given interaction, whereas in the case of the renormalized interaction, substituting Eqs. (1.37) and (1.39) into Eq. (1.13), we have

$$K_c(p,p') = K_1(p,p') + 2K_2(p,p') + 2K_3(p,p') + K_4(p,p') + K_5(p,p')$$

(1.41a)

and

$$K_s(p,p') = K_1(p,p') + K_4(p,p') + K_5(p,p'),$$ (1.41b)

where $K_1(p,p'), K_2(p,p'), K_3(p,p'), K_4(p,p')$ and $K_5(p,p')$ are given by Eqs. (1.37a), (1.37b), (1.37c), (1.40a) and (1.40b), respectively. It is seen from Eqs. (1.41a) and (1.41b) that the contributions from the particle-hole excitations to the effective kernel of interaction appear in the charge channel only and that there is no such contributions in the spin channel.

In analogy to Eq. (1.13), the similar relations can also be written for the vertex part and the irreducible response functions. That is, using Eq. (1.13) in Eq. (1.12) we can have $\Gamma_{c,s}(p,k) = \Gamma_{\uparrow\uparrow}(p,k) \pm \Gamma_{\uparrow\downarrow}(p,k)$ for the vertex part. In addition, in terms of $\Gamma_{c,s}(p,k)$ in Eq. (1.11) it also follows that $\tilde{\chi}_{c,s}(k) = \tilde{\chi}_{\uparrow\uparrow}(k) \pm \tilde{\chi}_{\uparrow\downarrow}(k)$. Thus, the vertex functions in the charge and spin channels, respectively, are determined from the solution of the equations

$$\Gamma_{c,s}(p,k) = 1 + i \int \frac{d^4 p'}{(2\pi)^4} K_{c,s}(p,p') G(p') G(p'-k) \Gamma_{c,s}(p',k). \qquad (1.42)$$

In analogy to Eq. (1.35), the irreducible response functions in the charge and spin channels can be obtained from the equation

$$\widetilde{\chi}_{c,s}(k) = -2i \int \frac{d^4 p'}{(2\pi)^4} G(p') G(p'-k) \Gamma_{c,s}(p',k). \qquad (1.43)$$

Next, we will consider the pairing instability and the effect of the vertex corrections on the irreducible response functions.

In analogy to the previous section, we are interested in the solution of Eqs. (1.42) and (1.43) with $K^{\sigma\sigma'}(p,p')$, the kernel of interaction given by Eq. (1.39). From Eq. (1.41) it follows that for the model with the renormalized interaction, $K_c(p,p') \neq K_s(p,p')$ due to the particle-hole excitations as discussed before. Thus, $\Gamma_c(p,k) \neq \Gamma_s(p,k)$ and $\widetilde{\chi}_c(k) \neq \widetilde{\chi}_s(k)$ in Eqs. (1.42) and (1.43).

First of all, we consider the solution to Eqs. (1.42) and (1.43) in the charge channel. To solve Eq. (1.42) for $\Gamma_c(p,k)$, the vertex correction in the charge channel, we consider $K_c(p,p')$, the kernel of interactions in the charge channel given by Eq. (1.41a). We can separate Eq. (1.41a) into two parts: one of them is the regular part $\widetilde{I}(p,p')$ that does not have the contributions with a logarithmic singularity and the other will be called the Cooper kernel $I_c(p,p')$ that only includes the terms with a logarithmic singularity. In the lowest order $\widetilde{I}(p,p') = K_1(p,p') + 2K_2(p,p') + K_4(p,p')$, where $K_1(p,p'), K_2(p,p')$ and $K_4(p,p')$ are given by Eqs. (1.37a), (1.37b) and (1.40a), respectively. In the lowest order $I_c(p,p') = 2K_3(p,p') + K_5(p,p')$ with $K_3(p,p')$ and $K_5(p,p')$ given by Eqs. (1.37c) and (1.40b), respectively. The diagrams for $\widetilde{I}(p,p')$ and $I_c(p,p')$ in the lowest order are shown in Figs. 1.9(a) and 1.9(b), respectively. The latter was obtained by twisting the upper hole-lines in Fig. 1.7(a) to represent interactions in the particle-particle channel. One can see that $I_c(p,p')$ includes both the ladder of the Cooper-like diagrams coming from the particle-hole channel and the ladder of the conventional Cooper diagrams. To evaluate $\widetilde{I}(p,p')$, we first

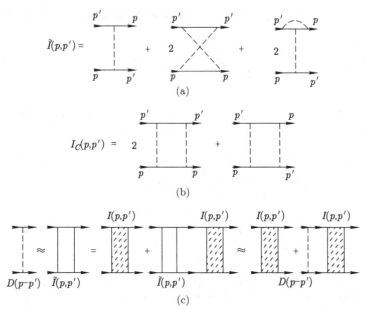

Fig. 1.9 (a) The diagram representation of the irreducible kernel of interactions $\tilde{I}(p, p')$ in the particle-particle channel in the lowest order in the model with the renormalized interaction; (b) the Cooper part of interactions in the lowest order $I_c(p, p')$; (c) the equation for $I(p, p')$ in the ladder approximation. The bold dashed lines stand for the boson Green's function in the RPA as in Fig. 1.8. The two vertical lines represent $\tilde{I}(p, p')$, the irreducible kernel of interactions and the hatched area stands for $I(p, p')$, the effective two-particle interaction or the effective two-particle vertex function.

consider $2K_2(p, p') + K_4(p, p')$ which can be written as

$$2K_2(p, p') + K_4(p, p')$$
$$= -2i \int \frac{d^4q}{(2\pi)^4} [D^2(p - q) - D(p - p')D(p - q)]G(q)G(p' - p + q).$$

$$(1.44)$$

Equation (1.44) represents a very small term in $\tilde{I}(p, p')$ because the expression in the bracket of Eq. (1.44) is almost zero. Thus, $|2K_2(p, p') + K_4(p, p')| \ll |K_1(p, p')|$. Here $2K_2(p, p')$, the contribution to $\tilde{I}(p, p')$ from the particle-hole channel, practically eliminates $K_4(p, p')$, the correction to the interaction. As a result, for the quantity $\tilde{I}(p, p')$ it suffices to have the simple first-order vertex, i.e. $\tilde{I}(p, p') \approx K_1(p, p') = D(p - p')$, namely, the vertex correction does not affect the electron-boson interaction. It should be mentioned that this conclusion will not be changed even when the characteristic boson cutoff frequency is comparable to the Fermi energy, $\omega_c \approx \varepsilon_F$. In contrast, for the model of a given interaction, the vertex correction essentially affects $\tilde{I}(p, p')$ when $\omega_c \approx \varepsilon_F$.

Next, let us consider $I_c(p, p')$, the Cooper part of interactions (Fig. 1.9(b)). In the lowest order, $I_c(p, p') = 2K_3(p, p') + K_5(p, p')$, which can be approximated as

$$2K_3(p, p') + K_5(p, p')$$

$$= -i \frac{d^4q}{(2\pi)^4} [2D^2(p - q) - D(p - q)D(q - p')]G(q)G(p' + p - q)$$

$$\approx K_3(p, p') \approx -K_5(p, p'). \tag{1.45}$$

Thus, in the lowest order $I_c(p, p') \approx -K_5(p, p')$, the Cooper part of interactions approximately coincides with the conventional Cooper diagram but with an opposite sign. Such a conclusion will remain valid in the higher orders, as can be proven. Furthermore, the irreducible kernel of interaction $\tilde{I}(p, p')$ can be found from the equation $\tilde{I}(p, p') = I(p, p') - I_c(p, p')$, where $I(p, p')$ is the effective two-particle vertex part. Thus, carrying out the ladder summation over all the diagrams shown in Figs. 1.9(a) and 1.9(b), we can get the equation for the irreducible kernel of interactions $\tilde{I}(p, p') \approx D(p - p')$. It satisfies in this case the following equation

$$D(p-p') = I(p, p') + i \int \frac{d^4q}{(2\pi)^4} D(p-q)G(q)G(p'+p-q)I(q, p'). \tag{1.46}$$

The analytic derivation of Eq. (1.46) is given in Appendix B. Thus, Eq. (1.46), in fact, corresponds to the inverted Eq. (1.14), in which $I(p, p') \leftrightarrows \tilde{I}(p, p')$, namely, the vertices should be interchanged. This means that the particle-hole excitations lead to the inversion of Eq. (1.14) for the effective two-particle vertex, Fig. 1.9(c). One can see that in Eq. (1.46) $I(p, p') = I(Q)$ is a function of $Q = p' + p$ only as before. Using this fact and carrying out the frequency summation in Eq. (1.46), we can obtain a solution given by

$$I(\boldsymbol{Q}, \omega_o) \approx \frac{\lambda}{1 + \lambda \displaystyle\sum_{q, |\xi_q| < \omega_c} \frac{1 - n_F(\xi_a) - n_F(\xi_{o-a})}{\omega_o - \xi_q - \xi_{o-a} + i\delta}}, \tag{1.47}$$

where ω_c is the characteristic boson cutoff frequency, $\lambda = \langle D(\boldsymbol{p} - \boldsymbol{p}', 0) \rangle$ (note here that $\lambda = -\langle D_0(\boldsymbol{p} - \boldsymbol{p}', 0) \rangle$ in the model of a given interaction) was introduced as usual with $\langle \cdots \rangle$ being the average over the angle between $\boldsymbol{p}$ and $\boldsymbol{p}'$,

$$\langle D \rangle N_F = \frac{1}{(2\pi)^3} \int \frac{d^2p}{v_F} \frac{d^2p'}{v_F'} D(\boldsymbol{p} - \boldsymbol{p}', 0) \int \frac{d^2p}{v_F},$$

and the use of the analytic continuation $(i\omega_Q \rightarrow \omega_Q + i\delta), \boldsymbol{Q} = \boldsymbol{p} + \boldsymbol{p}'$ and $\omega_Q = \omega' + \omega$ was made as before. Equation (1.47) leads to the result concerning the pairing instability caused by the particle-hole excitations if the interaction is repulsive, i.e., $\lambda > 0$ in Eq. (1.47). There is no pairing instability due to the particle-hole excitations in the case of an attractive interaction ($\lambda < 0$) as seen from Eqs. (1.47). We can examine Eq. (1.47) in a way similar to that used for Eq. (1.20). As an example, from Eq. (1.47) we have

$$I(\omega_Q) - \lambda \Big/ \left\{ 1 - \lambda N_F \left[\ln \left(\frac{2\omega_c}{\omega_Q} \right) + i\frac{\pi}{2} \right] \right\}$$

at $\boldsymbol{Q} = \boldsymbol{0}$ and $T = 0$, assuming that ω_Q is real and positive. Thus, the quantity $I(\omega_Q)$ has a pole in the case of a repulsive interaction ($\lambda > 0$) at the point $\omega_Q = i\Omega$, where $\Omega = 2\omega_c \exp(-1/\lambda N_F)$. In analogy to Eq. (1.20), the pole of $I(\boldsymbol{Q}, \omega_Q)$ as a function of $|\boldsymbol{Q}|$ exists for $\omega_Q = i\Omega(1 - v_F^2 |\boldsymbol{Q}|^2 / 6\Omega^2)$ when $\boldsymbol{Q}$ is non-zero ($\boldsymbol{Q} \neq 0$ and $v_F|\boldsymbol{Q}| \ll |\omega_Q|$). The absolute value of ω_Q hence decreases as $|\boldsymbol{Q}|$ increases. For the case of $T \neq 0$, Eq. (1.47) allows us to evaluate T_c, the temperature at which a pole first appears in $I(\boldsymbol{Q}, \omega_Q)$ at $|\boldsymbol{Q}| = \omega_Q = 0$. It follows from Eq. (1.47) that $I^{-1}(0, 0) = 0$ at $T = T_c = 1.13\omega_c \exp(-1/\lambda N_F)$ in the case of $\omega_c \ll \varepsilon_F$. Equation (1.47) gives rise to the maximum value of T_c as $T_c = 1.13\varepsilon_F \exp(-1/\lambda N_F)$ in the case of $\omega_c \approx \varepsilon_F$. Thus, the particle-hole excitations provide much higher transition temperature than a given attractive interaction because in the case of the particle-hole excitations there is no effect of the vertex correction on the electron-boson interaction constant λ as mentioned above after Eq. (1.31). In contrast, for the model of a given attractive interaction in the case of $\omega_c \approx \varepsilon_F$, the vertex correction significantly affects the electron-boson interaction constant as described above.

Now we can evaluate $\Gamma_{c,s}(p, k)$, the vertex functions, and $\tilde{\chi}_{c,s}(k)$, the irreducible response functions in both charge and spin channels. It is worthwhile to point out that in the case that the characteristic boson cutoff frequency is much less than the Fermi energy, $\omega_c \ll \varepsilon_F$, the effect of the vertex corrections on the self-energy and irreducible response is negligible. This is analogous to the case of a given interaction. Therefore, we only consider the case that the characteristic boson cutoff frequency is comparable to the Fermi energy, $\omega_c \approx \varepsilon_F$. First, we will focus on the charge channel. As done for Eq. (1.32), substituting Eq. (1.47) into Eq. (1.29) and then substituting Eq. (1.29) into Eq. (1.43), we obtain

$$\widetilde{\chi}_c(\mathbf{k},\omega) \approx \frac{\chi_0(\mathbf{k},\omega)}{1 + \dfrac{\lambda}{2}\chi_0(\mathbf{k},\omega)/\left[1 + \dfrac{\lambda}{2}B(\omega + i\delta)\right]}, \qquad (1.48)$$

where $B(\omega)$ is given by Eq. (1.31). Equation (1.51) is valid in the region of $v_F|\mathbf{k}| \ll \omega_c \approx \varepsilon_F$ and $v_F|\mathbf{k}| \gg \omega$.

In analogy to the treatment of Eq. (1.33), we can separate the real and imaginary parts of Eq. (1.48). The irreducible response in the charge channel can be expressed by

$$\widetilde{\chi}_c(\mathbf{k},\omega)|_{\omega < v_F k < \varepsilon_F} = \mathrm{Re}\widetilde{\chi}_c(\mathbf{k},\omega)|_{\omega < v_F k < \varepsilon_F} + i\mathrm{Im}\widetilde{\chi}_c(\mathbf{k},\omega)|_{\omega < v_F k < \varepsilon_F}, \qquad (1.49)$$

where

$$\mathrm{Re}\widetilde{\chi}_c(\mathbf{k},\omega)|_{\omega < v_F k < \varepsilon_F} \approx -\frac{2N_F}{1 - \lambda N_F/A_c(\omega)} \qquad (1.50a)$$

and

$$\mathrm{Im}\widetilde{\chi}_c(\mathbf{k},\omega)|_{\omega < v_F k < \varepsilon_F} \approx -\frac{\pi N_F}{[1 - \lambda N_F/A_c(\omega)]^2}\frac{\omega}{v_F k} + \widetilde{\lambda}_c^2 N_F^2 \tanh\frac{\omega}{2T}, \qquad (1.50b)$$

with $A_c(\omega) = 1 + (\lambda/2)\mathrm{Re}B(\omega) = 1 - \lambda N_F \ln(\varepsilon_F/\max\{\omega,T\})$ and $\widetilde{\lambda}_c = \lambda/A_c(\omega)$ is the effective electron-boson interaction constant in the charge channel for the model of the repulsive ($\lambda > 0$) interaction. Equations (1.50a) and (1.50b) are quite similar to and yet different from those equations for the model of a given interaction (see Eqs. (1.34a) and (1.34b)). However, in the static ($\omega = 0$) and long wavelength limit, Eq. (1.50) turns to the form of the result of Landau's Fermi liquid theory [86, 87], $\widetilde{\chi}(0,0) = -2N_F/(1+F_0^s)$, with the Landau's Fermi liquid parameter in the charge channel at the zero frequency $F_0^s = -\lambda N_F/A_c(0) - \lambda N_F/[1 - \lambda N_F \ln(\varepsilon_F/T)]$. Thus, $F_0^s = -\lambda N_F/A_c(\omega)$ is the frequency-dependent Landau's Fermi liquid parameter in the charge channel.

In addition, as we can see from Eq. (1.51), the multiple scattering (the ladder of both the Cooper-like diagram due to the particle-hole excitations and the conventional Cooper diagram) produces the strong frequency dependence of the response function. The irreducible response now contains the anomalous Fermi liquid term $\text{Im}\widetilde{\chi} \sim -\omega/T$ as well as the normal Fermi liquid response $\text{Im}\widetilde{\chi} \sim -\omega/v_F k$. The contribution of the anomalous Fermi liquid term $\text{Im}\widetilde{\chi} \sim -\omega/T$ to the irreducible response strongly depends on the value of the coupling constant $\lambda_c = \lambda/A_c(\omega)$ which is also frequency- and temperature-dependent. The scaling of ω/T in the anomalous term is in agreement with the marginal Fermi-liquid hypothesis proposed by Varna [49, 97, 98] *et al.* to explain the universal anomalies in the normal state of the cuprate high-T_c superconductors. Moreover, this anomalous Fermi liquid term appears to be negative in contrast to the model of a given interaction (see Eq. (1.34)). Thus, it is understood that the marginal Fermi-liquid behavior is related to the particle-hole excitations. It should be also noted that the anomalous Fermi liquid term does not depend on the sign of interaction because $\text{Im}\widetilde{\chi}_{\text{an}} \sim -\widetilde{\lambda}_c^2\omega/T$. Equations (1.50) and (1.51) for the irreducible response function correspond to the model of a repulsive interaction ($\lambda > 0$) and are still valid in the case of an attractive interaction ($\lambda < 0$). In the case of the attractive interaction ($\lambda < 0$) there is, however, no pairing instability due to the particle-hole excitations according to Eq. (1.47).

Now we can evaluate the irreducible response in the spin channel. In the spin channel the kernel of interactions up to the second order is given by Eq. (1.41b). This equation, in fact, does not have the contribution from the particle-hole excitations and is thus equivalent to the model of a given interaction. Therefore, to get the equation for the irreducible response in the spin channel, we can use Eq. (1.32) in which $\widetilde{\lambda}$ should be replaced by $-\widetilde{\lambda}_s$, where $\widetilde{\lambda}_s = \lambda(1 + 2\lambda N_F)$ is the effective interaction constant in the spin channel for the case of the repulsive interaction with $\lambda = <D(\boldsymbol{p}-\boldsymbol{p}', 0) \gg 0$. Thus, the irreducible response in the spin channel in this case takes the form

$$\widetilde{\chi}_s(\boldsymbol{k}, \omega) \approx \frac{\chi_0(\boldsymbol{k}, \omega)}{1 + \dfrac{\lambda_s}{2}\chi_0(\boldsymbol{k}, \omega)/\left[1 - \dfrac{\widetilde{\chi}_s}{2}\text{B}(\omega + i\delta)\right]}, \qquad (1.51)$$

where $\widetilde{\lambda}_s = \lambda(1 + 2\lambda N_F)$ is the effective interaction constant.

Further, Eq. (1.51) can also be turned into a form similar to Eq. (1.49). Thus, we have

$$\widetilde{\chi}_s(\boldsymbol{k}, \omega)|_{\omega < v_F k < \varepsilon_F} = \text{Re}\widetilde{\chi}_s(\boldsymbol{k}, \omega)|_{\omega < v_F k < \varepsilon_F} + i\text{Im}\widetilde{\chi}_s(\boldsymbol{k}, \omega)|_{\omega < v_F k < \varepsilon_F},$$

$$(1.52)$$

where

$$\mathrm{Re}\widetilde{\chi}_s(\boldsymbol{k},\omega)|_{\omega < v_F k < \varepsilon_F} \approx -\frac{2N_F}{1 - \widetilde{\lambda}_s N_F / A_s(\omega)} \qquad (1.53a)$$

and

$$\mathrm{Im}\widetilde{\chi}_s(\boldsymbol{k},\omega)|_{\omega < v_F k < \varepsilon_F} \approx -\frac{\pi N_F}{[1 - \widetilde{\lambda}_s N_F / A_s(\omega)]^2}\frac{\omega}{v_F k} - \widetilde{\lambda}_s^2 N_F^2 \tanh\frac{\omega}{2T},$$

$$(1.53b)$$

with $A_s(\omega) = 1 - (\widetilde{\lambda}_s/2)\mathrm{Re}B(\omega) = 1 + \widetilde{\lambda}_s N_F \ln[\varepsilon_F/\max(\omega,T)]$ and $\widetilde{\lambda}_s = \lambda_s/A_s(\omega)$. In the static ($\omega = 0$) and long wavelength limit, Eq. (1.41) turns to the form of the result of Landau's Fermi liquid theory [86, 87], $\widetilde{\chi}_s(0,0) = -2N_F/(1+F_0^a)$ with $F_0^a = -\widetilde{\lambda}_s N_F/A_s(0) - \widetilde{\lambda}_s N_F/[1+\widetilde{\lambda}_s N_F \ln(\varepsilon_F/T)]$ being Landau's Fermi liquid parameter in the spin channel at the zero frequency. Thus, $F_0^a(\omega) = -\widetilde{\lambda}_s N_F/A_s(\omega)$ is the frequency-dependent Landau's Fermi liquid parameter in the spin channel. As can be seen, Eqs. (1.52) and (1.53) are analogous to Eqs. (1.33) and (1.34). Therefore, all the conclusions about the behavior of the irreducible response in the spin channel are analogous to the behavior of the irreducible response in the case of a given interaction.

1.6.3.3 For a renormalized interaction in an anisotropic L2D electron gas

Now, as an example, we can make some estimates of the transition temperature T_c and other relevant properties due to the particle-hole excitations for a model system of a cuprate superconductor — a layered 2D (L2D) electron gas. In the model of an L2D electron gas, it is assumed that an electron has an isotropic 2D spectrum $\varepsilon_k = \boldsymbol{k}^2/2m^*$ with m^* being the effective mass. Thus, we consider a layered metal with the cylindrical topology of the Fermi surface. In this case the free-particle polarizability $\chi_0(\boldsymbol{k},\omega) \approx \mathrm{Re}\chi_0(\boldsymbol{k},\omega) = -2N_F[1 - \mathrm{Re}\sqrt{1 - 4p_F^2/k^2}] = \chi_0(\boldsymbol{k},0)$ for $\omega < v_F k$, where $N_F = m^*/2\pi$ is the 2D density of states per spin, while $\mathrm{Im}\chi_0(\boldsymbol{k},\omega) \approx -2N_F\omega/v_F k$. Therefore, for the non-retarded interaction like the pure Coulomb interaction, in which $D_0(\boldsymbol{k}) = V(\boldsymbol{k}), D(\boldsymbol{k},\omega) = V(\boldsymbol{k})/[1 - V(\boldsymbol{k})\chi_0(\boldsymbol{k},\omega)] \approx D(\boldsymbol{k},0)$ is almost frequency-independent for $\omega < v_F k$ because of a weak frequency-dependent part ($|\mathrm{Re}\chi_0(\boldsymbol{k},\omega)| \gg |\mathrm{Im}\chi_0(\boldsymbol{k},\omega)|$ for $\omega \ll v_F k$).

It deserves a mention that the bare Coulomb interaction $V(\boldsymbol{k})$ here in an L2D system [45–47, 113, 115] is given by Eq. (1.9c). We can rewrite Eq. (1.9c) as $V(k) = V_0(k_\parallel)f(\boldsymbol{k})$, in which $V_0(k_\parallel) = 2\pi e^2/\kappa k_\parallel$ is the bare

Coulomb interaction in a pure 2D case with κ being the dielectric constant of the lattice background and with $k_\parallel$ being the momentum components in the plane, and $f(\boldsymbol{k})$ is the form factor of an L2D electron gas with c being an interlayer spacing and k_z being the momentum component in the direction normal to the plane.

We evaluate the average of $D(\boldsymbol{p} - \boldsymbol{p}')$ in an L2D system as

$$\langle D(\boldsymbol{p} - \boldsymbol{p}')\rangle = \frac{c}{(2\pi)^2} \int_{-\pi/c}^{\pi/c} dp'z \int_0^{2\pi} d\varphi D(\boldsymbol{p} - \boldsymbol{p}')_{|\boldsymbol{p}|-|\boldsymbol{p}'|-p_F} = \frac{2e^2}{\kappa p_F} F(\alpha, \zeta), \tag{1.54}$$

where

$$F(\alpha, \zeta) = \int_0^1 dx[(1 - x^2)(x^2 + 2\alpha x \coth \zeta x + \alpha^2)]^{-1/2} \tag{1.55}$$

and $k_\parallel^2 = |\boldsymbol{p}_\parallel - \boldsymbol{p}'_\parallel|^2 = 2p_F^2(1 - \cos\varphi)$ was used. In the above equations $\alpha = e^2/\kappa v_F$ is the inter-electronic interaction constant with $v_F = p_F/m^*$ and is related to the dimensionless density parameter r_s by the relation $r_s = \sqrt{2}\alpha$ and $\zeta = 2p_F c$ with the 2D Fermi momentum $p_F = \sqrt{k\pi n_2}$ (n_s the 2D electron density) and the interlayer spacing c. From Eq. (1.55), $F(\alpha, \zeta)$ can be analytically evaluated in the limiting cases for an L2D electron gas. For example, in the case of the high density limit (or weak inter-electronic coupling), $\alpha \ll 1$, and large interlayer spacing, $\zeta = 2p_F c \gg 1, F(\alpha, \zeta) \approx \ln(2/\alpha)$. In the low density limit, $\alpha \gg 1$, and $\zeta = 2p_F c \gg 1, F(\alpha, \zeta) \approx \pi/(2\alpha)$. The value of $F(\alpha, \zeta)$ weakly depends on the interlayer spacing c when $c \gg a_B^*$ (where $a_B^* == \kappa/e^2 m^*$ is the effective Bohr radius). It decreases with the decrease of c when $c < a_B^*$ as $F(\alpha, \zeta) \approx (\pi/2\alpha)/(1 + a_B^*/c)^{1/2}$.

Using Eq. (1.54) in Eq. (1.47) with $\lambda = \langle D(\boldsymbol{p} - \boldsymbol{p}')\rangle$, we can derive the equation for T_c as

$$T_c \approx 1.13\varepsilon_F \exp[-\pi/(\alpha F(\alpha, \zeta))]. \tag{1.56}$$

Equation (1.56) with Eq. (1.55) allows us to obtain T_c in the limiting cases. In the case of the high density limit (or weak coupling), $\alpha \ll 1$, and a large interlayer spacing, $\zeta = 2p_F c \gg 1, T_c \sim \varepsilon_F \exp(-\pi/\alpha)$ is very low. In the low density limit or strong inter-electronic coupling ($\alpha > 1$), however, $T_c \approx 1.13\varepsilon_F \exp(-2) \approx 0.153\,\varepsilon_F \approx 0.48\,n_s/m^*$. T_c weakly depends on the interlayer spacing when $c > a_B^*$ and decays with the decrease of c when $c < a_B^*$. As a result, Eq. (1.56) tends to $T_c \sim \varepsilon_F \exp(-2\sqrt{a_B^*/c})$ in this limit. These results seem to be qualitatively in line with the experimental observations.

It is of interest to note that in the case of the high density ($\alpha \ll 1$), an estimate of T_c due to the particle-hole excitations for an isotropic 3D metal is $T_c \sim \varepsilon_F \exp(-2\pi/\alpha)$. T_c due to the particle-hole excitations for an isotropic 3D metal is much lower than that for the 2D and layered 2D cases. This implies that the particle-hole excitations do not lead to a high transition temperature in the case of high density and isotropic 3D. Nevertheless, the particle-hole excitations may also be important in understanding other phenomena of the conventional superconductivity.

It should be pointed out here that our estimate of T_c based on Eq. (1.47) is actually valid only in the case of the weak and possibly intermediate coupling because this approach in the consideration of the superconducting transition (pairing instability) mediated by the particle-hole excitations omits other contributions from the higher-order corrections in the particle-hole channel. Such contributions are not important in the case of the weak coupling and just lead to a minor change in the T_c estimate.

1.6.3.4 *Vertex corrections and transition temperature in a strongly coupled electron gas*

To see how the results will be changed in the strong coupling, we consider Eq. (1.10) for the single-particle self-energy containing the vertex part $\Gamma(p, k)$. It is understood that Eq. (1.10) is general and valid for both the strong coupling and the weak coupling limits. Let us assume that we have already evaluated the vertex part. Now, we can formally evaluate $K^{\sigma\sigma'}(p, p')$ by using Eq. (1.10) and Ward's identity in terms of the diagrammatic derivative method. The expression for the kernel of interaction can now be written as

$$K^{\sigma\sigma'}(p, p') = K_1^{\sigma\sigma'}(p, p') + K_2^{\sigma\sigma'}(p, p') + K_3^{\sigma\sigma'}(p, p')$$
$$+ K_4^{\sigma\sigma'}(p, p') + K_5^{\sigma\sigma'}(p, p'), \qquad (1.57)$$

where

$$K_1^{\sigma\sigma'}(p, p') = D(p - p')\Gamma(p, p - p')\delta_{\sigma\sigma'}, \qquad (1.58a)$$

$$K_2^{\sigma\sigma'}(p, p') = -i \int \frac{d^4q}{(2\pi)^4} D^2(p - q)$$
$$\Gamma(p, p - q)\Gamma(p', p' - q)G_\sigma(q)G_\sigma(q)G_{\sigma'}(p' - p + q), \qquad (1.58b)$$

and

$$K_3^{\sigma\sigma'}(p,p') = -i \int \frac{d^4q}{(2\pi)^4} D^2(p-q)\Gamma(p,p-q)$$

$$\times \Gamma(p'+p-q,p-q)G_\sigma(q)G_{\sigma'}(p'+p-q). \quad (1.58c)$$

Note here that we adopt the italic notations to distinguish this general case from the one described in Eq. (1.39). The diagrams for these interactions are shown in Fig. 1.10. The equations for kernels $K_4^{\sigma\sigma'}(p,p')$ and $K_5^{\sigma\sigma'}(p,p')$ will be discussed later. One can see that Eqs. (1.58a), (1.58b) and (1.58c)

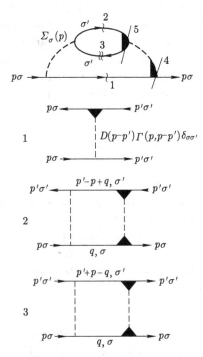

Fig. 1.10   Diagrammatic functional derivatives of the diagram of the self-energy with respect to the electron Green's function in the model of a renormalized interaction due to the particle-hole excitations in the case of strong coupling; sub-diagrams 1, 2, and 3 come from the procedures of cutting the electron lines at 1, 2 and 3 in $\Sigma_\sigma(p)$. Diagrams: 1 — the first-order screened exchange interaction corrected by the vertex ($K_1^{\sigma\sigma'}(p,p')$); 2 — the second-order screened exchange interaction corrected by the vertex ($K_2^{\sigma\sigma'}(p,p')$); 3 — the Cooper-like diagram ($K_3^{\sigma\sigma'}(p,p')$). The procedures of cutting the vertex $\Gamma(p,k)$ at 4 and 5 leads to $K_4^{\sigma\sigma'}(p,p')$ and $K_5^{\sigma\sigma'}(p,p')$, respectively (see Eqs. (1.59a) and (1.59b)). The thick dashed line represents the boson Green's function beyond the RPA.

will have the same forms as Eqs. (1.37a), (1.37b) and (1.37c), respectively, only if $D(k)\Gamma(p,k)$ is replaced by $\widetilde{D}(p,k)$, the corrected boson propagator beyond the RPA. Also, such a correction of $D(k)$ by $\Gamma(p,k)$ can be done by using either the Hubbard-like correction to $D(k)$ in the RPA [49, 114] or the self-consistent approach. In addition, with the replacement of $D(k)$ in Eq. (1.36) by $\widetilde{D}(p,k) \approx \widetilde{D}(p,k)_{|\mathbf{p}|\approx p_F} = \widetilde{D}(k)$, Eqs. (1.37a), (1.37b) and (1.37c) will include much larger contributions than the original kernel given by Eq. (1.36).

Now we consider two other contributions in Eq. (1.57)—the kernels $K_4^{\sigma\sigma'}(p,p')$ and $K_5^{\sigma\sigma'}(p,p')$. They come from the derivatives (diagrammatic cutting) of the vertex part $\Gamma(p,k)$ and the irreducible response $\widetilde{\chi}(k)$, respectively, and can be written as

$$K_4^{\sigma\sigma'}(p,p') = \int \frac{d^4k}{(2\pi)^4} G_\sigma(p-k)D(k)\frac{\delta\Gamma(p,k)}{\delta G_{\sigma'}(p')},\qquad (1.59a)$$

and

$$K_5^{\sigma\sigma'}(p,p') = -i\int \frac{d^4k}{(2\pi)^4} G_\sigma(p-k)D^2(k)\Gamma(p,k)$$
$$\cdot \left[\sum_{\sigma''}\int \frac{d^4q}{(2\pi)^4} G_{\sigma''}(q)G_{\sigma''}(q-k)\frac{\delta\Gamma(q,k)}{\delta G_{\sigma'}(p')}\right]. \qquad (1.59b)$$

Thus, the contributions of $K_4^{\sigma\sigma'}(p,p')$ and $K_5^{\sigma\sigma'}(p,p')$ to the effective kernel $K^{\sigma\sigma'}(p,p')$ given by Eq. (1.57) are indeed of higher order than $K_1^{\sigma\sigma'}(p,p'), K_2^{\sigma\sigma'}(p,p')$ and $K_3^{\sigma\sigma'}(p,p')$ because the contributions of both $K_4^{\sigma\sigma'}(p,p')$ and $K_5^{\sigma\sigma'}(p,p')$ are proportional to $\sim \delta\Gamma/\delta G_{\sigma'}(p')$. Using Eq. (1.12) for $\Gamma(p,k)$, with the kernel $K_1^{\sigma\sigma'}(p,p')$ given by Eq. (1.37a), in Eq. (1.59a), we can show that kernels $K_4^{\sigma\sigma'}(p,p')$ and $K_5^{\sigma\sigma'}(p,p')$ in Eq. (1.39) follow from Eq. (1.59a). Thus, the well-known Cooper diagram appears as a result of Eq. (1.59a). Now, we consider Eq. (1.59b) which appears in Eq. (1.57) due to the functional derivative of $\widetilde{\chi}(k)$, the irreducible response, with respect to the Green function, i.e., $K_5^{\sigma\sigma'}(p,p') \sim \delta\widetilde{\chi}(k)/\delta G_{\sigma'}(p')$. Such a functional derivative $(\delta\widetilde{\chi}(k)/\delta G_{\sigma'}(p'))$ appears because the renormalized boson propagator satisfies the Dyson equation $D(k) = D_0(k) + D_0(k)\widetilde{\chi}D(k)$. Thus, the contribution of $K_5^{\sigma\sigma'}(p,p')$ to Eq. (1.57) comes from the vertex part in the particle-hole bubble. This contribution has not been accounted for in our previous treatment of the effect of the particle-hole excitations on the pairing instability. In other words, Eq. (1.46) and, hence, Eq. (1.47) does not include the additional effect of the particle-hole excitations related to $\delta\widetilde{\chi}(k)/\delta G_{\sigma'}(p')$. Therefore, to estimate the effect of the term $K_5^{\sigma\sigma'}(p,p')$ on the effective kernel of interaction $K^{\sigma\sigma'}(p,p')$, let us consider the operator

$$\widehat{\zeta}(p,k) = iD(k)\Gamma(p,k)\int \frac{d^4q}{(2\pi)^4}G(q)G(q-k)$$

$$= -\frac{1}{2}D(k)\Gamma(p,k)\chi_0(k) \approx \widetilde{\zeta} \approx \langle \widetilde{D}(k)\rangle N_F. \tag{1.60}$$

As seen from Eq. (1.60), in the weak coupling limit, $\lambda N_F = \langle D(p - p')\rangle N_F \ll 1$, and $\widetilde{\zeta}(p,k) - \widetilde{\zeta} \approx \lambda N_F \ll 1(\Gamma(p,k) \approx 1)$. whereas in the strong coupling, $\lambda N_F \approx 1/2$, we have $\widetilde{\zeta} \approx \lambda N_F = \langle \widetilde{D}\rangle N_F = \langle D(p-p')\rangle\Gamma(p,p-p')N_F \approx 1/2$. In addition, $\widetilde{\zeta} \approx 1/2$ is valid even in the case of the intermediate coupling because $\widetilde{\zeta}$ includes $\Gamma(p,k) > 1$ which, in this case, cannot be ignored. Using Eq. (1.60) in Eq. (1.59b) and assuming $q \approx p \sim p_F$ in $\Gamma(q,k)$, we can roughly estimate $K_5^{\sigma\sigma'}(p,p')$ as

$$K_5^{\sigma\sigma'}(p,p') = -2\int \frac{d^4k}{(2\pi)^4}G_\sigma(p-k)D(k)\widetilde{\zeta}(p,k)\frac{\delta\Gamma(q,k)}{\delta G_{\sigma'}(p')} \approx -2\widetilde{\zeta}K_4^{\sigma\sigma'}(p,p'), \tag{1.61}$$

where the factor 2 appears due to the summation over spins σ'' in Eq. (1.59b). It follows from Eq. (1.49) that the contribution of kernels $K_4^{\sigma\sigma'}(p,p')$ and $K_5^{\sigma\sigma'}(p,p')$ to the effective kernel of interaction $K^{\sigma\sigma'}(p,p')$ given by Eq. (1.57) can be estimated as $K_4^{\sigma\sigma'}(p,p')+K_5^{\sigma\sigma'}(p,p') \approx (1-2\widetilde{\zeta})K_4^{\sigma\sigma'}(p,p')$. In the case of weak coupling ($\widetilde{\zeta} \approx \lambda N_F \ll 1$), $|K_5^{\sigma\sigma'}(p,p')| \ll |K_4^{\sigma\sigma'}(p,p')|$. Therefore, our treatment of the Cooper instability leads to Eqs. (1.58a)–(1.58c) and (1.59a)–(1.59b). In the case of the strong coupling and repulsive interaction, i.e., when $\lambda =< D(p-p') \gg 0$, we have $2\widetilde{\zeta} \approx 1$, and the contribution of $K_4^{\sigma\sigma'}(p,p')+K_5^{\sigma\sigma'}(p,p') \approx (1-2\widetilde{\zeta})K_4^{\sigma\sigma'}(p,p') \sim 0$ to the effective kernel of interaction is small. Such a conclusion is valid even in the case of the intermediate coupling because $\widetilde{\zeta}\widetilde{\sigma}\widetilde{\lambda} \approx \langle \widetilde{D}\rangle \approx \langle D(p-p')\rangle\Gamma(p,p-p')\rangle$. Γ in this case is not negligible. Thus, the contribution of the particle-hole excitations coming from the renormalized particle-hole bubble practically eliminates the higher-order contributions in the strong or intermediate coupling. It is understood that such a compensation of the higher-order contributions by the particle-hole bubble may not exist in the case of an attractive ($\lambda < 0$) original interaction because $\widetilde{\zeta} < 0$ in such a case. Therefore, in the case of the strong or intermediate coupling we can use Eq. (1.36) with Eqs. (1.37a), (1.37b) and (1.37c) where $D(k)$, the boson propagator in the RPA, should be replaced by

$\tilde{D}(p, k) = D(k)\Gamma(p, k)|_{|\boldsymbol{p}|-p_F} \approx \tilde{D}(k)$. Using Eqs. (1.13a) and (1.13b), the kernels of interaction in the charge and spin channels, respectively, can be represented by

$$K_c(p, p') = \tilde{D}(p-p') - 2i \int \frac{d^4q}{(2\pi)^4} \tilde{D}^2(p-q)G(q)[G(p'-p+q)+G(p'+p-q)],$$

(1.62a)

and

$$K_s(p, p') = D(p - p')\Gamma_c(p, p - p'),$$ (1.62b)

where $\tilde{D}(k) \approx D(k)\Gamma(p, k)|_{|\boldsymbol{p}|-p_F}$ was assumed, and $\Gamma_c(p, k)$ is the three-point vertex function in the charge channel.

Moreover, to further simplify our consideration it is possible to assume that in Eq. (1.62a), $\tilde{D}(p, k) = D(k)\Gamma(p, k)|_{|\boldsymbol{p}|-p_F} \approx D(k)$, where $D(k)$ is the boson propagator in the framework of the self-consistent RPA, i.e., $D(k) = D_0(k) + D_0(k)\chi_0(k)D(k)$. Here $\chi_0(k)$ is the "free-particle" polarizability that includes the self-energy corrections. The self-energy $\Sigma(p)$ is calculated by using Eq. (1.10) incorporating the vertex correction $\Gamma(p, k)$. Such an approximation means that $\chi_0(k)$ and, hence, $D(k)$, includes the vertex corrections through the self-energy dependence of $\Gamma(p)$. In Eq. (1.62b) for $K_s(p, p')$ we keep $\Gamma_c(p, k)$, the vertex correction in the charge channel. The reason for this is as follows. In the charge channel the major contribution comes from the last term in Eq. (1.62a) that is related to the particle-hole excitations and has a logarithmic singularity. Such a logarithmic singularity in the last term in Eq. (1.62a) exists in any case no matter the kind of interaction: $D^2(k)$ or $D^2(k)\Gamma^2(p, k)$, in which the vertex correction will only affect the value of the effective interaction constant $\lambda^2 = \langle D^2\Gamma^2 \rangle \approx \langle D^2 \rangle$. However, such a correction to λ^2 can self-consistently be accounted for in the RPA as described before. In the spin channel in the case of the strong or intermediate coupling, the term, $D(p - p')\Gamma_c(p, p - p')$, survives only in $K_s(p, p')$ as shown before. In this case the singularity of $\Gamma_s(p, k)$ and, therefore, $\tilde{\chi}_s(k)$, the irreducible response in the spin channel, are related to the singularity of $\Gamma_c(p, k)$. However, such a singularity of $\Gamma_c(p, k)$ should directly be included in the evaluation of $\Gamma_s(p, k)$ because it is extremely difficult to separate the singularity of $\Gamma_c(p, k)$ in calculation of $\chi_0(k)$ and, hence, $D(k)$ including the self-energy corrections. That is the reason why it is necessary to keep $\Gamma_c(p, k)$ in $K_s(p, p')$.

Using Eqs. (1.62a) and (1.62b) in Eqs. (1.42) and (1.43), and performing the same procedures we employed in the case of a given interaction, we can get the expressions for $\Gamma_{c,s}(p,k)$ and $\widetilde{\chi}_{c,s}(k)$ in the charge and spin channels, respectively, as described in the paper of the authors published in a set of Proceedings [47]. In addition, it is also understood that in the framework of our approximations we do not need to write down equations for $I_{c,s}(p,p')$, the effective four-point vertex parts in the charge and spin channels, respectively, because $I_c(p,p') \approx K_c(p,p')$, and $I_s(p,p') \approx K_s(p,p')$, where $K_c(p,p')$ and $K_s(p,p')$ are given by Eqs. (1.62a) and (1.62b), respectively. Therefore, we write down the final equations for $\widetilde{\chi}_c(k)$ and $\widetilde{\chi}_x(k)$, respectively, as

$$\widetilde{\chi}_c(\boldsymbol{k},\omega) = \frac{\chi_0(\boldsymbol{k},\omega)}{1 + [A_c - B_c(\omega)]\chi_0(\boldsymbol{k},\omega)}, \qquad (1.63)$$

and

$$\widetilde{\chi}_s(\boldsymbol{k},\omega) = \frac{\chi_0(\boldsymbol{k},\omega)}{1 + [A_s - B_s(\omega)]\chi_0(\boldsymbol{k},\omega)}, \qquad (1.64)$$

where

$$A_c = \frac{1}{2}[\langle D(\boldsymbol{p}-\boldsymbol{q})\rangle - \langle D^2(\boldsymbol{p}-\boldsymbol{q})\rangle N_F], \qquad (1.65a)$$

$$B_c(\omega) = \frac{1}{2}\langle D^2(\boldsymbol{p}-\boldsymbol{q})\rangle \sum_{q,|\xi_q|<\varepsilon_F} \frac{1 - 2n_F(\xi_q)}{\omega - \xi_q + i\delta}, \qquad (1.65b)$$

and

$$A_s = \frac{1}{2}[\langle D_c(\boldsymbol{p}-\boldsymbol{p}',0)\rangle + \langle \widetilde{D}_c(\boldsymbol{p}-\boldsymbol{p}',0)\rangle N_F A_c], \qquad (1.66a)$$

$$B_s(\omega) = \frac{1}{2}\langle \widetilde{D}_c(\boldsymbol{p}-\boldsymbol{p}',0)\rangle\langle D^2(\boldsymbol{p}-\boldsymbol{q})\rangle N_F \sum_{q,|\xi_q|<\varepsilon_F} \frac{1 - 2n_F(\xi_q)}{\omega - 2\xi_q + i\delta}, \qquad (1.66b)$$

where N_F is the quasiparticle density of states including the summation over spins. In the case that the original interaction is of Coulomb type, $D(k) = D(\boldsymbol{k},0) = V(\boldsymbol{k})/[1+V(\boldsymbol{k})N_F]$ is the screened Coulomb in the framework of the self-consistent RPA and $D_c(\boldsymbol{k},\omega) = V(\boldsymbol{k})/[1 - V(\boldsymbol{k})\widetilde{\chi}_c(\boldsymbol{k},\omega)]$ is the renormalized screened Coulomb with $V(k)$ given by Eq. (1.9c) in an L2D system, where $\widetilde{\chi}_c(\boldsymbol{k},\omega)$ is given by Eq. (1.63). Further, in Eqs. (1.66a) and (1.66b), $\widetilde{D}_c(\boldsymbol{k}, \omega)$ is the renormalized screened Coulomb interaction corrected by the vertex. It can be written as

$$\widetilde{D}_c(\boldsymbol{k}, \omega) = \frac{V(k)}{1 - V(k)[1 - G_c(\boldsymbol{k}, \omega)]\chi_0(\boldsymbol{k}, \omega)}, \tag{1.67}$$

where $G_c(\boldsymbol{k}, \omega) = V^{-1}(k)[A_c - B_c(\omega)]$ is introduced as the frequency-dependent local field factor in the charge channel with A_c and $B_c(\omega)$ given by Eqs. (1.65a) and (1.65b).

Further, using Eqs. (1.65a), (1.65b), (1.66a), (1.66b) and (1.67), we can also represent Eqs. (1.63) and (1.64) in the form similar to Eq. (1.49) (or Eq. (1.52)). Thus, for the charge and spin response functions, respectively, we have

$$\widetilde{\chi}_c(\boldsymbol{k}, \omega)|_{\omega < v_F k < \varepsilon_F} = \mathrm{Re}\widetilde{\chi}_c(\boldsymbol{k}, \omega)|_{\omega < v_F k < \varepsilon_F} + i\mathrm{Im}\widetilde{\chi}_c(\boldsymbol{k}, \omega)|_{\omega < v_F k < \varepsilon_F}, \tag{1.68}$$

and

$$\widetilde{\chi}_s(\boldsymbol{k}, \omega)|_{\omega < v_F k < \varepsilon_F} = \mathrm{Re}\widetilde{\chi}_s(\boldsymbol{k}, \omega)|_{\omega < v_F k < \varepsilon_F} + i\mathrm{Im}\widetilde{\chi}_s(\boldsymbol{k}, \omega)|_{\omega < v_F k < \varepsilon_F}, \tag{1.69}$$

where

$$\mathrm{Re}\widetilde{\chi}_c(\boldsymbol{k}, \omega)|_{\omega < v_F k} \approx -\frac{N_F}{1 + F_0^s(\omega)}, \tag{1.70a}$$

and

$$\mathrm{Im}\widetilde{\chi}_c(\boldsymbol{k}, \omega)|_{\omega < v_F k} \approx -\frac{N_F}{1 + F_0^s(\omega)^2}\left[\frac{\omega}{v_F k} + \frac{\pi}{4}N_F^2\langle\widetilde{D}_c\rangle\langle D^2\rangle \tanh\frac{\omega}{2T}\right] \tag{1.70b}$$

for the irreducible charge density response function,

$$\mathrm{Re}\widetilde{\chi}_s(\boldsymbol{k}, \omega)|_{\omega < v_F k} \approx -\frac{N_F}{1 + F_0^a(\omega)}, \tag{1.71a}$$

and

$$\mathrm{Im}\widetilde{\chi}_s(\boldsymbol{k}, \omega)|_{\omega < v_F k} \approx -\frac{N_F}{1 + F_0^a(\omega)^2}\left[\frac{\omega}{v_F k} + \frac{\pi}{8}N_F^3\langle D^2\rangle \tanh\frac{\omega}{4T}\right] \tag{1.71b}$$

for the irreducible spin density response function. In Eqs. (1.70a) and
(1.70b), $F_0^s(\omega) = -N_F[A_c - \mathrm{Re}B_c(\omega)]$ is the Landau Fermi liquid parameter in the charge channel, where $\mathrm{Re}B_c(\omega) \approx -(1/2)N_F\langle D^2\rangle \ln(\varepsilon_F/$
$\max\{\omega, T\})$. In Eqs. (1.71a) and (1.71b), $F_0^a(\omega) = -N_F[A_s - \mathrm{Re}B_s(\omega)]$ is
the Landau Fermi liquid parameter in the spin channel, where $\mathrm{Re}B_s(\omega) \approx$
$-(1/4)N_F^2\langle\widetilde{D}_c\rangle\langle D^2\rangle \ln(\varepsilon_F/\max\{\omega/2, T\})$, In the static limit are reduced,
Eqs. (1.70a) and (1.71a) to the results of Landau's Fermi liquid theory
[86, 87] with $F_0^s(0)$ and $F_0^a(0)$ being the Landau Fermi liquid parameters in the charge and spin channels at the zero frequency. Thus, $F_0^s(\omega)$
and $F_0^a(\omega)$ are the frequency-dependent Landau's Fermi liquid parameters in the charge and spin channels, respectively. In addition, Eq. (1.70b)
for the irreducible charge density response function incorporates both
Varma's behavior [49], i.e., $\mathrm{Im}\chi \sim -\omega/T$, and the standard Fermi liquid result, $\mathrm{Im}\chi \sim -\omega/v_F k$. It is of interest to note that the irreducible
response in the spin channel, which is given by Eq. (1.71b), also con-
tains both Varma's behavior, $\mathrm{Im}\chi \sim -\omega/T$, and the standard Fermi liquid result, $\mathrm{Im}\chi \sim -\omega/v_F k$, while the irreducible response in the spin
channel in the case of the weak coupling given by Eq. (1.53b) pos-
sesses such an anomalous Fermi liquid with a positive sign in addition
to the normal Fermi liquid response. It is concluded from Eq. (1.71b)
that the contribution of the term $\sim -\omega/T$ to the response function in
the spin channel is much smaller than that in the charge channel because
$\mathrm{Im}\widetilde{\chi}_s \sim \lambda^3$, where $\lambda \sim N_F\langle D\rangle$, whereas in the charge density channel it is
$\mathrm{Im}\widetilde{\chi}_c \sim \lambda^2$.

Now, we can see that the irreducible charge density response function
given by Eq. (1.70a) has a singularity because $\Gamma_c(p, k)$ does. Substituting
Eq. (1.25) into Eq. (1.29) and replacing the effective two-particle vertex
function (effective two-particle interaction) $I(p, p')$ in Eq. (1.25) by $K_c(p, p')$
given by Eq. (1.62a), we can express $\Gamma_c(p, k)$ as

$$\Gamma_c(p_F, \omega; \boldsymbol{k}, \omega')_{v_F k \gg \omega', \omega}$$

$$\approx 1 + \frac{C}{1 - N_F[\langle D\rangle - 2\langle D^2\rangle N_F] + \langle D^2\rangle N_F \sum_{q, |\xi_q| < \varepsilon_F} \dfrac{1 - 2n_F(\xi_q)}{\omega' - \xi_q + i\delta}},$$
$$\tag{1.72}$$

where N_F is the density of states per spin, and $C \approx 1 + N_F\langle K_c(\boldsymbol{p}_F, 0; \boldsymbol{p}'_F, 0)\rangle$
is a constant with $K_c(p, p')$ given by Eq. (1.62a). It is obvious that Eq. (1.72)

has a pole in the upper half of the complex plane. As an example, we can examine Eq. (1.72) at $T = 0$. If we assume that ω in Eq. (1.56) is real and positive, we have

$$\Gamma_c(p_F\omega; 0, 0) \approx 1 + \frac{\widetilde{C}}{1 - 2\lambda_{\text{eff}}N_F \left[\ln\dfrac{\varepsilon_F}{\omega} + i\dfrac{\pi}{2}\right]}, \tag{1.73}$$

where $\widetilde{C} = C/\{1 - N_F[\langle D\rangle - 2\langle D^2\rangle N_F]\}$ is a constant, and

$$\lambda_{\text{eff}} = \frac{\langle D^2\rangle N_F}{1 - N_F[\langle D\rangle - 2\langle D^2\rangle N_F]} \tag{1.74}$$

is the effective coupling constant. As seen from Eq. (1.73), $\Gamma_c(\omega)$ has a pole at the point $\omega = i\Omega$, where $\Omega = \varepsilon_F \exp[-1/(2\lambda_{\text{eff}}N_F)]$. For the case $T \neq 0$, Eq. (1.72) allows us to evaluate the transition temperature T_c at which a pole first appears in $\Gamma_c(0)$. It follows from Eq. (1.72) that $\Gamma_c^{-1}(0) = 0$ at $T = T_c = 1.13\varepsilon_F \exp[-1/(2\lambda_{\text{eff}}N_F)]$. Equation (1.73) can be used to estimate T_c in an L2D electron gas. In analogy to Eq. (1.54), we can calculate $\langle D^2\rangle$ as

$$\langle D^2(\boldsymbol{p} - \boldsymbol{p}')\rangle N_F = \alpha\frac{e^2}{\kappa p_F}\widetilde{F}(\alpha, \zeta), \tag{1.75}$$

where

$$\widetilde{F}(\alpha, \zeta) = \int_0^1 \frac{dx}{\sqrt{1 - x^2}} \frac{x\coth\zeta x + \alpha}{(x^2 + 2\alpha x \coth\zeta x + \alpha^2)^{3/2}}. \tag{1.76}$$

$\alpha = e^2/\kappa v_F$ is the original inter-electronic interaction constant, $N_F = m^*/2\pi$ is the 2D density of states per spin with the quasiparticle mass m* that should be determined self-consistently from Eq. (1.10) and $\zeta = 2p_F c$ with the interlayer distance c.

It is worthwhile to show the limiting behaviors of $\widetilde{F}(\alpha, \zeta)$. In the case of the large interlayer spacing, $\zeta \gg 1$ and weak inter-electronic coupling or high density limit, $\alpha \ll 1, \widetilde{F}(\alpha, \zeta) \approx 1/\alpha$, while in the case of the strong coupling or low density limit, $\alpha \gg 1, \widetilde{F}(\alpha, \zeta) \approx \pi/(2\alpha^2)$. The value of $\widetilde{F}(\alpha, \zeta)$ weakly depends on the interlayer spacing when c $\gg a_B^*$. It decreases with the decrease of c when $c \ll a_B^*$ as $\widetilde{F}(\alpha, \zeta) \approx (\pi/2\alpha^2)(1 + a_B^*/2c)/(1 + a_B^*/c)^{3/2}$. This is similar to the c-dependence of $F(\alpha, \zeta)$ in Eq. (1.55). Using

Eqs. (1.54), (1.75) and (1.76) in Eq. (1.74), we can finally represent the equation for T_c in an L2D electron gas as

$$T_c = 1.13\varepsilon_F \exp[-1/\lambda_{\text{eff}}(\alpha, \zeta)] = 1.13 E_B^* \alpha^{-2} \exp[-1/\lambda_{\text{eff}}(\alpha, \zeta)], \quad (1.77)$$

where

$$\lambda_{\text{eff}}^{-1}(\alpha, \zeta) = \frac{1}{\alpha \widetilde{F}(\alpha, \zeta)} \left[\frac{\pi}{\alpha} - F(\alpha, \zeta) + \alpha \widetilde{F}(\alpha, \zeta) \right], \quad (1.78)$$

where $F(\alpha, \zeta)$ and $\widetilde{F}(\alpha, \zeta)$ are given by Eqs. (1.55) and (1.76), respectively. $\lambda_{\text{eff}}(\alpha, \zeta) = 2\lambda_{\text{eff}} N_F$ is the dimensionless effective coupling constant in the particle-hole channel with λ_{eff} given by Eq. (1.74) and $E_B^* = e^4 m^*/2\kappa^2$ is the effective Bohr energy.

From Eqs. (1.77) and (1.78) we can analytically discuss the limiting behaviors of T_c

(1) In the case of the high density limit (or weak coupling), $\alpha \ll 1$, and large interlayer spacing, $\zeta = 2p_F c \gg 1, T_c \sim \varepsilon_F \exp(-\pi/\alpha)$ is very low.
(2) In the low density limit or strong coupling, $\alpha \gg 1, T_c \approx 1.13\varepsilon_F \exp(-2)$ $\approx 0.153\varepsilon_F \approx 0.48\, n_s/m^*$.
(3) Equation (1.77) weakly depends on the interlayer spacing when $c \gg a_B^*$. T_c decays with the decrease of c when $c < a_B^*$. As a result, $T_c \sim \varepsilon_F \exp(-4\sqrt{a_B^*/c})$. Thus, in the limiting case Eq. (1.77) in fact gives the same results as that from Eq. (1.56).
(4) Equation (1.77) is more appropriate for the case of the intermediate ($\alpha \gg 1$) and strong ($\alpha \sim 1$) coupling cases, whereas Eq. (1.56), strictly speaking, is valid only in the case of the weak ($\alpha \gg 1$) and possibly intermediate coupling.
(5) As discussed above after Eq. (1.55), the effective Bohr radius a_B^* is a criterion. When $c/a_B^* \gg 1$, T_c does not appreciably depend on the interlayer distance whereas it decreases with the reduction of c when $c/a_B^* < 1$. This implies that any superconductor may present either positive or negative pressure effect on T_c, contingent upon the condition at which one is detecting it. This is in good agreement with experimental observations. Experimentally, one claims the sample that shows a positive pressure effect on T_c (T_c increases with an increases of pressure) to be of BCS-type, while those, such as MgB_2, that present a negative pressure effect on T_c belong to non-BCS-type [21].

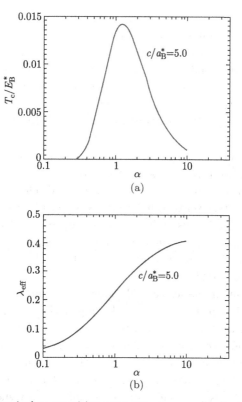

Fig. 1.11   (a) Dimensionless transition temperature normalized with respect to the effective Bohr energy, T_c/E_B^*; and (b) the effective interelectron coupling constant $\lambda_{\text{eff}}(\alpha, \varsigma)$ vs. α for the value of $c/a_B^* = 5$.

1.6.3.5   Numerical calculations of the transition temperature

Figure 1.11 shows the numerical results of T_c/E_B^*, the superconducting transition temperature normalized with respect to the effective Bohr energy, and $\lambda_{\text{eff}}(\alpha, \zeta)$, the effective coupling constant, vs. $\alpha = e^2/\kappa v_F$, the original interaction constant. As seen from Fig. 1.11, T_c/E_B^* as a function of α has a bell-shape, whereas $\lambda_{\text{eff}}(\alpha, \zeta)$ monotonically increases with the increase of α. In our opinion, the present results make it evident that the particle-hole excitations with higher-order vertex corrections can give rise to a high T_c and can be responsible for the anomalous behavior of the normal state of the correlated Fermi liquid.

(1) The bell-shape that has been experimentally observed is repro-
duced from Eq. (1.77) as shown in Fig. 1.11(a), whereas $\lambda_{\text{eff}}(\alpha, \zeta)$

monotonically increases with the increase of λ, as seen in Fig. 1.11(b). Since different samples have a variety of Bohr energy $E_B^* = e^4 m_b / 2\kappa^2$, the absolute transition temperature T_c is quite diverse.

(2) Under extreme conditions, T_c can reach as high as the room temperature 300 K. It deserves a mention that this predicted maximum transition temperature 300 K may not be really reached because the extreme condition for the dielectric constant κ, etc. (see Fig. 1.12) may never be realized, but T_c close to the refrigerator or freezing temperature seems to be realistic.

(3) Meanwhile, Eq. (1.77) presents a way of raising the effective electron-coupling constant α, which has a complicated relation to the dielectric constant κ, interlayer distance c and band mass of electron m_b, so does T_c. Equation (1.77) also tells us that the optimum of T_c does not occur at the maximum of the effective coupling constant. Instead, it happens at the original coupling constant $\alpha \sim 1.3$ and the effective coupling constant $\lambda_{\text{eff}} \sim 0.25$.

(4) Moreover, our numerical calculations show that the relation of T_c to the dielectric constant is also bell-shaped (Fig. 1.12), which has not been experimentally reported, to our best knowledge. This prediction may serve as a keystone to examine the validity of the theory.

In addition, the difference between the two curves in Fig. 1.12 may also be used to explain why the LaSrCuO cuprate superconductor, with $T_c \sim 35$ K, can be easily reproduced everywhere, whereas the

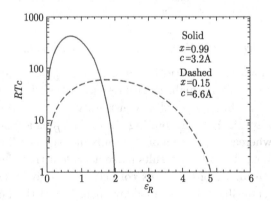

Fig. 1.12   Comparison of transition temperatures T_c vs. $\varepsilon_R = \kappa / R$ at optimal doping, where $R = m_b / m_e$ with m_b and m_e being the band mass of electron and electron mass, respectively, and κ the dielectric constant of the cuprate under investigation. E_B^* is the effective Bohr energy (see the text) and x is the ratio of carriers relative to Cu atoms.

once-reported samples with $T_c > 200$ K could not be reproduced even by the same research group. This is perhaps due to the sensitivity of T_c to the dielectric constant of the superconductors with higher T_c. The macroscopically similar conditions controlled in a laboratory may not ensure the microscopically similar conditions for the sample under synthesis, especially, while it is out of one's attention, to receive the exact dielectric constant.

(5) A numerical calculation shows that there is a singularity in T_c when c keeps decreasing — T_c first increases and then suddenly drops to 0 at a critical point. This implies that the sample has become an insulator after it [49].

1.6.3.6 Predictions of the DIA Theory

Based on the above theory developed from the DIA, we have finally derived Eq. (1.77), which is the result for a model system of a cuprate superconductor — L2D electron gas and is much more informative for experimentalists to synthesize a new sample superconductor with a higher transition temperature. We are now summarizing the predictions from it for the consideration of experimentalists in the field of superconductivity.

(1) Equation (1.77) suggests that T_c should be closely related to physical, chemical and structural parameters that are measurable. This is very informative to experimentalists searching for new samples with higher T_c.

The point here is to comprehensively control the parameters experimentally. Clearly, it will take many efforts to produce a best match of those parameters. This needs close cooperation between experimentalists and theorists.

(2) Under high pressure, the layered distance in a cuprate superconductor is reduced due to compression, and the superconductor may change to an insulator [46]. Therefore, a possible new phenomenon of superconductor-insulator phase transition under high pressure is expected in some cases. However, this kind of transition may not happen at all in some cases even if the sample is crushed.

(3) Moreover, there is the prediction in Fig. 1.12 that transition temperature varies with changes of the dielectric constant of the sample background. This can be used to examine the validity of our theory based on the DIA.

(4) In addition, from Fig. 1.12 it can be concluded as well that the heavy fermion superconductors can never present high T_c because of their large band mass.

1.6.4   *Physics of superconductivity*

Superconductivity is a collective consequence of many-electron correlation effects in some specifically structured materials. Therefore, its correct understanding must be based on many-particle theory. Any theory based on the mean field approximation (MFA), for instance, the BCS theory, may furnish wrong information in some cases, while it provides a rough description in other aspects. Strong coupling plays the major role in high-temperature superconductors (HTS) and the MFA is thus definitely inadequate for HTS. That is why the BCS remains valid for conventional superconductors only if its concept of Cooper pairing mediated by phonons is replaced by the reversed Coulomb interaction, but fails to provide self-consistent and reasonable interpretations for the anomalies at the normal state of HTS, not to mention any prediction. There must be a more general, or simply, a unified theory that can lead to both high- and low-temperature superconductivity. That is the theory described above on the basis of the DIA. Conventional (low-T_c) superconductivity should result from this general theory as a natural consequence under a limiting condition when the interlayer spacing c tends to zero.

Superconductivity is a physical phenomenon of charged particles undergoing a second-order phase transition with such general formalisms of transition temperature as $T_c \sim \varepsilon \exp(-1/\lambda)$, where ε may be the Fermi energy ε_F or phonon cutoff frequency ω_c and λ is the coupling constant. However, a metal-insulator transition pertains also to a second-order phase transition and has the same type of formalisms for the transition temperature. Therefore, this type of formula is just a necessary condition for a superconducting transition, but insufficient. To identify the feature of the transited phase, one has to then examine conductivity σ, whether tending to zero or infinity, as temperature approaches T_c. Obviously, the BCS theory is unable to do so, whereas our theory based on the DIA is able to, but not yet to calculate it. We plan to do this in the future studies.

As seen from the above discussions, the instability condition of the effective two-particle interaction $I^{-1} = 0$ is its pole condition and implies that the interacting electron system under investigation reaches its "resonance" and undergoes a phase transition. This resonating condition is realized with repulsive Coulombic interaction between electrons only. Knowing this may help one understand how important the repulsion between electrons is in such a second-order phase transition, with conductivity σ tending to infinity as the temperature T approaches T_c. Repulsive interaction (scattering) can also help explain why resistivity goes to zero after the resonance. Otherwise, for an attractive interaction between two electrons the resonance promotes localization of electrons, and thus conductivity σ will go to zero, leading to a metal-insulator transition.

However, as indicated above, Cooper pairing is a physical process in momentum space where one is dealing with "quasiparticles" and realized due to the sign reversal of Coulomb interaction between two quasiparticles (more exactly speaking here, quasielectrons). So, there should be no confusion here between the repulsive bare interaction for a pair of electrons in real space and the negative reversed interaction of Coulomb type for two quasielectrons in momentum space.

Finally, we are led to this cognition that the issue in superconductivity mechanism is in fact an issue that has not been completely resolved in physics. In order to really unveil the mystery of superconductivity there must be, so to speak, a development of many-particle physics. That is the DIA introduced above.

1.7   Summary and conclusions

In summary, we have reinterpreted and examined the BCS Hamiltonian and indicated several issues existing in it, which leads to a demand of developing the theory that was established based on the concept of a mean field approximation in the case of weak coupling. Especially, our emphasis is focused on the concept of Cooper pairing and its reinterpretation as well as a suggestion of the ground state of quartets. Moreover, we have introduced a new approach — diagrammatic iteration approach (DIA) that was developed more than ten years ago towards the unification of low- and high-T_c superconductivity. The DIA did make a development of the diagrammatic methodology and an extension of the diagrammatic approach to the region beyond RPA on the basis of Feynman diagrams and the field theory incorporating Migdal's expressions. The development of the theory exhibits its logic and rationality. The results deduced from its applications to a superconductor exhibit a great success in complying with experimental data in a single model and in a unified theory without invoking other hypotheses except for reasonable and necessary simplifications. We believe that the DIA is the best approach at present in describing a strongly coupled electron gas and thus fits to HTS. One of the objectives of this work is to draw more attention to and discussions of it. Of course, the DIA is devoted to a unified description of superconductivity and is of course not perfect; instead, it is just one more step to approach the truth of superconductivity. Nevertheless, it at least paves a path towards the unification of superconductivity theory. Some of the predicted results obtained from the approach may be used to examine the validity of the theory.

Moreover, it is informative to recall that thirty years ago Kohn and Luttinger [123] proposed a non-phonon mechanism of superconductivity caused by the sharpness of the Fermi surface. They found out that the Cooper instability should happen in a weakly interacting system of fermions down to the zero temperature, no matter the form of the interaction. It was also indicated there that the instability of the normal state appears even in the case of purely repulsive forces between particles. Thus, our model, taking into account the particle-hole excitations, in fact, complies with Kohn and Luttinger's idea. Moreover, the electron pairing is analogous to the resonance valence binding (exchange interaction). Therefore, it resembles, but is different from, the RVB model proposed by Anderson [123, 124] to explain such unusual high values of T_c and anomalous properties of the Cu-O-based layered 2D metals.

As shown above, using the DIA approach, we have examined the pairing instability induced by either a given interaction or the particle-hole excitations. In the case of a given interaction (which is similar to the electron-phonon or electron-plasmon interactions, etc.) the consideration of the pairing instability is equivalent to the Cooper instability issue [115, 118–120] and followed by the conventional scheme — the BCS theory. The instability of the normal state due to the particle (electron) pairing appears only if the given interaction is attractive and leads to the metal-insulator transition. However, for the particle-hole excitations, it happens due to the repulsive interaction between two particles (electrons), leading to both low and high transition temperature.

It is concluded as summaries that

(1) The vertex corrections to any one of the three: the interaction constant, the self-energy or response function becomes unimportant when

$\omega_c \ll \varepsilon_F$. This conclusion is in agreement with the well-known Migdal theorem.

(2) When $\omega_c \approx \varepsilon_F$, however, the vertex corrections to the interaction constant greatly limit the value of T_c [111].

(3) In this case ($\omega_c \approx \varepsilon_F$) the vertex corrections are not negligible and essentially affect the irreducible response. In the irreducible response appears the anomalous Fermi liquid term $\mathrm{Im}\chi \sim \omega/T$ in addition to the normal response $\sim -\omega/v_F k$.

(4) The contribution of such an anomalous term to the response strongly depends on the value of the coupling constant which is dependent on frequency and temperature.

(5) Nevertheless, this anomalous term possesses an opposite sign with respect to the normal response and is negligible in the case of weak coupling ($\lambda \ll 1$).

It has been shown that the incorporation of the particle-hole excitations into the pairing instability, in fact, corresponds to the replacement of the given interaction by the renormalized one which satisfies the Dyson equation, i.e., $D_0(k) \to D(k) = D_0(k) + D_0(k)\tilde{\chi}(k)D(k)$, where $\tilde{\chi}(k)$ is the renormalized particle-hole bubble or the so-called irreducible response incorporating the three-point vertex function.

(1) First of all, the particle-hole excitations give rise to the Cooper-like diagram which is similar to, but different from the conventional Cooper diagram because of its identical initial and final momenta in scattering. The Cooper-like diagram appears to be in the iterative series with a factor 2 due to the spin summation and with a negative sign with respect to the conventional Cooper's one.

(2) Thus, the contribution of the particle-hole channel to the pairing instability definitely exceeds the conventional Cooper's contributions as shown.

(3) We have also shown that the incorporation of the particle-hole excitations reverses the equation for the effective two-particle interaction (in which $I(p, p') \leftrightarrows \tilde{I}(p, p')$, namely, the vertices should be interchanged) with respect to the case of a given interaction.

(4) Therefore, the pairing instability caused by the particle-hole excitations appears only if the interaction is repulsive, as proven.

(5) In addition, it is shown that the particle-hole excitations practically eliminate the effect of the vertex correction on the coupling constant

even when $\omega_c \approx \varepsilon_F$ and thereby can provide much higher T_c than that due to a given attractive interaction.

(6) Our evaluation of the irreducible response shows that the particle-hole excitations also lead to the appearance of an anomalous Fermi liquid term $\text{Im}\chi \sim -\omega/T$ in both the charge and spin responses of the Fermi liquid in addition to the normal response $\sim -\omega/v_F k$.

(7) The anomalous Fermi liquid term exhibits a negative sign in contrast to the positive sign in the case of a given interaction. The scaling ω/T of the anomalous term in the response is in agreement with the marginal Fermi liquid hypothesis proposed by Varma *et al.* [49, 97, 98, 121, 122] to explain the universal anomalies in the normal state of the cuprate high-T_c superconductors.

(8) It is also shown that in the case of intermediate or strong coupling it is sufficient to include only the first step of iteration beyond the RPA in the consideration of the vertex corrections to the responses and in the evaluation of T_c of the transition mediated by the particle-hole excitations.

(9) We have also made some estimates of T_c in a layered 2D system of electrons which is a more suitable model for the Cu-O superconductors. It has shown that a layered structure of a superconductor favors the transition temperature T_c due to the enhancement of correlation.

(10) Even if in a layered 2D structure, T_c remains not very high as can be seen from Eq. (1.56), unless the higher-order vertex correction in the case of strong coupling has been taken into account in each step beyond RPA, as evidenced by Eq. (1.77).

References

[1] J. Bardeen, L. N. Cooper and J. R. Schrieffer, *Phys. Rev.* **108**, 1175 (1957).

[2] J. G. Bednorz and K. A. Mueller, *Z. Phys.* **B**-*Cond. Matt.* **64**, 189 (1986).

[3] J. Nagamatsu, N. Nakagawa, T. Muranaka, Y. Zenitani, and J. Akimitsu, *Nature (London)* **410**, 63 (2001).

[4] D. Larbalestier, A. Gurevich, D. M. Feldmann, and A. Polyanskii, *Nature* **414**, 368 (2001).

[5] B. G. Levi, *Physics Today Oct.*, 19 (2001).

[6] B. Lorenz, R. L. Meng, Y. Y. Xue, and C. W. Chu, *Phys. Rev. B.* **64**, 052513 (2001).

[7] L. Bud'ko, G. Lapertot, C. Petrovic, C. E. Cunningman, N. Anderson, and P. C. Canfeld, *Phys. Rev. Lett.* **86**, 1877 (2001).

[8] A. Knigavko and E. Marsiglio, *Phys. Rev. B.* **64**, 172513 (2001).

[9]   H. D. Yang, J. -Y. Lin, H. H. Li, H. H. Hsu, C. J. Liu, S.-C. Li, R. C. Yu, and C.-Q. Jin, *Phys. Rev. Lett.* **87**, 167003 (2001).

[10] X. K. Chen, M. J. Konstantinovite, and J. C. Irwin, *Phys. Rev. Lett.* **87**, 157002 (2001).

[11] F. Giubileo, D. Roditchev, W. Sacks, R. Lamy, and J. Klein, Cond-Mat/0105146.

[12] S. Tsuda, T. Yokoya, T. Kiss, Y. Takano, K. Togano, H. Kito, H. Ihara, and S. Shin, *Phys. Rev. Lett.* **87**, 177006 (2001).

[13] G. Rubio-Bollinger, H. Suderow, and S. Vieira, Cond-Mat/0102242.

[14] H. Schmidt, J. F. Zasazinski, K. E. Gray, and D. G. Hinks, *Phys. Rev. B.* **63**, R220504 (2001).

[15] T. Takahashi *et al.*, *Phys. Rev. Lett.* **86**, 4915 (2001).

[16] T. Takahashi, T. Sato, S. Souma, T. Muranaka, and J. Akimitsu, Cond-Mat/0103079.

[17] H. Kotegawa, K. Ishida, Y. Kitaoka, T. Maranaka, and J. Akimitsu, Cond-Mat/0102334.

[18] F. Bouquet R. A. Fisher, N. E. Phillips, D. G. Hinks, and J. D. Jorgensen, Cond-Mat/0104206.

[19] B. Gorshunov, C. A. Kuntscher P. Haas, M. Dressel, F. P. mena, A. B. Kuz'menko, D. van der Marel, T. Maranaka, and J. Akimitsu, Cond-Mat/0103164.

[20] G. Karapetrov, M. Iavarone, W. K. Kwok, G. W. Crabtree, and D. G. Hinks, *Phys. Rev. Lett.* **86**, 4374 (2001).

[21] B. Lorenz, R. L. Meng, and C. W. Chu, *Phys. Rev. B.* **64**, 012507 (2001).

[22] E. Satio, T. Taknenobu, T. Ito, Y. Iwasa, K. Prassides, and T. Arima, *J. Phys.: Condensed Matter,* **13**, L267 (2001).

[23] R. Prozorov, R. W. Giannetta, S. L. Bud'ko, and P. C. Canfield, *Phys. Rev. B.* **64**, 180501 (2001).

[24] C. Panagopoulos, B. D. Rainfold, T. Xiang, C. A. Scott, M. Kambara, and I. H. Inone, *Phys. Rev. B.* **64**, 094514 (2001).

[25] A. V. Pronin, A. Pimenov, A. Loidl, and S. I. Krasnosvobodtsev, *Phys. Rev. Lett.* **87**, 097003 (2001).

[26] X. H. Chen, Y. Y. Xue, R. L. Meng, and C. W. Chu, *Phys. Rev. B.* **64**, 172501 (2001).

[27] C. Walti, E. Felder, C. Degen, G. Wigger, R. Monnier, B. Delley, and H. R. Ott, *Phys. Rev. B.* **64**, 172515 (2001).

[28] J. Kotus, I. J. Mazin, K. D. Belashchenko, V. P. Antropov, and L. L. Boyer, *Phys. Rev. Lett.* **86**, 4656 (2001).

[29] J. E. Hirsch, *Physica C* **364–365**, 37–42 (2001).

[30] H. Takahashi, K. Igawa, K. Arii, Y. Kamihara, M. Hirano and H. Hosono, *Nature* **453**, 376 (2008).

[31] X. H. Chen, T. Wu, G. Wu, R. H. Liu, H. Chen and D. F. Fang, *Nature (London)* **453**, 761 (2008).

[32] H. H. Wen, C. Mu, L. Fang, H. Yang, and X. Zhu, *Europhys. Lett.* **82**, 17009 (2008).

[33] Z. A. Ren, J. Yang, W. Lu, X. L. Shen, Z. C. Li, G. C. Che, X. L. Dong, L. L. Sun, F. Zhou and Z. X. Zhao, *Europhys. Lett.* **82**, 57002 (2008).

[34] K. Sasmal, B. Lv, B. Lorenz, A. M. Guloy, F. Chen, Y. Y. Xue and C. W. Chu, *Phys. Rev. Lett.* **101**, 107007 (2008).

[35] M. Rotter, M. Tegel and D. Johrendt, *Phys. Rev. Lett.* **101**, 107006 (2008).

[36] A. S. Sefat, R. Y. Jin, M. A. McGuire, B. C. Sales, D. J. Singh, and D. Mandrus, *Phys. Rev. Lett.* **101**, 117004 (2008).

[37] J. H. Tapp, Z. Tang, B. Lv, K. Sasmal, B. Lorenz, C. W. Chu, and A. M. Guloy, *Phys. Rev. B* **78**, 060505(R) (2008).

[38] P. W. Anderson, Science **235**, 1196 (1987).

[39] R. B. Laughlin, *Phys. Rev. Lett.* **60**, 2677 (1988).

[40] S. A. Kivelson, D. S. Rokhsa, and J. P. Sethna, *Phys. Rev. B.* **35**, 8865 (1987).

[41] S. Chakravarty and S. A. Kivelson, *Phys. Rev. B* **64**, 064511 (2001).

[42] S. Chakravarty and S. A. Kiverlson, *Europhys. Lett.* **16**, 751 (1991).

[43] S. Chakravararty, M. Gelfand, and S. A. Kivelson, *Science* **254**, 970 (1991).

[44] G. Baskaran, and E. Tosarit, *Cur. Sci.* **61**, 22 (1991).

[45] Y. M. Malozovsky and J. D. Fan, *Phys. Rev. B* **49**, 4334 (1994).

[46] Y. M. Malozovsky, J. D. Fan, *Physica C* **231**, 63 (1994).

[47] Y. M. Malozovsky and J.D. Fan, *Supercond. Sci. and Technol.* **10**, 5 (1997). 259–277; as well as Y. M. Malozovsky and J. D. Fan, "A Diagrammatic Approach to the Pairing Instability in an Interacting Fermi Gas," Proceedings of SPIE - The International Society for Optical Engineering, ed. by Davor Povuna, Vol. 2697:129-140, January 30 - February 2, San Jose, California, USA (1996).

[48] J. D. Fan and Y. M. Malozovsky, *Inter. J. of Mod. Phys. B* **13**, 3505 (1999).

[49] C. M. Varma, P. B. Littlewood, S. Schmitt-Rink, E. Abrahams and A. E. Ruckenstein, *Phys. Rev. Lett.* **63**, 1996 (1989).

[50] J. Ruvalds, *Supercond. Sci. Technol.* **9**, 905 (1996).

[51] V. L. Ginzburg, *Physics — Uspekhi* **43** (6) 573 (2000).

[52] J. R. Schrieffer, *Theory of Superconductivity*, New York: Benjamin Inc. (1964).

[53] V. L. Ginzburg and L. D. Landau, *Zh. Eksp. Teor. Fiz.* **20**, 1064 (1950), English translation in: L. D. Landau, Collected papers (Oxford: Pergamon Press, 1965) 546.

[54] J. P. Blaizot and G. Ripka: Quantum Theory of Finite Systems, MIT Press (1985). A. Fetter and J. Walecka: Quantum Theory of Many-Particle Systems, Dover (2003).

[55] G. M. Eliashberg, *Zh. Fksp. Teor. Fiz.* **38**, 966; **39**, 147 [*Sov; Physics JETP* **11**, 696; **12**, 1000] (1960).

[56] J. W. Lynn, "high-Temperature Superconductivity," Graduate texts in Contemporary Physics, Springer-Verlag, (1990).

[57] L. N. Cooper, *Phys. Rev.* **104**, 1189 (1956).

[58] G. D. Mahan, *Many-Particle Physics*, Second Edition, New York: Plenum (1991).

[59] M. Tinkham, *Introduction to Superconductivity*, New York: McGraw Hill (1975).

[60] P. de Gennes, *Superconductivity of Metals and Alloys*, Menlo Park, C. A.: Benjamin, (1966).

[61] J. D. Fan and Y. M. Malozovsky, *Inter. J. of Mod. Phys. B* **17**, 3242 (2003).

[62] J. D. Fan and Y. M. Malozovsky, *Inter. J. of Mod. Phys. B* **27**, 1362035 (2013).

[63] Z. X. Shen and D. S. Dessau, *Phys. Rep.* **253**, 1 (1995).

[64] R. Doll and M. Nabauer, *Phys. Rev. Lett.* **7**, 51 (1961).

[65] B. S. Deaver Jr. and W. M. Fairbank, *Phys. Rev. Lett.* **7**, 43 (1961).

[66] S. Shapiro, *Phys. Rev. Lett.* **11**, 80 (1963).

[67] M. Tinkham, *Rev. of Mod. Phys.* January, 268 (1964).

[68] J.M. Blatt, *Superconductivity*, New York: Academic Press (1964).

[69] L. D. Landau, *Zh. Eksp. Teor. Fiz.* **11**, 592 (1941) [J. Phys. USSR 5, 71, (1941)].

[70] V. L. Ginzburg and L. D. Landau, *Zh. Eksp. Teor. Fiz.* **20**, 1064 (1950); a discussion in English is given by J. Bvardeen, in Handbook der Physik, etited by S. Fligge, (Springer-Valag, Berlin, 1956) Vol. 15, p. 324.

[71] V. L. Ginzburg, *Usp. Fiz. Nauk* **167**, 429 (1997) [Phys. Usp. 40, 407 (1997)].

[72] L. P. Gor'kov, *Zh. Eksp. Teor. Fiz.* **36**, 1918; **37**, 1407 (1959); Sov. Phys. JETP **9**, 1364 (1959); **10**, 998 (1960).

[73] N. N. Bogoliubov, *Zh. Eksp. Teor. Fiz.* **34**, 41 (1958) [Sov. Phys. JETP 7, 51(1958)].

[74] J. Valatin, *Nuovo Cimento* **7**, 843 (1958).

[75] C. Kuper, *Adv. Phys.* 8(29), 1 (1959).

[76] K. J. Cheng, *Study on Mechanism of Superconductivity*, Beijing: New Times Press, (1991).

[77] K. J. Cheng and S. Y. Cheng, *Inter J. of Mod. Phys. B*, **29–31**, 2894 (1998).

[78] Y. M. Malozovsky and J. D. Fan, *Phys. Lett. A* **257**, 332 (1999).

[79] J. D. Fan and Y. M. Malozovsky, *Physica C* **341–348**, 197 (2000).

[80] J. J. Sakurai, *Modern Quantum Mechanics*, Revised edition, New Jersey: Addison-Wesley Publishing, 1994.

[81] J. D. Fan and Y. M. Malozovsky, *J of Supercond. and Novel Magnetism*: Vol. **23**, Issue 5, 655 (2010).

[82] J. D. Fan and Y. M. Malozovsky, *Inter. J. of Mod. Phys. B* **17**, 3458 (2003).

[83] L. D. Landau, *Zh. Eksp. Teor. Fiz.* **30**, 1058 (1956) [English transl.: Sov. Phys.-JETP 3, 920 (1957)].

[84] L. D. Landau, *Zh. Eksp. Teor. Fiz.* **32**, 59 (1957) [English transl.: Sov. Phys.-JETP 5, 101 (1957)].

[85] D. Pines, P. Nozieres, *The Theory of Quantum Liquids*, Vol. **1**, New York: Benjamin, (1966).

[86] G. Baym, C. Pethick, Landau Fermi Liquid Theory, New York: Wiley (1991).

[87] A. A. Abrikosov, L. P. Gorkov and I. E. Dzyaloshinski, Method of Quantum Field Theory in Statistical Physics, New Jersey: Prentice-Hall (1964).

[88] A. B. Migdal, Theory of finite Fermi Systems and Application to Atomic Nuclei, New York: Wiley Interscience (1967).

[89] M. Gel-Mann, and K. Brueckner, *Phys. Rev.* **106**, 364 (1957).

[90] K. K. Sawada, K. N. Fukada, and R. Brout, *Rev. A. Brueckner, Phys. Rev.* **108**, 507 (1957).

[91] J. Hubbard, *Phys. R. Soc. London Ser. A.* **243**, 336 (1957).

[92] P. Nozieres and D. Pines, *Phys. Rev.* **111**, 442 (1958).

[93] J. J. Quinn and A. Ferrell, *Phys. Rev.* **112**, 812 (1958).

[94] A. Isihara, Solid State Physics **42**, 271 (1984), (Academic Press, 1989) and references therein.

[95] L. Swierkowski and D, Nelson, *Phys. Rev. Lett.* **67**, 240 (1991).

[96] C. W. J. Beenakker and H. van Houten, *Solid State Physics* **44**, 1 (1991), (Academic press) and references therein.

[97] P. B. Littlewood and C. M. Varma, *Phys. Rev. B* **46**, 405 (1992).

[98] C. M. Varma, *J. Phys. Chem. Solid* **54**, 1081(1993).

[99] E. Dagotto, *Rev. Mod. Phys.* **66**, 763 (1994).

[100] D. J. van Harlingen, *Rev. Mod. Phys.* **67**, 515 (1995).

[101] D. J. Scallapino, *Phys. Rep.* **250**, 329 (1995).

[102] P. W. Anderson, *Phys. Rev. B* **64**, 1839 (1994).

[103] G. M. Eliashberg, *Sov. Phys. JETP* **11**, 696 (1960); *Sov. Phys. JETP* **12**, 1000 (1960).

[104] A. B. Migdal, *Sov. Phys. JETP* **34**, 996 (1958), or A. B. Migdal, *Soviet Phys. J. Exper. Theor. Phys.* **7**, 996 (1958).

[105] G. M. Eliashberg, *Sov. Phys. JETP* **11**, 696 (1960); *Sov. Phys. JETP* **12**, 1000 (1960).

[106] S. Engelsberg and J. R. Schrieffer, *Phys. Rev.* **131**, 993 (1963).

[107] P. Nozieres, *Theory of interacting Fermi system*, New York: W. A. Benjamin, Inc. (1964).

[108] P. Nozieres and D. Pines, *The Theory of Quantum Liquids*, New York: W. A. Benjamin, Inc. (1966).

[109] Y. M. Malozovsky and J. D. Fan, *J. Phys,: Condens. Matter.* **8**, 10435 (1996).

[110] L. P. Gor'kov and T. K. Melik-Barkhudarov, *Sov. Phys.-JETP* **12**, 1018 (1961).

[111] M. Grabowski and L. J. Sham, *Phys. Rev. B* **29**, 6132 (1984).

[112] C. G. Olson, R. Liu, A. Yang, D. M. Lynch, A. J. Arko, R. S. List, B. Veal, Y. Chang, P. Jiang, A. Paulikas, *Science* **245**, 731 (1989); *Phys. Rev. B* **42**, 381 (1990); D. S. Dessau, Z. X. Shen, D. M. King, D. S. Marshall, L. W. Lombardo, P. H. Pichinson, A. G. Loeser, J. DiCarlo, C. H. Park, A. Kapitulnik, W. E. Spicer, *Phys. Rev. Lett.* **71**, 2781 (1993), references therein.

[113] A. Fetter, *Ann. Phys.* (N. Y.) **88**, 1 (1974).

[114] Y. M. Malozovsky and J. D. Fan, *Phys. Rev. B* **49**, 4334 (1994).

[115] Y. M. Malozovsky, S. M. Bose and P. Longe, *Phys. Rev. B* **47**, 15242 (1993).

[116] C. W. Chu, L. Gao, F. Chen, Z. J. Huang, R. L. Meng and Y. Y. Xue, *Science* **365**, 226 (1993).
[117] J. E. Crow and Nai-Phuan Ong, Ed. by J. W. Lynn, High Temperature Superconductivity, Springer-Verlag, p. 208 (1990).
[118] Y. M. Malozovsky, S. M. Bose and P. Longe, *Phys. Rev.* B **47**, 10504 (1993).
[119] W. Kohn and J. M. Luttinger, *Phys. Rev. Lett.* **15**, 524 (1965).
[120] D. Allender, J. Bray and J. Bardeen, *Phys. Rev.* B **7**, 1020 (1973).
[121] C. Sire, C. M. Varma, A. E. Ruckenstein and T. Giamarachi; *Phys. Rev. Lett.* **72**, 2478 (1994).
[122] C. M. Varma, *Phys. Rev. Lett.* **75**, 898 (1995).
[123] P. W. Anderson, in Frontiers and Borderlines in Many Particle Physics, International School of Physics "Enrico Fermi", Course CIV, edited by R. Broglia and R. Schrieffer (North-Holland, Amsterdam, 1988).
[124] P. W. Anderson, *Science* **256**, 1526 (1992).

Appendix A

Taking account of the spin indices in Eq. (1.1) and assuming $(p, k) = 1$ in it, we can derive the expression for the kernel of interaction in the RPA based on Ward's identity as shown below:

$$K^{\sigma\sigma'}(p, p') = -i\frac{\delta \Sigma_\sigma^{\mathrm{RPA}}(p)}{\delta G_{\sigma'}(p')}$$

$$= \frac{\delta}{\delta G_{\sigma'}(p')}\left[\int \frac{d^4q}{(2\pi)^4}\frac{D_0(p-q)}{1 - D_0(p-q)\chi_0(p-q)}G_\sigma(q)\right], \quad (A1)$$

where

$$\chi_0(k) = -i\sum_\sigma \int \frac{d^4p}{(2\pi)^4}G_\sigma(p)G_\sigma(p-k) \qquad (A2)$$

is the "free-particle" response function, and $D_0(k)$ is the un-renormalized boson propagator. Using the identity $\delta G(p)/\delta G(p') = \delta(p-p')\delta(\sigma-\sigma')$ and the relation

$$\frac{\delta\chi_0(k)}{\delta G_{\sigma'}(p')} = -i\sum_\sigma \int \frac{d^4p}{(2\pi)^4}[G_\sigma(p-k)\delta(p-p')\delta_{\sigma\sigma'}$$

$$+ G_\sigma(p)\delta(p-k-p')\delta_{\sigma\sigma'}]$$

$$= -i[G_{\sigma'}(p'-k) + G_{\sigma'}(p'+k)] \qquad (A3)$$

in Eq. (A1), we have

$$K_{\text{RPA}}^{\sigma\sigma'}(p,p') = D(p-p')\delta_{\sigma\sigma'} - i\int \frac{d^4q}{(2\pi)^4}D^2(p-q)$$
$$\cdot\, G_\sigma(q)[G_{\sigma'}(p'-p+q)+G_{\sigma'}(p'+p-q)], \qquad (A4)$$

where $D(k) = D_0(k)/[1-D_0(k)\chi_0(k)]$ is the boson propagator in the RPA. One can see that Eq. (A4) is identical to Eq. (1.7).

Appendix B

In terms of Eqs. (1.31), and (1.32), the kernel of interaction in the charge channel up to the second order can be written as follows:

$$K_c(p,p') = D(p-p') - i\int \frac{d^4q}{(2\pi)^4}D(p-q)G(q)G(p'+p-q)D(q-p').$$
$$(B1)$$

One can see, from Eq. (B1) that the second term in Eq. (B1), in fact, represents the second-order correction to the first-order interaction $D(p-p')$. Further, $D(q-p')$ in the right hand side of Eq. (B1) stands for the first-order interaction in the charge channel and should also be corrected by the higher order in the interaction which, according to Eq. (B1), should have the same form as the second term in Eq. (B1). Thus, the effective kernel of interaction in the charge channel including the higher-order corrections can be written as below:

$$K_c(p,p') = D(p-p') - i\int \frac{d^4q}{(2\pi)^4}D(p-q)G(q)G(p'+p-q)\Big\{D(q-p')$$

$$-i\int \frac{d^4q_1}{(2\pi)^4}D(q-q_1)G(q_1)G(p'+p-q_1)\Big\{D(q_1-p')$$

$$-i\int \frac{d^4q_2}{(2\pi)^4}D(q_1-q_2)G(q_2)G(p'+p-q_2)[D(q_2-p')$$

$$-\ldots]\ldots\Big\}\Big\}. \qquad (B2)$$

From Eq. (B2) one can see that the correction to $D(q-p')$ in the right hand side of Eq. (B2) appears with a minus sign, instead of plus, as might be thought at first sight. The series in Eq. (B2) corresponds to a ladder summation of all the graphs shown in Figs. 1.6(a) and 1.6(b). Replacing $K_c(p, p')$ in the left hand side of Eq. (B2) by $I(p, p')$, the effective two-particle vertex part, the equation for it can be written as below

$$I(p, p') = D(p - p') - i \int \frac{d^4 q}{(2\pi)^4} D(p - q) G(q) G(p' + p - q) I(q, p'),$$

(B3)

which is the same as Eq. (1.33).

In the case of a given interaction, the kernel of interaction up to the second order can be written as follows:

$$K_c(p, p') = D(p - p') + i \int \frac{d^4 q}{(2\pi)^4} D(p - q) G(q) G(p' + p - q) D(q - p'),$$

(B4)

where for simplicity we used $D(p - p')$, the first-order vertex in the interaction, instead of $\tilde{I}(p, p')$, the irreducible kernel of interaction. From Eq. (B4), one can see that the second term in Eq. (B1) appears with a positive sign in contrast to Eq. (B1). Further, $D(q - p')$ in the right hand side of Eq. (B4), in analogy to Eq. (B2), should be corrected by the higher order in the interaction. Thus, the effective kernel of interaction in the charge channel, including the higher-order corrections in the case of a given interaction, can be written as below:

$$K_c(p, p') = D(p - p') - i \int \frac{d^4 q}{(2\pi)^4} D(p - q) G(q) G(p' + p - q) \left\{ D(q - p') \right.$$

$$- i \int \frac{d^4 q_1}{(2\pi)^4} D(q - q_1) G(q_1) G(p' + p - q_1) \left\{ D(q_1 - p') \right.$$

$$- i \int \frac{d^4 q_2}{(2\pi)^4} D(q_1 - q_2) G(q_2) G(p' + p - q_2) [D(q_2 - p')$$

$$\left. \left. + \ldots \right] \ldots \right\} \right\}.$$

(B5)

The series in Eq. (B5) corresponds to the well-known ladder summation of the exchange graphs. Replacing $K_c(p, p')$ in the left hand side of Eq. (B5) by $I(p, p')$, the effective two-particle vertex part, the equation for it can be written as

$$I(p, p') = D(p - p') + i \int \frac{d^4q}{(2\pi)^4} D(p - q)G(q)G(p' + p - q)I(q, p'),$$

$$(B6)$$

which has the same form as Eq. (1.15) if $D(p - p')$ in Eq. (B6) is replaced by $\tilde{I}(p, p')$. A comparison of Eq. (B6) with Eq. (B3) indeed shows that the incorporation of the particle-hole excitations leads to the inversion of Eq. (B6) for the effective two-particle vertex part, namely $I(p, p') \leftrightarrows \tilde{I}(p, p')$, i.e., the vertex parts should be interchanged.

Part II
Experiments

2

Spectroscopic Studies of Quasiparticle Low-Energy Excitations in Cuprate and Iron-Based High-Temperature Superconductors

Nai-Chang Yeh[1]

Abstract

Recent development in the physics of high-temperature superconductivity (SC) is reviewed, with special emphasis on the studies of the low-energy excitations of cuprate and iron-based superconductors. For cuprate superconductors, a phenomenology based on coexisting competing orders with superconductivity in the ground state of these doped Mott insulators is shown to provide a consistent account for a wide range of experimental findings. In the case of iron-based superconductors, studies of the low-energy excitations reveal interesting similarities and differences when compared with cuprate superconductors. In contrast to the single-band cuprate superconductivity with an insulating parent state, the ferrous superconductors are multi-band materials with a semi-metallic parent state and exhibit two-gap superconductivity when doped. On the other hand, both systems exhibit strong antiferromagnetic correlation and Fermi-surface distortion, leading to unconventional pairing symmetries with sign-changing order parameters on different parts of the Fermi surface. These findings suggest that the pairing potentials in both the cuprate and the ferrous superconductors are generally repulsive, thus favoring a pairing mechanism that is electronically driven and a pairing strength that is closely related to the electronic correlation. The physical implications of the unified phenomenology based on antiferromagnetic correlations and remaining open issues associated with the cuprate and ferrous superconductivity are discussed.

[1]Department of Physics and Kavli Nanoscience Institute, California Institute of Technology, USA.

2.1   Introduction

Since the discovery of superconductivity about 100 years ago, the subject of superconductivity has remained one of the most intellectually challenging topics in condensed matter physics. Although much progress has been made in the physics and applications of superconductivity, to date it is still impossible to predict from first principles whether a specific type of material or compound would become superconducting below a given temperature. Moreover, the discovery of high-temperature superconducting (high-T_c) cuprates in 1986 [1] and the subsequent discovery of the iron-based superconductors in 2008 [2] have completely defied the conventional wisdom to avoid oxides and magnetic materials in search of high-T_c superconductors. Despite intense research efforts worldwide, the pairing mechanism for high-T_c superconductivity remains elusive to date. On the other hand, the discovery of iron-based high-T_c superconductors provides interesting comparisons with the cuprate superconductors, which help shed new light on the mystery of the high-T_c pairing mechanism.

The objective of this review is to survey the up-to-date status of experimental manifestations of unconventional low-energy excitations in the cuprates, explore the feasibility of a unified phenomenology for all cuprates, compare the findings in the cuprates with those of the iron-based superconductors, and then discuss the implications of these studies on the microscopic pairing mechanism of high-T_c superconductivity.

This article is structured as follows. In Sec. 2.2 an overview is given for the representative empirical findings of unconventional low-energy excitations in the cuprate superconductors, followed by discussions of the underlying physics associated with these phenomena in various states, including the zero-field pairing state below the superconducting transition, the zero-field normal state above the superconducting transition, and the vortex state in the presence of finite magnetic fields. In Sec. 2.3 we review the basic properties of iron-based superconductors and the characteristics of low-energy quasiparticles and spin excitations, and then discuss the theoretical implications associated with these empirical findings. In Sec. 2.4 we compare the low-energy excitations of the cuprate and ferrous superconductors, and discuss the physical implications on the mechanism for high-temperature superconductivity. Additionally, important open issues and possible clues in the quest for high-temperature superconductivity are summarized. Finally, Sec. 2.5 concludes the status of our current understanding of cuprate and ferrous superconductivity and the challenges required to unravel the mystery of the high-temperature superconducting mechanism.

2.2 Unconventional low-energy excitations in the cuprate superconductors

2.2.1 Overview

Cuprate superconductors are doped antiferromagnetic (AFM) Mott insulators with strong electronic correlation [3–7]. Mott insulators differ from conventional "band insulators" in that the latter are dictated by the Pauli exclusion principle when the highest occupied band contains two electrons per unit cell, whereas the former are influenced by the strong on-site Coulomb repulsion such that double occupancy of electrons per unit cell is energetically unfavorable and the electronic system behaves like an insulator rather than a good conductor at half filling. An important signature of doped Mott insulators is the strong electronic correlation among the carriers due to poor screening and the sensitivity of their ground state to the doping level. In the cuprates, the ground state of the undoped perovskite oxide is an antiferromagnetic (AFM) Mott insulator, with nearest-neighbor Cu^{2+}-Cu^{2+} AFM exchange interaction in the CuO_2 planes [8]. Depending on doping with either electrons or holes into the CuO_2 planes [8, 9], the Néel temperature (T_N) for the AFM-to-paramagnetic transition decreases with increasing doping level. Upon further doping of carriers, long-range AFM vanishes, spin fluctuations become important, and various competing orders (COs) begin to appear in the ground state, followed by the occurrence of superconductivity (SC). As schematically illustrated in the phase diagrams for the hole- and electron-type cuprates in Fig. 2.1(a), the superconducting transition temperature (T_c) first increases with increasing doping level (δ), reaching a maximum T_c at an optimal doping level, then decreases and finally vanishes with further increase of doping.

Although much similarity exists between the phase diagrams for the hole- and electron-type cuprates, closer inspection indicates asymmetric characteristics: For hole-type cuprates in the under- and optimally-doped regime, the physical properties that occur above T_c but below a crossover temperature T^*, known as the low-energy pseudogap (PG) temperature, are significantly different from those of Fermi liquids, including slightly suppressed electronic density of states (DOS) referred to as the low-energy

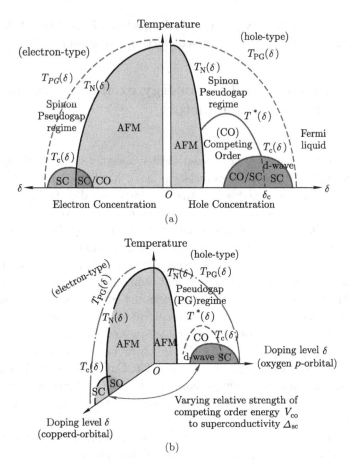

Fig. 2.1 (a) A schematic zero-field temperature (T) versus doping level (δ) generic phase diagram for electron- and hole-type cuprates [7] T^*: low-energy PG temperature, T_{PG}: high-energy pseudogap temperature). (b) A feasible explanation for the asymmetric hole- and electron-type phase diagrams may be attributed to the differences in the ratio of the SC energy gap (Δ_{SC}) relative to the competing order energy gap (V_{CO}): For hole- type cuprates revealing low-energy PG phenomena, empirical evidences suggest $V_{CO} > \Delta_{SC}$ [7, 10–16]. In contrast, tunneling experiments indicate a "hidden CO gap" only revealed in the vortex state of the electron-type cuprate superconductors with $V_{CO} < \Delta_{SC}$ [7, 10, 13, 17].

pseudogap (PG) phenomenon [18–20] and incompletely recovered Fermi surfaces in the momentum space known as the Fermi arc phenomenon [16, 19, 21]. Here the PG phenomenon refers to the observation of a soft gap without coherence peaks in the quasiparticle excitation spectra above T_c

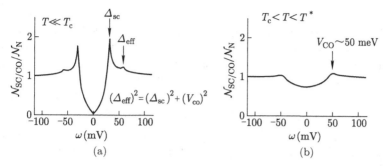

(a)   (b)

Fig. 2.2   Manifestations of the PG phenomena in quasiparticle tunneling spectra based on the assumptions of coexisting SC with a CO in the ground state of a cuprate superconductor [7, 10–13] and $V_{CO} > \Delta_{SC}$, where $V_{CO}(\Delta_{SC})$ denotes the CO (SC) energy gap. (a) The quasiparticle density of states ($\mathcal{N}_{SC/CO}$) at $T \ll T_c$, as calculated from the CO scenario elaborated in Sec. 2.2.2.2, is normalized relative to the normal-state density of states ($\mathcal{N}_N$) as a function of the quasiparticle energy ω. Here it is assumed $V_{CO} = 50$ meV and $\Delta_{SC} = 33$ meV, which correspond to the spectral parameters obtained from fitting the data of optimally doped Bi-2212 system with $T_c \sim 92$ K [12, 13]. The calculated spectrum exhibits sharp coherence peaks at energy $\omega = \pm\Delta_{SC}$ and small "humps" at $\omega = \pm\Delta_{\text{eff}}$, where the effective gap energy Δ_{eff} is defined by $(\Delta_{\text{eff}})^2 \equiv (\Delta_{SC})^2 + (V_{CO})^2$. This finding is in general agreement with the tunneling spectra obtained from under and optimally-hole-doped cuprate superconductors [20]. (b) For temperatures in the range $T_c < T \ll T^*$, a spectral PG feature remains at $\omega <\sim \pm V_{CO}$, which is consistent with the fact that Δ_{SC} vanishes above T_c [12, 13].

in hole-type cuprates and below a PG temperature T^*, as exemplified in Fig. 2.2. Evidence for the PG in hole-type cuprates has been reported in tunneling measurements [7, 10–15, 20, 22–25], [63]Cu spin-lattice relaxation rate and [63]Cu Knight shift in nuclear magnetic resonance (NMR) experiments [18, 26–31], optical conductivity experiments [18, 32], Raman scattering experiments [33, 34] and angle resolved photoemission spectroscopy (ARPES) measurements [19, 21]. Intimately related to the PG is the observation of Fermi arcs in the hole-type cuprates, which refers to an incomplete recovery of the full Fermi surface for temperature in the range $T_c < T < T^*$ [16, 19, 21], and will be discussed in more detail in Sec. 2.2.3. On the other hand, neither low-energy PG [7, 17, 35] nor Fermi arc phenomena [36] can be found in the electron-type cuprate superconductors in the absence of magnetic fields. Interestingly, however, break-junction tunneling spectra of a one-layer electron-type cuprate revealed PG phenomena at $T < T_c$ in the vortex state [37, 38]. Similarly, spatially resolved scanning tunneling spectroscopic studies of the vortex state of the infinite-layer electron-type cuprate $Sr_{0.9}La_{0.1}CuO_2$ revealed PG features with a characteristic energy

$\Delta_{PG} < \Delta_{SC}$ inside the vortex core [7, 10, 17]. Moreover, electronic Raman scattering experiments on $Nd_{2-x}Ce_xCuO_4$ also revealed contributions of an additional small energy gap in the SC state [39]. Thus, many of the seemingly puzzling asymmetric properties between the hole- and electron-type cuprates may be explained by the differences in the ratio of the SC energy gap (Δ_{SC}) relative to a CO energy gap (V_{CO}) and by attributing the origin of the low-energy PG phenomena to the presence of a CO energy gap so that $V_{CO} \sim \Delta_{PG}$ [7, 10–17]. The presence (absence) of the zero-field low-energy PG phenomena in the hole-type (electron-type) cuprate superconductors may be considered as the result of $V_{CO} > \Delta_{SC}(V_{CO} < \Delta_{SC})$, as schematically illustrated in Fig. 2.1(b). A feasible physical cause for such differences will be discussed later.

In addition to the contrasting low-energy charge excitations among the hole- and electron-type cuprates, the low-energy spin excitations also exhibit electron-hole asymmetry: Neutron scattering experiments on the hole-type $La_{2-x}Sr_xCuO_{4-y}$ reveal incommensurate spin correlations in the superconducting state, with a temperature independent spin gap observed both below and above T_c [40–42]. On the other hand, the one-layer electron-type cuprate $Nd_{2-x}Ce_xCuO_{4-y}$ (NCCO) [43] displayed commensurate spin correlations and a temperature-dependent spin gap in neutron scattering experiments [43]. Furthermore, the spin gap of NCCO was observed to reach a maximum as $T \to 0$ and disappeared as $T \to T_c$ [43]. Therefore, the spin gap in both hole- and electron-type cuprates may be related to the charge PG, as the spin gap and PG are both absent above T_c in the electron-doped cuprates and present above T_c in hole-doped cuprates. Moreover, enhancement of a static commensurate magnetic order by application of a magnetic field up to 9 T in electron-doped $Pr_{0.89}LaCe_{0.11}CuO_4$ (PLCCO) is observed [44], and a commensurate quasi-2D spin density wave (SDW) enhanced by the application of a magnetic field equal to 5 T in under-doped $Pr_{0.88}LaCe_{0.12}CuO_4$ is also reported [45].

In addition to the low-energy PG phenomenon that is correlated with the Fermi arcs and only found in hole-type cuprates slightly above T_c, there is a high-energy PG (denoted by T_{PG} in Fig. 2.1(a)) that is present in both electron- and hole-type cuprates according to optical [46, 47] and neutron scattering [48] experiments. As further elaborated in this review, the low-energy PG may be associated with the onset of COs. In contrast, the high-energy PG appears to be related to the short-range magnetic exchange coupling in the cuprates [3].

Physically, the existence of various COs besides SC in the ground state of the cuprates may be attributed to the complexity of the cuprates and the strong electronic correlation, which is in stark contrast to conventional superconductors where SC is the sole ground state. The presence of COs has been manifested by a wide range of experiments, including X-ray and neutron scattering [40–45, 48–55], muon spin resonance (μ SR) [56], NMR [57, 58], optical conductivity [46, 47] and Raman scattering [33, 34, 39], ARPES [21, 59, 60] and scanning tunneling microscopy/spectroscopy (STM/STS) [7, 10–17, 24, 61–64]. Moreover, theoretical evidences for COs have been provided by analytical modeling and numerical simulations [3–6, 65–75]. The occurrence of a specific type of CO such as the spin density wave (SDW) [5, 6, 68–70], pair density wave (PDW) [71, 72], d-density wave (DDW) [73, 74] or charge density wave (CDW) [4, 75] depends on the microscopic properties of a given cuprate, such as electron- or hole-doping, the doping level (δ), and the number of CuO_2 layers per unit cell (n) [7, 10–17, 76–78].

Although the relevance of competing orders to cuprate superconductivity remains unclear to date, the existence of COs has a number of important physical consequences. In addition to the aforementioned non-universal phenomena among different cuprates, quantum criticality emerges naturally as the result of competing phases in the ground state [5–7, 66, 67]. Moreover, strong quantum fluctuations are expected due to the proximity to quantum criticality [7, 14, 15, 66, 77, 78], and the low-energy excitations from the ground state become unconventional due to redistributions of the spectral weight between SC and COs in the ground state [7, 10–17]. Macroscopically, the presence of COs and strong quantum fluctuations naturally lead to weakened superconducting stiffness upon increasing T and magnetic field H [7, 14, 77–80], which contributes to the extreme type-II nature and the novel vortex dynamics of cuprate superconductors [81–97]. Additionally, a novel vortex liquid may exist in the presence of high magnetic fields at low temperatures [77, 78], giving rise to quantum oscillations in the "strange metallic state" of the cuprates when the applied magnetic fields (H) are still much smaller than the upper critical field H_{c2} [98–102]. Indeed, theoretical analysis of the experimental data taken on under-doped hole-type cuprates $YB_2Cu_3O_{6+x}$ (Y-123) reveals that the commonly held assumption that the oscillation period is given by the underlying Fermi-surface area via the Onsager relation becomes invalid [103], prompting conjectures for reconstructed Fermi surfaces due to incommensurate SDW

[102, 104]. However, the observation of negative Hall effects [105] in the low-temperature high-field limit where quantum oscillations appear is similar to the commonly observed anomalous sign-reversal Hall conductivity of both hole- and electron-type cuprates in the flux flow limit [106–111]. These findings suggest that quantum fluctuations and COs may both be relevant to the appearance of quantum oscillations.

Given that the manifestation of unconventional low-energy excitations is one of the natural consequences of COs in the ground state of cuprates, investigation of the low-energy excitations can help reveal the characteristics of relevant COs. Here we summarize the best known unconventional phenomena to be discussed in this section: the appearance of satellite features [7, 10–15, 20] and energy-independent local density of states (LDOS) modulations in the quasiparticle spectra below T_c [7, 10, 11, 24, 61–64]; the existence of a low-energy PG [23–25] and the appearance of the Fermi arcs for $T_c < T < T^*$ in hole-type cuprates [16, 19, 21]; "dichotomy" in the momentum-dependence of quasiparticle coherence in hole-type cuprates [12, 19, 112, 113]; PG-like vortex-core states in both electron- and hole-type cuprate superconductors [10, 11, 17, 114]; strong quantum fluctuations found in all cuprates for $H^* < H \ll H_{c2}$ at $T \to 0$, where the crossover field H^* is dependent on the doping level, the electronic anisotropy, and the number of CuO_2 layers per unit cell [14, 15, 77, 78]; the occurrence of quantum oscillations in the vortex liquid phase of hole-type cuprate superconductors for magnetic fields applied perpendicular to the CuO_2 planes [98–102]; the anomalous sign-reversal of the Hall conductivity in the vortex liquid state [106–111]; and the anomalous Nernst effect appearing in the normal state of hole-type cuprate superconductors [115, 116].

Various theoretical attempts have been made to explain the aforementioned anomalous behaviors among the cuprates, which may be largely categorized into two types of scenarios known as the "one-gap" [117] and "two-gap" models [7]. The former is associated with the "pre-formed pair" conjecture that asserts strong phase fluctuations in the cuprates so that formation of Cooper pairs may occur at a temperature well above the superconducting transition [117]. The latter considers coexistence of COs and SC with different energy scales in the ground state of the cuprates [7], as described in this overview. While the unconventional and asymmetric low-energy excitations amongst the hole- and electron-type cuprates have puzzled researchers and derailed the successful development of a microscopic theory that consistently accounts for all the differences found in cuprates of varying doping levels, there appears to be a converging consensus recently based on the two-gap model to consistently account for most experimental phenomena [7, 10, 21, 33, 34, 57, 58, 60, 64]. Moreover, the occurrence of COs does not exclude the possibility of pre-formed Cooper pairs [25] above the superconducting transition, because there is no apparent reason for a CO that coexists with coherent Cooper pairs below T_c to be incompatible with incoherent Cooper pairs above T_c and below the PG temperature T^*. Nonetheless, several open issues remain. Further, whether the presence of COs is relevant or even devastating to the occurrence of cuprate superconductivity is still inconclusive.

In this section we survey the up-to-date status of experimental manifestations of unconventional low-energy excitations in the cuprates, explore the feasibility of a unified phenomenology for all cuprates based on the CO scenario, and then discuss the implications of the phenomenology in the context of microscopic pairing mechanism of cuprate superconductivity. The survey is divided into three sub-sections for the pairing state and normal state in zero fields, and for the vortex state in finite fields.

2.2.2   The zero-field pairing state

As mentioned in the overview, the proximity of the cuprates to Mott insulators implies strong electronic correlation in the parent compound and in the small doping limit. To understand the formation of holes in the strongly correlated cuprates, an effective one-band t-J model was formulated by Zhang and Rice [118], in which the eigenstate of a single CuO$_4$ cluster is considered. That is, when a hole is introduced into the CuO$_4$ cluster, an orbital with $d_{x^2-y^2}$-symmetry among the four oxygen atoms can form, which will hybridize with the $3d^9$-state of the central copper site, whereas the $3d^8$-state of the copper site (which corresponds to a double occupancy of holes at the same site) is forbidden due to the large (up to ~ 8 eV) on-site Coulomb repulsion energy. More specifically, the lowest-energy hybrid configuration with Coulomb interaction included would involve a singlet combination of the two states $[(1p^2 2p^6)_3 (1p^2 2p^5_\sigma)(3d^9_{-\sigma})]$ and $[(1p^2 2p^6)_4 (3d^{10})]$, which is known as the "Zhang–Rice singlet". Here, σ denotes that spin state of an electron in the p- or d-orbital. By considering the tunneling among these singlets in a half-filled CuO_2 plane, Zhang and Rice were able to derive an effective one-band t-J model [118]. Such singlet pairing for delocalized holes could reduce the strong on-site Coulomb repulsion, and would favor the unconventional $d_{x^2-y^2}$-wave pairing symmetry.

2.2.2.1 Unconventional pairing symmetry

Historically, one of the most heated debates in the first decade of cuprate superconductivity research was the pairing symmetry [119–122]. From the symmetry point of view, the pairing channels of a singlet superconductor with the square-lattice symmetry must be consistent with even orbital quantum numbers (such as s, d, g, ... for $l = 0, 2, 4, ...$ or their linear combinations). Given the quasi-two dimensional nature of most cuprates and the strong on-site Coulomb repulsion, it is feasible that the pairing symmetry is predominantly associated with the d-channel rather than the s-channel in order to minimize the Coulomb repulsion and to accommodate the quasi-two dimensional nature of the cuprates at the price of a higher kinetic energy.

The issue of the pairing symmetry in cuprate superconductors was eventually settled by means of phase-sensitive measurements [119–122]. It was also realized later on that directional quasiparticle tunneling spectra by means of STS studies of the cuprate single crystals along different crystalline planes were also good experimental verifications of the pairing symmetry [123–127], because different pairing symmetries would result in distinctly different characteristics in the directional quasiparticle tunneling spectra due to the phase-changing pairing potential with varying quasiparticle momentum k [128–130], as schematically exemplified in Fig. 2.3 for the s-wave and $d_{x^2-y^2}$-wave pairing potentials. Indeed, overwhelming experimental evidences [119–127] have revealed signatures for predominantly $d_{x^2-y^2}$ pairing symmetry in all hole-type cuprate superconductors that are under-doped or optimally-doped.

The principle for verifying the pairing symmetry of a superconductor with a pairing potential Δ_k is based on the generalized Blonder–Tinkham–Klapwijk (BTK) model [131] first derived by Hu, Tanaka and Kashiwaya [128–130]. Specifically, for an N-I-S junction (N: normal metal, I: insulator, S: superconductor) such as in the case of an STS experiment with a metallic tip, the tunneling conductance (dI_{NS}/dV) under a biased voltage V and

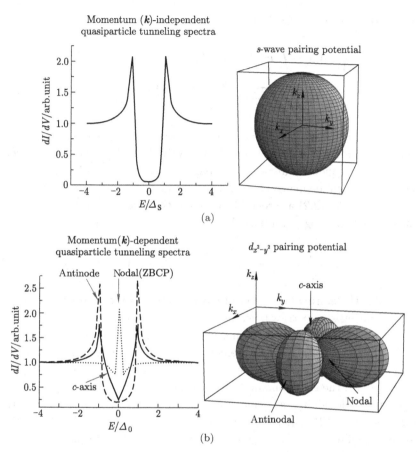

(a)

(b)

Fig. 2.3   Comparison of the quasiparticle tunneling spectra obtained from the s-wave and $d_{x^2-y^2}$-wave pairing symmetries [128–130, 132]: (a) Momentum ($\boldsymbol{k}$)-independent quasiparticle tunneling conductance (dI/dV) vs. energy (E) normalized to the isotropic s-wave superconducting gap (Δ_s). The right panel depicts the isotropic s-wave superconducting gap in the $\boldsymbol{k}$-space. (b) Momentum ($\boldsymbol{k}$)-dependent quasiparticle tunneling conductance spectra for a $d_{x^2-y^2}$-wave pairing superconductor, where the $d_{x^2-y^2}$-wave pairing potential is approximated by the relation $\Delta_{\boldsymbol{k}} = \Delta_0 \cos(2\theta_{\boldsymbol{k}})$, and $\theta_{\boldsymbol{k}}$ denotes the incident angle of the quasiparticle momentum $\boldsymbol{k}$ relative to the anti-nodal direction of the $d_{x^2-y^2}$-wave pairing potential. The right panel depicts the anisotropic $d_{x^2-y^2}$-wave superconducting gap in the $\boldsymbol{k}$-space. We note strong $\boldsymbol{k}$-dependent spectral characteristics. In particular, for $\boldsymbol{k}$ along the nodal direction, a sharp peak known as the zero-bias conductance peak (ZBCP) [128–130] appears due to the presence of Andreev bound states. For more details, see Refs. [123–130].

Thus, given a pairing potential of a superconductor, the incident angle of the quasiparticles relative to the N-I-S junction and the tunneling cone for the incident quasiparticles, the resulting tunneling spectra can be derived from Eqs. (2.1)–(2.3), provided that SC is the only ground state and that the sample is relatively clean so that spectral broadening due to the finite quasiparticle lifetime may be neglected.

at $T = 0$ is given by [128–130]:

$$\frac{dI_{NS}}{dV}(V) \propto \int d\theta[1 + |a(E,\theta)|^2 - |b(E,\theta)|^2]e^{-(\theta^2/\beta^2)}, \qquad (2.1)$$

$$\sim \frac{1 + \sigma_N|\Gamma_+|^2 + (\sigma_N - 1)|\Gamma_+\Gamma_-|^2}{|1 + (\sigma_N - 1)\Gamma_+\Gamma_-e^{i(\phi_- - \phi_+)}|^2}, \qquad (2.2)$$

where $|a(E,\theta)|^2$ and $|b(E,\theta)|^2$ in Eq. (2.1) refer to the Andreev reflection and normal reflection coefficients, respectively, $E = eV$ is the quasiparticle energy, θ is the incident angle of quasiparticles relative to the N-I-S junction, β denotes the tunneling cone, σ_N is the normal state conductance, and $\Gamma_{+,-}$ and $\phi_{+,-}$ in Eq. (2.2) are related to the electron-like (+) and hole-like (−) quasiparticle pairing potentials $\Delta_{+,-}$ as follows [128–130]:

$$\Gamma_{+,-} = \frac{E - \sqrt{E^2 - |\Delta_{+,-}|^2}}{|\Delta_{+,-}|} \ (\Delta_{+,-} \equiv |\Delta_{+,-}|e^{i\,\phi_{+,-}}). \qquad (2.3)$$

While the pairing symmetry of most cuprates is predominantly $d_{x^2-y^2}$, mixed pairing symmetries (such as $d_{x^2-y^2} + s$) have been widely reported in a number of cuprates, including in the tunneling junction studies [124, 133, 134], phase-sensitive measurements [135, 136], microwave spectra [137], optical spectra [138], and μ SR penetration depth measurements [139, 140]. In particular, the subdominant s-wave component appears to increase with increasing hole-doping, as demonstrated by both STS and Raman spectroscopic studies [124, 138]. As exemplified in Fig. 2.4(a) for the theoretical c-axis tunneling spectra associated with different pairing symmetries, and in Fig. 2.4(b) and Fig. 2.4(c) for representative experimental tunneling spectra of an optimally-doped hole-type cuprate superconductor $YBa_2Cu_3O_{7-\delta}$ (with $T_c = 93$ K) and an over-doped cuprate $(Y_{0.7}Ca_{0.3})Ba_2Cu_3O_{7-\delta}$ (with $T_c \sim 78$ K), it is apparent that a subdominant s-wave component becomes non-negligible with increasing hole-doping. This finding is consistent with the notion that the $d_{x^2-y^2}$-wave pairing is more favorable when on-site Coulomb repulsion is significant near the Mott insulator limit, whereas the s-wave pairing component may become energetically preferred in the over-doped limit when cuprate superconductors become more like conventional superconductors. We further note that the $(d_{x^2-y^2} + s)$-pairing leads to a reduction in the rotation symmetry from four- to two-fold symmetry while maintaining nodes as well as changing signs in the pairing potential as a function of the quasiparticle momentum.

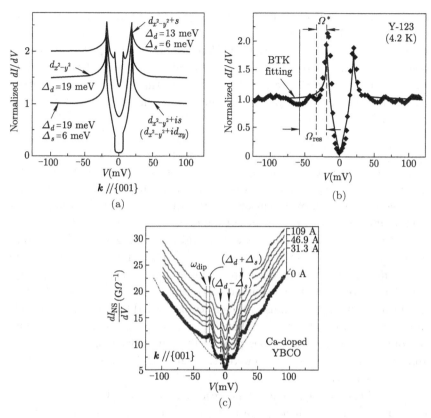

Fig. 2.4   Evidence for increasing subdominant s-wave component with increasing doping in hole-type cuprate superconductors with $(d_{x^2-y^2}+s)$-wave pairing potentials [124,
125]: (a) Theoretically tunneling spectra for different pairing potentials of pure $d_{x^2-y^2}$,
$(d_{x^2-y^2}+s)$ and $(d_{x^2-y^2}+is)$. (b) A representative c-axis $(k//\{001\})$ tunneling spectrum
of an optimally-doped cuprate superconductor $YBa_2Cu_3O_{7-\delta}$ (Y-123, with $T_c = 93k$),
showing predominantly a $d_{x^2-y^2}$-wave pairing potential. (c) A series of c-axis $(k//\{001\})$
tunneling spectra of an over-doped cuprate $(Y_{0.7}Ca_{0.3})Ba_2Cu_3O_{7-\delta}$ (with $T_c \sim 78$ K),
showing spatially homogeneous tunneling spectra with tunneling characteristics consistent with the theoretical curve in (a) for a $(d_{x^2-y^2}+s)$-wave pairing potential. For more
details, see Refs. [124, 125].

An alternative approach to revealing the underlying pairing symmetry
of cuprate superconductors is to investigate the spatial evolution of
the quasiparticle low-energy excitations near quantum impurities. It is
well-known that magnetic quantum impurities can suppress conventional
superconductivity effectively [141–146], whereas non-magnetic impurities

in the dilute limit appear to inflict negligible effects on conventional superconductivity, as explained by the Anderson theory for dirty superconductors [147]. However, the findings of strong effects of spinless quantum impurities on the hole-type cuprate superconductors [124–126, 148–159] and related theoretical studies suggest that the effects of quantum impurities depend on the pairing symmetry and the existence of magnetic correlation in cuprate superconductors [160–168]. For instance, fermionic nodal quasiparticles in the cuprates with either $d_{x^2-y^2}$- or $(d_{x^2-y^2} + s)$-pairing symmetry can interact strongly with the quantum impurities in the CuO_2 planes and incur significant suppression of superconductivity regardless of the spin configuration of the impurity [160–164]. Moreover, the spatial evolution of the quasiparticle spectra near quantum impurities would differ significantly if a small component of complex order parameter existed in the cuprate. For instance, should the pairing symmetry contain a complex component such as $(d_{x^2-y^2} + id_{xy})$ that broke the time-reversal $(\mathcal{T})$ symmetry, the quasiparticle spectrum at a non-magnetic impurity site would have revealed two resonant scattering peaks at energies of equal magnitude but opposite signs in the electron-like and hole-like quasiparticle branches [24]. In contrast, for either $d_{x^2-y^2}$ or $(d_{x^2-y^2} + s)$ pairing symmetry, only one resonant scattering peak at the impurity site is expected [160, 162–164]. All empirical data to date [124–126, 148–159] are consistent with the latter scenario.

In addition, the existence of nearest neighbor AFM Cu^{2+}-Cu^{2+} correlation in the superconducting state of the cuprates can result in an unusual Kondo-like behavior near a spinless impurity [165–167] due to an induced spin-1/2 $(S = 1/2)$ moment when one of the Cu^{2+} ions is substituted with a spinless ion such as Zn^{2+}, Mg^{2+}, Al^{3+} and Li^+ [124–126, 148–159]. Indeed, the Kondo-like behavior associated with isolated spinless impurities in hole-type cuprates has been confirmed from the nuclear magnetic resonance (NMR) [148, 149, 157] and the inelastic neutron scattering (INS) experiments [154, 155], and the spinless impurities are found to have more significant effects on broadening the NMR linewidth, damping the collective magnetic excitations and reducing the superfluid density than the magnetic impurities such as Ni^{2+} with $S = 1$ [124–126, 148–159]. On the other hand, both types of impurities exhibit similar effects on suppressing superconducting transition temperature (T_c), increasing the microwave surface resistance in the superconducting state and increasing the normal state resistivity [124–126, 148–159].

The overall stronger suppression of superconductivity due to non-magnetic impurities in d-wave cuprates has been attributed to the slower spatial relaxation of spin polarization near the spinless impurities than that near the $S = 1$ impurities, the latter being partially screened by the surrounding antiferromagnetically coupled Cu^{2+} spins [167, 168]. The detailed spatial evolution of the quasiparticle tunneling spectra near these quantum impurities in the cuprates can further provide useful insights into the pairing state of the cuprates, and has been investigated in impurity-substituted $YBa_2Cu_3O_{7-\delta}$ and $Bi_2Sr_2CaCu_2O_{8+\delta}$ systems using the low-temperature scanning tunneling microscopy (STM) techniques [124–126, 158, 159]. As exemplified in Fig. 2.5 for an optimally-doped $YBa_2Cu_3O_{7-d}$ (Y123) single crystal with 0.26% Zn and 0.4% Mg substituted into the Cu sites in the CuO_2 planes, hereafter denoted as (Zn, Mg)-Y123. The superconducting transition temperature of the sample is $T_c = 82.0$ K, which is substantially lower than that of the pure optimally-doped Y123 with $T_c = 93.0$ K. For an STM tip significantly far away from any impurities, the tunneling spectra were similar to the typical c-axis quasiparticle tunneling spectra in pure Y123, as shown in the upper panel of Fig. 2.5(a). However, the global superconducting energy gap Δ_d was suppressed to (25 ± 2) meV from the value $\Delta_d = (29 \pm 1)$ meV in pure Y123 [124–126]. Moreover, the energy w_{dip} associated with the "dip-hump" satellite features had also shifted substantially relative to that in pure Y123. We note that the dip-hump feature has been attributed to the effects of quasiparticle damping by the background many-body excitations such as spin fluctuations [169, 170] or phonons [171], and the resonant energy of the many-body excitation may be empirically given by $|\Omega_{res}| = |w_{dip} - \Delta_d|$. The finding that $|\Omega_{res}|$ in the (Zn, Mg)-Y123 sample decreased significantly to (7 ± 1) meV from the value $|\Omega_{res}| = (17 \pm 1)$ meV in the pure Y123 with a very small impurity concentration in our (Zn, Mg)-Y123 clearly rules out phonons as the relevant many-body excitations associated with the satellite features. Therefore, one may associate the w_{dip} satellite features with the inelastic scattering of quasiparticles by spin fluctuations. Further, the strong suppression of w_{dip} by a small amount of non-magnetic impurities implies that the Cu^{2+} spin fluctuations in the optimally-doped cuprates are susceptible to the type of quantum impurities that interrupt the coherence of pair formation [124–126].

Interestingly, detailed studies on the surface of (Zn, Mg)-Y123 revealed apparent impurity scattering spectra that could be associated with two types of impurities, with maximum scattering intensity occurring at either

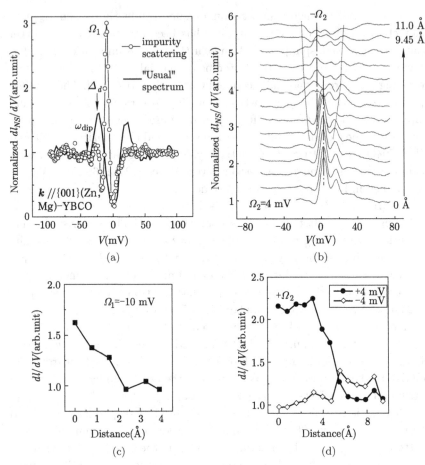

Fig. 2.5   Normalized c-axis differential conductance (dI/dV) versus bias voltage (V) quasiparticle tunneling spectra of the (Zn, Mg)-Y123 single crystal near impurity sites at 4.2 K [124–126]. (a) Upper panel: A representative impurity scattering spectrum with a resonant peak at $\Omega_1 \sim -10$ meV and a typical spectrum away from impurities. Lower panel: Spatial variation of the impurity-induced resonant peak intensity. (b) Upper panel: Representative spectra revealing spatial variations in the quasiparticle spectra along the Cu-O bonding direction from an impurity with a maximum scattering at $\Omega_2 \sim +4$ meV. We note the spatially alternating resonant peak energies between $+4$ meV and -4 meV and the particle-hole asymmetry in the degrees of suppression of the SC coherence peaks. Lower panel: Spatial variation of the impurity-induced resonant peak intensity, showing alternating peak intensities at energies $+4$ meV and -4 meV with distance from an impurity.

$\Omega_1 \sim -10$ meV or $\Omega_2 \sim -4$ meV [124–126]. Further, regardless of the impurity species, the intensity of each resonant peak decreased rapidly within approximately one Fermi wavelength along the Cu-O bonding direction, as exemplified in the lower panels of Figs. 2.5(a) and 2.5(b), while the coherence peaks associated with the SC gap were significantly suppressed near the quantum impurities, and the degree of suppression was asymmetric between the electron-like and hole-like branches. Moreover, as the STM tip was moved away from the impurity site, the resonant scattering peak appeared to alternate between energies of the same magnitude and opposite signs, as exemplified in the upper panel of Fig. 2.5(b). Such spatial variations are expected for both Kondo-like and charge-like impurities [160–168]. Finally, the impurity effects on the variations in the quasiparticle spectra appeared to have completely diminished at approximately two coherence lengths ($\sim$3 nm) away from the impurity, as shown in the lower panel of Fig. 2.5(b).

Theoretically, the consideration for the effect of quantum impurities has been limited to a perturbative and one-band approximation without self-consistently solving for the spatially varying pairing potential in the presence of impurities [160–167]. Further, the existence of COs and the interaction among impurities have been neglected, which may be justifiable in the limit of dilute impurities and for over-doped cuprates. In this simplified approximation, the Hamiltonian $\mathcal{H}$ is approximated by $\mathcal{H} = \mathcal{H}_{\mathrm{BCS}} + \mathcal{H}_{\mathrm{imp}}$, where $\mathcal{H}_{\mathrm{BCS}}$ is the $d_{x^2-y^2}$-wave BCS Hamiltonian that contains the normal (diagonal) one-band single-particle eigen-energy and anomalous (off-diagonal) $d_{x^2-y^2}$-wave pairing potential $\Delta_k (= \Delta_d \cos 2\theta_k, \theta_k$ being the angle relative to the anti-node of the order parameter in the momentum space) of the unperturbed host, and

$$\mathcal{H}_{\mathrm{imp}} = \mathcal{H}_{\mathrm{pot}} + \mathcal{H}_{\mathrm{mag}} = U \sum_{\sigma} c_{0\sigma}^{\dagger} c_{0\sigma} + \sum_{R} J_R \boldsymbol{S} \cdot \boldsymbol{\sigma}_R \qquad (2.4)$$

denotes the impurity perturbation due to both the localized potential scattering term $\mathcal{H}_{\mathrm{pot}} (= U \sum_{\sigma} c_{0\sigma}^{\dagger} c_{0\sigma}$; U: the on-site Coulomb scattering potential) and the Kondo-like magnetic exchange interaction term $\mathcal{H}_{\mathrm{mag}} (= \sum_{R} J_R \boldsymbol{S} \cdot \boldsymbol{\sigma}_R)$ between the spins of the conduction carriers on the R sites ($\boldsymbol{\sigma}_R$) and those of the localized magnetic moments (S), with J_R being the exchange coupling constant. Assuming the aforementioned model Hamiltonian $\mathcal{H}$, one can obtain the quasiparticle spectra due to impurities from the Green function derived from $\mathcal{H}$. If the effects of the

tunneling matrix are further neglected for simplicity, one obtains, in the
pure potential scattering limit (where $\mathcal{H}_{\text{imp}} = \mathcal{H}_{\text{pot}}$), a resonant energy at
Ω on the impurity site that satisfies the following relation [160, 161]:

$$|\Omega/\Delta_d| \approx \{(\pi/2)\cot\delta_0/\ln[8/(\pi\cot\delta_0)]\}, \qquad (2.5)$$

where δ_0 is the impurity-induced phase shift in the quasiparticle wavefunc-
tion at a long distance. Generally $\delta_0 \to (\pi/2)$ is true in the strong potential scattering (unitary) limit. On the other hand, in the case of magnetic
impurities with both contributions from $\mathcal{H}_{\text{pot}}$ and $\mathcal{H}_{\text{mag}}$, one expects two
spin-polarized impurity states at energies $\Omega_\pm$, which are given by [163]:

$$|\Omega_\pm/\Delta_d| = 1/[2\mathcal{N}_F(U \pm W)\ln|8\mathcal{N}_F(U \pm W)|], \qquad (2.6)$$

where $\mathcal{N}_F$ is the density of states at the Fermi level, $W \equiv J(\boldsymbol{S} \cdot \boldsymbol{\sigma})$ implies
that magnetic impurities are isolated and equivalent at all sites, and the
two energies Ω_+ and Ω_- are associated with the upper and lower signs in
Eq. (2.6), respectively.

Comparing the STS studies of (Zn, Mg)-Y123 with those of a 0.6%
Zn-substituted $Bi_2Sr_2CaCu_2O_{8-x}$ (Bi-2212) [158], it is found that the primary features such as the appearance of a single resonant scattering peak
and strong suppression of the superconducting coherence peaks at the impurity site, as well as the rapidly decreasing intensity of the resonant peak with
the displacement from the impurity site, are generally comparable in both
systems. These findings are consistent with the simplified theoretical model
outlined above in the unitary limit, and further imply the preservation of
time-reversal $(\mathcal{T})$ symmetry in both systems, suggesting the absence of any
discernible complex order parameter in the pairing symmetry. The agreement of experimental findings with the simplified model Hamiltonian may
be understood in terms of the presence of nodal quasiparticles in hole-type
cuprates. As elaborated later in Sec. 2.2.2.2, the relevant competing orders
in the hole-type cuprates are primarily associated with the CDW or PDW
orders with a wave-vector parallel to the Cu-O bonding directions, and
therefore the effective energy gap of the hole-type cuprates in the pairing
state always vanishes at $\boldsymbol{k} = (\pm\pi, \pm\pi)$. Thus, nodal quasiparticles most
responsible for the quantum impurity-induced low-energy excitations are
always present in the hole-type cuprates, leading to experimental observations qualitatively consistent with the model Hamiltonian. In fact, the
model Hamiltonian may be generalized to include the existence of competing orders by replacing the superconducting d-wave gap Δ_d in Eqs. (2.2)

and (2.3) by the effective gap $\Delta_{\mathrm{eff}} = [(\Delta_d)^2 + (V_{\mathrm{CO}})^2]^{1/2}$, where V_{CO} denotes the energy gap associated with a given competing order.

On the other hand, several spectral differences are noteworthy between the (Zn, Mg)-Y123 and the Zn-substituted Bi-2212. First, the strength of impurity scattering appears weaker and longer-ranged in Y123. Second, various phenomena that are more consistent with the Kondo effect, such as alternating resonant peak energies between $+\Omega$ and $-\Omega$ with the distance from a non-magnetic impurity and temporal variations of the resonant peak, have only been observed in Y123. Third, global suppression of the SC energy gap and of the collective magnetic excitation energy has only been revealed in Y123. Such a difference may be attributed to the fact that pure Y123 generally exhibits long-range spectral homogeneity [124, 125, 172], whereas nano-scale spectral variations commonly observed in nominally pure Bi-2212 samples [173, 174] (probably due to the inherent non-stoichiometric nature of Bi-2212) yield difficulties in identifying the global effect of impurity substitutions.

While the hole-type cuprates are shown to be strongly affected by both magnetic and non-magnetic quantum impurities in the CuO_2 planes, in the case of an electron-type cuprate superconductor, the infinite-layer system $Sr_{0.9}La_{0.1}CuO_2$ (La-112), it is found to be insensitive to a small concentration of non-magnetic impurities so that the bulk magnetization studies [175] revealed no suppression in the bulk T_c up to 3% substitutions of Zn into the Cu sites, beyond which the compound became inhomogeneous and phase segregated. On the other hand, the substitution of magnetic Ni-impurities into the Cu sites of La-112 yielded significant T_c suppression: With 1% Ni, T_c already decreased from 43 K to 32 K; 2% Ni dropped T_c to below 4 K, and 3% Ni completely suppressed the bulk superconductivity although the sample was still stoichiometrically homogeneous from X-ray diffraction. The quasiparticle spectra near quantum impurities in La-112 [35] also differ significantly from those of hole-type cuprates [124–126, 148–159]. Specifically, the substitutions of either Zn or Ni do not result in strong resonant peaks near the impurities [35]. Rather, the quasiparticle spectra only exhibit modifications to the height of the coherence peaks and an increase in the spectral weight within the superconducting gap, with symmetric suppression of both coherence peaks in the case of Zn impurities (see Fig. 2.6(a)) [35] and asymmetric suppression of the particle-like and hole-like coherence peaks in the case of Ni impurities (see Fig. 2.6(b)) [35]. Moreover, substantially increased sub-gap spectral weight is found in the case of Zn impurities, suggesting

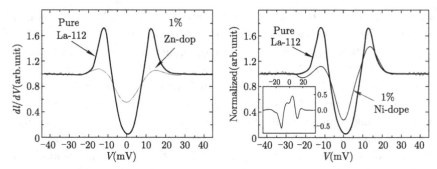

Fig. 2.6   Comparison of the quasiparticle tunneling spectrum (thick solid line) of the pure electron-type optimally-doped cuprate $Sr_{0.9}La_{0.1}CuO_2$ (La-112) with those of the quantum impurity-substituted La-112 at $T = 4.2$ K [35]: (a) A representative tunneling spectrum of 1% Zn-substituted La-112, $Sr_{0.9}La_{0.1}(Cu_{0.99}Zn_{0.01})O_2$ shown by the dashed line reveals symmetrically suppressed superconducting coherence peaks and a substantially increased sub-gap spectral weight, suggesting strong enhancement of low-energy quasiparticle excitations induced by the Zn impurities. (b) A representative tunneling spectrum of 1% Ni-substituted La-112, $Sr_{0.9}La_{0.1}(Cu_{0.99}Ni_{0.01})O_2$ (shown in the main panel by the thin line), reveals asymmetrically suppressed superconducting coherence peaks and moderately increased sub-gap spectral weight. This finding suggests weaker effects induced by the Ni impurities. The inset shows the spectral difference due to Ni-impurities. The spectral difference for impurity bound states is found to extend over a long range [35], similar to the Shiba impurity bands [142].

strong enhancement of low-energy quasiparticle excitations induced by the Zn substitution. In contrast, the spectral difference due to Ni impurities reveals long-range impurity bound states [35] that are similar to the Shiba impurity bands for magnetic impurities in fully gapped conventional superconductors [142]. The assumption of impurity bands may be justified by noting that the average Ni-Ni separation for 1% Ni substitutions (~ 1.8 nm) is shorter than the in-plane superconducting coherence lengths of the pure La-112 sample (~ 4.8 nm) [176] so that substantial overlap of the impurity wavefunctions may be expected.

The puzzling spectral response of the electron-type La-112 to Zn and Ni impurities was initially considered as supporting evidence for s-wave pairing symmetry in the La-112 system [35]. However, further investigations of the electron-type cuprates [7, 10, 16, 17] (see Sec. 2.2.2.2 for more details) suggest that the co-existence of a SDW competing order with a (π, π) wave-vector and an energy gap smaller than the SC gap ($\Delta_{SDW} < \Delta_{SC}$) could qualitatively account for the experimental observation. That is, the effective gap of the electron-type cuprate superconductors is fully gapped and anisotropic for all k-values even though the $d_{x^2-y^2}$-wave SC gap Δ_{SC} vanishes at $(\pm\pi/4, \pm\pi4)$ [7, 10, 16], which is consistent with the ARPES finding [177]. Hence, there are effectively no nodal quasiparticles interacting with quantum impurities so that no sharp spectral resonances could be found at quantum impurities in the electron-type cuprate superconductors. In this context, the characteristics of the low-energy excitations in electron-type cuprates due to either non-magnetic or magnetic impurities are analogous to those of fully gapped, anisotropic conventional superconductors. However, to date there has not been theoretical investigation of the spectral effects of quantum impurities in electron-type cuprates so that no quantitative comparison can be made with experimental observation.

2.2.2.2 Unconventional low-energy excitations in zero fields

As mentioned in the introduction, one of the most widely debated issues in cuprate superconductivity is the possibility of pre-formed Cooper pairs and the origin of the PG phenomenon. In this section the zero-field spectral characteristics of various types of hole- and electron-type cuprates are examined, with special emphasis on the analysis of the quasiparticle local density of state (LDOS) spectra taken by means of STM/STS, and the momentum-dependent quasparticle spectra taken by means of the ARPES. We shall show below that many spectral details as a function of the energy, momentum and temperature cannot be explained by considering Bogoliubov quasiparticles as the sole low-energy excitations in the cuprate superconductors. In contrast, the incorporation of COs coexisting with SC can provide consistent descriptions for all data.

Our theoretical analysis begins with a mean-field Hamiltonian $\mathcal{H}_{MF} = \mathcal{H}_{SC} + \mathcal{H}_{CO}$ that consists of coexisting SC and a CO at $T = 0$ [10–17]. We further assume that the SC gap Δ_{SC} vanishes at T_c and the CO order parameter vanishes at T^*, and that both T_c and T^* are second-order phase transitions. The SC Hamiltonian is given by:

$$\mathcal{H}_{SC} = \sum_{k,\alpha} \xi_k c_{k,\alpha}^\dagger c_{k,\alpha} - \sum_k \Delta_{SC}(k)(c_{k,\uparrow}^\dagger c_{-k,\downarrow}^\dagger + c_{-k,\downarrow} c_{k,\uparrow}), \qquad (2.7)$$

where $\Delta_{SC}(k) = \Delta_{SC}(\cos k_x - \cos k_y)/2$ for $d_{x^2-y^2}$-wave pairing, k denotes the quasiparticle momentum, ξ_k is the normal-state eigen-energy relative to the Fermi energy, $c^\dagger$ and c are the creation and annihilation operators, and $\alpha = \uparrow, \downarrow$ refers to the spin states. The CO Hamiltonian is specified by the energy V_{CO}, a wave-vector Q, and a momentum distribution δQ that depends on a form factor, the correlation length of the CO, and also on the degree of disorder [10–17]. We have previously considered the effect of various types of COs on the quasiparticle spectral density function $A(\boldsymbol{k}, \omega)$ and the density of states $\mathcal{N}(\omega)$. For instance, in the case that CDW is the relevant CO, we have a wave-vector $\boldsymbol{Q}_1$ parallel to the CuO$_2$ bonding direction $(\pi, 0)$ or $(0, \pi)$ in the CO Hamiltonian [10–17]:

$$\mathcal{H}_{\text{CDW}} = -\sum_{\boldsymbol{k},\alpha} V_{\text{CDW}}(\boldsymbol{k})(c_{\boldsymbol{k},\alpha}^{\dagger} c_{\boldsymbol{k}+\boldsymbol{Q}_1,\alpha} + c_{\boldsymbol{k}+\boldsymbol{Q}_1,\alpha}^{\dagger} c_{\boldsymbol{k},\alpha}). \qquad (2.8)$$

On the other hand, for commensurate SDW being the relevant CO, the SDW wave-vector becomes $\boldsymbol{Q}_2 = (\pi, \pi)$, and the corresponding CO Hamiltonian is [178]:

$$\mathcal{H}_{\text{SDW}} = -\sum_{\boldsymbol{k},\alpha,\beta} V_{\text{SDW}}(\boldsymbol{k})(c_{\boldsymbol{k}+\boldsymbol{Q}_2,\alpha}^{\dagger} \sigma_{\alpha,\beta}^3 c_{\boldsymbol{k},\beta} + c_{\boldsymbol{k},\alpha}^{\dagger} \sigma_{\alpha,\beta}^3 c_{\boldsymbol{k}+\boldsymbol{Q}_2,\beta}), \qquad (2.9)$$

where $\sigma_{\alpha\beta}^3$ denotes the matrix element $\alpha\beta$ of the Pauli matrix σ^3. Similarly, other types of COs such as the disorder pinned SDW [179] and the DDW [73] may be considered by using the following CO Hamiltonians [7, 10]:

$$\mathcal{H}_{\text{SDW}}^{\text{pinned}} = -g^2 \sum_{\boldsymbol{k},\alpha} V_{\text{SDW}}(\boldsymbol{k})(c_{\boldsymbol{k},\alpha}^{\dagger} c_{\boldsymbol{k}+\boldsymbol{Q}_3,\alpha} + c_{\boldsymbol{k}+\boldsymbol{Q}_3,\alpha}^{\dagger} c_{\boldsymbol{k},\alpha}), \qquad (2.10)$$

$$\mathcal{H}_{\text{DDW}} = -\sum_{\boldsymbol{k},\alpha} \frac{1}{2} V_{\text{DDW}}(\cos k_x - \cos k_y)$$

$$\times (ic_{\boldsymbol{k}+\boldsymbol{Q}_2,\alpha}^{\dagger} c_{\boldsymbol{k},\alpha} - ic_{\boldsymbol{k},\alpha}^{\dagger} c_{\boldsymbol{k}+\boldsymbol{Q}_2,\alpha}), \qquad (2.11)$$

where $\boldsymbol{Q}_3 = \boldsymbol{Q}_1/2$ for disorder pinned SDW [178], and g denotes the coupling strength between disorder and SDW. Thus, by incorporating realistic bandstructures and Fermi energies for different families of cuprates with given doping and by specifying the SC pairing symmetry and the form factor for the CO, $\mathcal{H}_{MF}$ can be diagonalized to obtain the bare Green function $G_0(\boldsymbol{k}, \omega)$ for momentum $\boldsymbol{k}$ and energy ω. Further, quantum phase fluctuations between the CO and SC may be included by solving the Dyson's equation self-consistently for the full Green function $G(\boldsymbol{k}, \omega)$ [10–17], which gives the quasiparticle spectral density function $A(\boldsymbol{k}, \omega) = -\text{Im}[G(\boldsymbol{k}, \omega)]/\pi$ for comparison with ARPES [16] and the quasiparticle density of states $\mathcal{N}(\omega) = \sum_{\boldsymbol{k}} A(\boldsymbol{k}, \omega)$ for comparison with STM spectroscopy [10–17].

Based on the Green function analysis outlined above for coexisting $d_{x^2-y^2}$-wave SC and a specific CO, the zero-field quasiparticle spectra

$\mathcal{N}(\omega)$ and $A(\mathbf{k}, \omega)$ at $T = 0$ can be fully determined by the parameters $\Delta_{\mathrm{SC}}, V_{\mathrm{CO}}, \mathbf{Q}, \delta \mathbf{Q}, \Gamma_{\mathbf{k}}$ (the quasiparticle linewidth), and η (the magnitude of quantum phase fluctuations), which is proportional to the mean value of the velocity-velocity correlation function [12, 13]. For finite temperatures, the temperature Green function is employed to account for the thermal distributions of quasiparticles.

Using the aforementioned theoretical analysis we have been able to consistently account for the T-dependent quasiparticle tunneling spectra in both hole- and electron-type cuprates if we assume the Fermi-surface nested CDW as the CO in the hole-type cuprates such as Y-123 and Bi-2212, and the commensurate SDW as the CO in the electron-type La-112 and PCCO [10–17], which are consistent with findings from neutron scattering experiments [180, 181]. On the other hand, it is found that quasiparticle spectra obtained from considering the DDW scenario generally do not agree with experimental observations.

Specifically for hole-type cuprates, in the spectra of Y-123 and Bi-2212 for example, the sharp peaks and satellite "hump" features at $T \ll T_c$ in Fig. 2.7(a) and also in Fig. 2.2(a) can be associated with $\omega = \pm \Delta_{\mathrm{SC}}$ and $\omega = \pm \Delta_{\mathrm{eff}}$, respectively, where $\Delta_{\mathrm{eff}} \equiv [(\Delta_{\mathrm{SC}})^2 + (V_{\mathrm{CO}})^2]^{1/2}$ is an effective excitation gap. Hence, the condition $V_{\mathrm{CO}} > \Delta_{\mathrm{SC}}$ in hole-type cuprates is responsible for the appearance of the satellite features at $T \ll T_c$ and the PG phenomena at $T^* > T > T_c$ [10–16]. In contrast, the condition $V_{\mathrm{CO}} < \Delta_{\mathrm{SC}}$ in electron-type cuprates, as exemplified in Fig. 2.7(b), is responsible for only one set of characteristic features at $\pm \Delta_{\mathrm{eff}}$ and the absence of PG above T_c [17].

By extending the above analysis to the tunneling spectra of different doping levels (δ) associated with the hole-type cuprates, it is found that $\Delta_{\mathrm{SC}}(\delta)$ generally follows the same non-monotonic dependence of $T_c(\delta)$ [7, 13]. In contrast, $V_{CO}(\delta)$ increases with decreasing δ, which is consistent with the general trend of the zero-field PG temperature in hole-type cuprates [7, 13]. Moreover, analysis of the FT-LDOS of hole-type cuprates of Y-123 [10, 11] and $\mathrm{Bi_2Sr_2CuO_{6+x}}$ (Bi-2201) [64] reveals energy-independent scattering wave-vectors that are in stark contrast to the strongly energy-dependent wave-vectors due to quasiparticle interferences (QPI) from elastic impurity scattering of Bogoliubov quasiparticles, as exemplified in Fig. 2.8(b) for the optimally-doped Y-123 single crystal [10, 11]. Additionally, the energy-independent scattering wave-vector found in Bi-2201 reveals a strong doping dependence [64], which is consistent with a CDW nesting

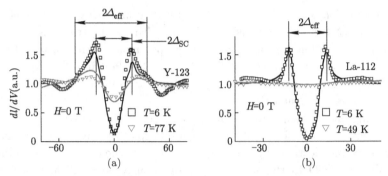

Fig. 2.7   Implication of CO from zero- and finite-field STS in Y-123 and La-112 [10–17]:
(a) Normalized zero-field tunneling spectra of Y-123 taken at $T = 6$ K (dark) and 77 K
(light). The solid lines represent fittings to the $T = 6$ K and 77 K spectra by assuming
coexisting SC and CDW, with fitting parameters of $\Delta_{SC} = 20$ meV, $V_{CDW} = 32$ meV and
$Q_{CDW} = (0.25\pi \pm 0.05\pi, 0)/(0, 0.25\pi \pm 0.05\pi)$, following Refs. [10–16]. (b) Normalized
zero-field tunneling spectra of La-112 taken at $T = 6$ K (dark) and 49 K (light). The solid
lines represent fittings to the $T = 6$ and 49 K spectra by assuming coexisting SC and
SDW, with fitting parameters $\Delta_{SC} = 12$ meV, $V_{SDW} = 8$ meV, and $Q_{SDW} = (\pm\pi, \pm\pi)$,
following Refs. [10–17]. We further note that the CO energies are consistent with those
obtained from neutron scattering experiments [180, 181].

wave-vector on the Fermi surface so that the wave-vector decreases with
increasing hole-doping, as schematically sketched in Fig. 2.8(a). Finally, in
the optimally-doped Bi-2212 [61, 62] and under-doped $Ca_{2-x}Na_xCuO_2Cl_2$
[63], the observed strong FT-LDOS intensity associated with an energy-
independent wave-vector along the $(\pm\pi, 0)/(0, \pm\pi)$ directions at $T \ll T_c$
and the remaining finite intensity of these four spots for $T_c < T < T^*$ [23]
are all in contradiction to the "one-gap" scenario, while they are naturally
accounted for if the energy-independent wave-vector is attributed to the
CDW nesting wave-vector that coexists with SC at $T \ll T_c$ [61–63] and
still remains for $T_c < T < T^*$ [23]. These viewpoints have been discussed
in detail through quantitative analysis of the experimental data [7, 10–16].

2.2.2.3   Dichotomy of quasiparticle coherence

Another empirical finding of unconventional low-energy excitations associated with the cuprates is the dichotomy of quasiparticle coherence revealed
by spectroscopic studies of the pairing state of hole-type cuprates [19, 112,
113, 182]. This finding can be naturally explained by considering the broadening of the spectral density function $A(\mathbf{k}, \omega)$ by the increasing magnitude of quantum phase fluctuations η [12, 13], and both the quasiparticle

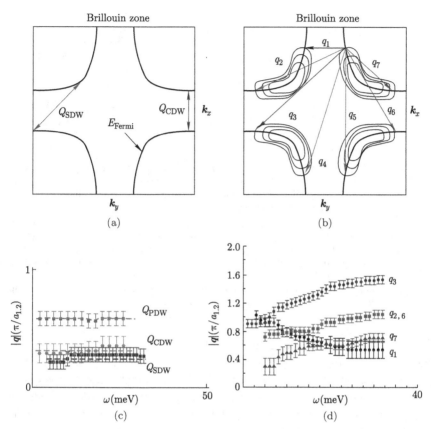

Fig. 2.8 Scattering wave-vectors obtained from the FT-LDOS of quasiparticle tunneling spectra [10, 11]: (a) Illustration of the wave-vectors associated with SDW and CDW excitations. (b) Illustration of the wave-vectors associated with elastic quasiparticle interferences (QPI) between pairs of points on equal energy contours with maximum joint density of states. (c) The nearly energy-independent collective modes $|\mathbf{Q}_{\mathrm{PDW}}|$, $|\mathbf{Q}_{\mathrm{CDW}}|$ and $|\mathbf{Q}_{\mathrm{SDW}}|$ obtained from the FT-LDOS data of Y-123 [10, 11]. (d) The QPI momentum ($|\mathbf{q}_i|$) versus energy (ω) dispersion relations derived from the FT-LDOS data of Y-123 [10, 11], which are found to be in excellent agreement with the QPI results obtained from the optimally doped Bi-2212 [112].

coherence, as manifested by the inverse linewidth (Γ^{-1}) of $A(\mathbf{k}, \omega)$, and the renormalized effective gap $\Delta_{\mathrm{eff}}(\mathbf{k})$ exhibit "dichotomy" in the momentum space [12, 13], showing different evolution in the Cu-O bonding direction $(0, \pi)/(\pi, 0)$ from that in the (π, π) nodal direction, as confirmed by recent ARPES results [113]. Specifically, dichotomy in the quasiparticle coherence can be manifested by comparing the linewidth of fluctuation-renormalized

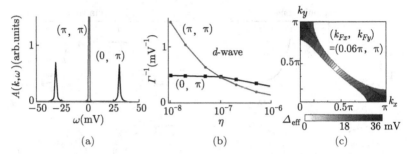

Fig. 2.9 Dichotomy in the spectral density function and excitation gap due to coexisting
SC/CO and quantum phase fluctuations [12, 13]: (a) Contrasts in the fluctuation renor-
malized $A(\mathbf{k}, \omega)$ at the Fermi level for $\mathbf{k}$ along $(\pi, 0)/(0, \pi)$ [darker (black) lines] and along
(π, π) [lighter (grey) lines] for coexisting $d_{x^2-y^2}$-wave SC and disorder pinned SDW with
$\eta = 3 \times 10^{-8}$, showing a narrower linewidth in $A(\mathbf{k}, \omega)$ along (π, π). (b) Dichotomy in
the quasiparticle lifetime ($\propto \Gamma^{-1}$), showing better quasiparticle coherence along (π, π)
than along $(\pi, 0)/(0, \pi)$ if the quantum fluctuations are sufficiently small, as manifested
by the Γ^{-1} vs. η plot. The latter finding is consistent with the empirical observation
of more coherent nodal quasiparticles in hole-type cuprate superconductors [112, 113,
182]. (c) Competing order-induced dichotomy in the momentum-dependent effective gap
$\Delta_{\text{eff}}(\mathbf{k})$ is illustrated in the first quadrant of the first BZ.

$A(\mathbf{k}, \omega \sim \Delta_{\text{eff}})$ for $\mathbf{k}$ along (π, π) with that for $\mathbf{k}$ along $(0, \pi)/(\pi, 0)$, as
exemplified in Figs. 2.9(a) and 2.9(b) for coexisting $d_{x^2-y^2}$-wave SC and
disorder pinned SDW or CDW [12]. The degree of dichotomy in the quasiparticle coherence decreases with increasing quantum fluctuations, as shown
in the inverse linewidth Γ^{-1}-vs.-η plots in Figs. 2.9(a) and 2.9(b). In particular, we note that quasiparticles of $d_{x^2-y^2}$-wave SC with disorder pinned
SDW or CDW exhibit better coherence along (π, π) than along $(0, \pi)/(\pi, 0)$
for small η, which is consistent with the ARPES and STS data [112, 113,
182]. The CO-induced dichotomy in the momentum-dependent effective gap
$\Delta_{\text{eff}}(\mathbf{k})$ is illustrated in the first quadrant of the first Brillouin zone (BZ)
in Fig. 2.9(c) [12, 13].

2.2.3 Normal state

2.2.3.1 Pseudogap phenomena

The low-energy PG phenomena described in the overview are most notably
observed in $Bi_2Sr_2CaCu_2O_x$ (Bi-2212) and $Bi_2Sr_2CuO_x$ (Bi-2201) as a
function of hole-doping [18, 19, 21–25]. The persistence of gapped quasiparticle spectral density functions near the $(\pi, 0)$ and $(0, \pi)$ portions of

the Brillouin zone above T_c in hole-type cuprates are the source of the incomplete recovery of the Fermi surface [16, 19, 21, 60]. In contrast, electron-type cuprates exhibit neither the low-energy PG nor the Fermi arc above T_c [35–37], although "hidden pseudogap" features in the quasiparticle excitation spectra have been observed under the superconducting dome in doping-dependent grain-boundary tunneling experiments on $Pr_{2-x}Ce_xCuO_{4-y}$ (PCCO) and $La_{2-x}Ce_xCuO_{4-y}$ (LCCO) when a magnetic field $H > H_{c2}$ is applied to suppress superconductivity [37, 38]. Further, a gap enhancement near the "hot spots" (π, π) of the Fermi surface in the SC state of the electron-type cuprate $Nd_{2-x}Ce_xCuO_{4-y}$ (NCCO) [39] is also consistent with the notion of a hidden PG with $\Delta_{PG} < \Delta_{SC}$ at $T < T_c$.

In addition to the PG phenomena manifested in the energy dependence of the LDOS spectra as exemplified in Fig. 2.2(b), STS studies further suggest that the PG phase in the hole-type cuprates in fact stems from lattice translational symmetry breaking, as represented by the energy-independent wave-vector obtained from the FT-LDOS for $T_c < T < T^*$ [23] and exemplified in Fig. 2.10 for theoretically calculated FT-LDOS as a

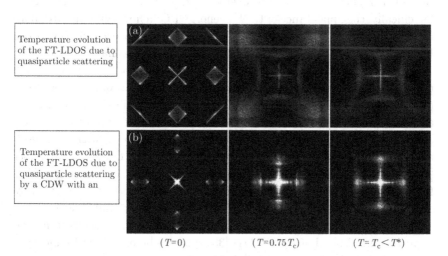

Temperature evolution of the FT-LDOS due to quasiparticle scattering

Temperature evolution of the FT-LDOS due to quasiparticle scattering by a CDW with an

$(T=0)$ $(T=0.75\,T_c)$ $(T=T_c<T^*)$

Fig. 2.10 The temperature evolution of the FT-LDOS maps in the first Brillouin zone at $T = 0, 0.75T_c, T_c$ (from left to right) for quasiparticle scattering in a $d_{x^2-y^2}$–wave superconductor [183] by (a) point impurities and (b) a competing order such as a CDW or a disorder pinned SDW with an incommensurate wave-vector parallel to the $(\pi, 0)/(0, \pi)$ direction. Here we have assumed that the CO has an energy gap V_{CO} larger than the SC gap and that V_{CO} does not vanish until the PG temperature $T^* > T_c$ [183]. The non-vanishing FT-LDOS intensities for a constant $|\mathbf{k}|$ value above T_c are consistent with experimental observation [23].

function of temperature in the presence (absence) of a CDW-like competing order [183]. Moreover, the low-energy excitation spectra for $T_c < T < T^*$ reveal particle-hole asymmetric spectral characteristics that differ fundamentally from the particle-hole symmetric Bogoliubov quasiparticle spectra for superconductivity [60]. Further, Raman scattering spectroscopic studies of the doping dependence of $YBa_2Cu_3O_{6+x}$ and $Bi_2Sr_2(Ca_xY_{1-x})Cu_2O_8$ and ARPES studies of $Bi_2Sr_2CuO_{6+x}$ (Bi-2201) also confirmed particle-hole symmetry breaking and pronounced spectral broadening, indicative of spatial symmetry breaking without long-range order at the opening of the PG, in agreement with the STS findings that the PG state is a broken-symmetry state that is distinct from homogeneous superconductivity [60].

2.2.3.2   Fermi arcs

Intimately related to the PG phenomena found in the quasiparticle tunneling spectra is the Fermi arc observed from ARPES studies of hole-type cuprate superconductors [18, 21], where the Fermi arc refers to the truncated Fermi surface not fully recovered at $T_c < T < T^*$ [18, 21, 182, 184]. Specifically, the appearance of the low-energy PG in hole-type cuprates may be correlated with the appearance of the Fermi arc above T_c and below the PG temperature T^* [18, 21, 182, 184] within the CO scenario [7, 10, 16].

As exemplified in Fig. 2.11 and detailed elsewhere [16], we find that the Fermi arc as a function of the quasiparticle momentum k, temperature (T) and doping level (δ) in Bi-2212 [21] may be explained consistently by assuming a CDW (or disorder pinned incommensurate SDW) as the relevant CO, which occurs at the PG temperature $T^* > T_c$ with an energy gap $V_{CDW} > \Delta_{SC}$ and an incommensurate wave-vector $Q_{CDW}||(\pi,0)/(0,\pi)$ [7, 10–16]. Thus, the PG gap is associated with the CO energy gap V_{CO} for $T_c < T < T^*$, whereas the effective gap $\Delta_{eff}(k)$ for $T < T_c$ is given by $\Delta_{eff}(k) \equiv \{[\Delta_{SC}(k)]^2 + [V_{CO}(k)]^2\}^{1/2}$. In contrast, the k- and T-dependence of the effective gap $\Delta_{eff}(k)$ and the absence of Fermi arcs in electron-type cuprates (e.g. $Pr_{0.89}LaCe_{0.11}CuO_4$) [35] can also be explained by incorporating a SDW with $V_{SDW} < \Delta_{SC}$ [39, 181] and a commensurate wave-vector $Q_{SDW} = (\pi, \pi)$ (see Eq. (2.9)) into spectral characteristics [16]. Hence, the CO scenario is shown to provide adequate phenomenology for a wide variety of experimental findings from electron-type to hole-type cuprate superconductors and as a function of temperature, doping level and quasiparticle momentum k.

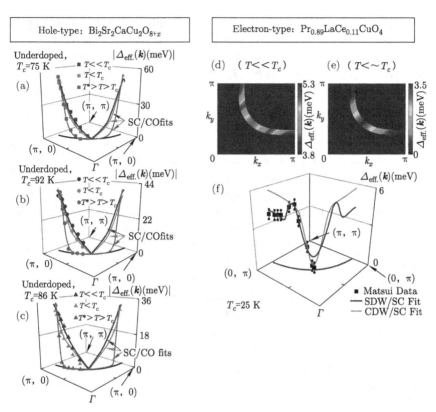

Fig. 2.11 (a)–(c) (color online) Theoretical fittings (solid color curves) [16] to the momentum ($\boldsymbol{k}$) dependent effective excitation gap Δ_{eff} (color symbols) determined from ARPES spectra on Bi-2212 for three different doping levels and at three temperatures [21]. By considering the scenario of coexisting $d_{x^2-y^2}$-SC and CDW [16], the calculated ARPES results agree well with experimental findings [21]: Below T_C, $\Delta_{\mathrm{eff}}(\boldsymbol{k})$ only vanishes at the nodal point, whereas above T_C and below T^*, $\Delta_{\mathrm{eff}}(\boldsymbol{k})$ vanishes over an "arc"-shaped of wave-vectors, as indicated by the momentum distribution curves on the zero effective gap plane shown in (a)–(c). In contrast, the $\boldsymbol{k}$- and T-dependences of $\Delta_{\mathrm{eff}}(\boldsymbol{k})$ in electron-type cuprates $Pr_{0.89}LaCe_{0.11}CuO_4$ differ from that in the hole-type cuprate, as exemplified in (d)–(f) [35]. The CO scenario can account for both the absence of Fermi arcs above T_C and the $\Delta_{\mathrm{eff}}(\boldsymbol{k})$-vs.-$\boldsymbol{k}$ behavior below T_C by assuming a SDW with a (π, π) wave-vector as the relevant CO and $V_{\mathrm{SDW}} < \Delta_{\mathrm{SC}}$ [16, 39, 181]. On the other hand, fitting with a CDW does not agree with the k-dependence of $\Delta_{\mathrm{eff}}(\boldsymbol{k})$.

2.2.3.3 Phase fluctuations above T_C

One of the major experimental signatures that strongly favor the "two-gap" scenario over the "one-gap" model is the finding of vanishing SC coherence at temperatures well below the PG temperature T^* [80]. Specifically, the one-gap conjecture [185–188] suggests that in under-doped cuprate super-conductors, Cooper pairs could form at T^* while significantly SC phase fluctuations prevent Cooper pairs from condensing into a true SC state until the temperature is lowered below T_c. Hence, one would expect remnants of SC phase coherence for $T_c < T < T^*$, which may be captured by means of high-frequency optical conductivity measurements if the relaxation time of the "pre-formed" Cooper pairs becomes sufficiently short as $T \to T^*$ from below. However, optical conductivity measurements for under-doped hole-type cuprates Bi-2212 over a range of doping levels [80] reveal that the complex paraconductivity associated with the SC coherence generally follows the Kosterlitz–Thouless–Berezinskii (KTB) theory for thermally generated vortices and only survives over a small temperature window above T_c and much below T^* [80]. The finding of rapidly vanishing superconducting phase coherence above T_c together with the breaking of particle-hole symmetry at $T_c < T < T^*$, and the quantitative studies outlined above for both the pairing state and the normal state properties of a variety of cuprate superconductors strongly suggest that the physical origin of the PG differs from the SC gap, and hence favors the two-gap scenario. As further discussed in the following section for the vortex-state properties, it is found that the PG phenomena observed above T_c in zero magnetic fields may be revealed by suppressing SC using an external magnetic field at $T \ll T_c$, again confirming the scenario of the PG physical origin being associated with COs.

2.2.4   Vortex state

High-temperature superconducting cuprates are extreme type-II supercon-ductors that exhibit strong thermal, disorder, and quantum fluctuations in their vortex states [77, 78, 81–97]. While much research has focused on the macroscopic vortex dynamics of cuprate superconductors with phe-nomenological descriptions [81–97], little effort has been made to address the microscopic physical origin of their extreme type-II nature until recently when spatially resolved vortex-state quasiparticle tunneling spectra became available [7, 10, 11, 17]. As discussed in previous sections, COs can coexist with superconductivity (SC) in the ground state of cuprate superconduc-tors [7, 10], which lead to the occurrence of quantum criticality [5, 6, 67, 68, 189]. The proximity to quantum criticality and the existence of COs can significantly affect the low-energy excitations of the cuprates due to strong quantum fluctuations [77, 78] and the redistribution of quasiparticle spectral weight among SC and COs [7, 12–15]. Moreover, external variables such as temperature (T) and applied magnetic field (H) can vary the interplay of SC and COs, such as inducing or enhancing [41, 52, 68] the COs at the price of more rapid suppression of SC, thereby leading to weakened SC stiffness and strong thermal and field-induced fluctuations [7, 77, 78]. On the other hand, the quasi-two dimensional nature of the cuprates can also result in quantum criticality in the limit of decoupling of CuO_2 planes [190]. In this section we review experimental studies of the unconventional low-energy excitations of the cuprates in the vortex state from both microscopic and macroscopic viewpoints.

2.2.4.1 Intra-vortex pseudogap and energy-independent wave-vectors in the quasiparticle tunneling spectra

In conventional type-II superconductors, superconductivity is suppressed inside periodic Abrikosov vortices [191], leading to continuous quasiparticle bound states and a peak of local density of states (LDOS) at zero energy [192–194]. In contrast, the effect of magnetic field on high-T_c superconductors is much more complicated than that on conventional type-II superconductors. Microscopically, neutron scattering experiments on the hole-doped cuprate $La_{1.84}Sr_{0.16}CuO_4$ reported an effective radius of vortices substantially larger than the superconducting coherence length ξ_{SC} [41, 52]. Scanning tunneling spectroscopic (STS) studies of optimally-doped Bi-2212 found PG-like features rather than zero-bias conductance peaks inside vortices [20, 195]. Further detailed spatially resolved STS studies of Bi-2212 in one magnetic field $H = 5$ T revealed a field-induced $(4a_0 \times 4a_0)$ conductance modulation inside each vortex, where $a_0 = 0.385$ nm is the planar lattice constant of Bi-2212 [61]. The latter finding has been attributed to the presence of a coexisting CO such as pair-density waves (PDW) [71, 72], pinned SDW [5, 6, 68–70], or CDW [4, 75] upon suppression of SC inside the vortices.

More recently, spatially resolved STS studies of the optimally-doped hole-type cuprate superconductor Y-123 and the optimally-doped electron-type cuprate superconductor La-112 in the vortex state have been carried out as a function of applied magnetic fields [10, 11, 17], which reveal rich information and interesting contrasts between the hole- and electron-type cuprate superconductors. In the case of optimally-doped Y-123, [10, 11] while the zero-field LDOS revealed highly homogeneous spectral characteristics, the vortex-state STS studies suggest strongly disordered vortices as

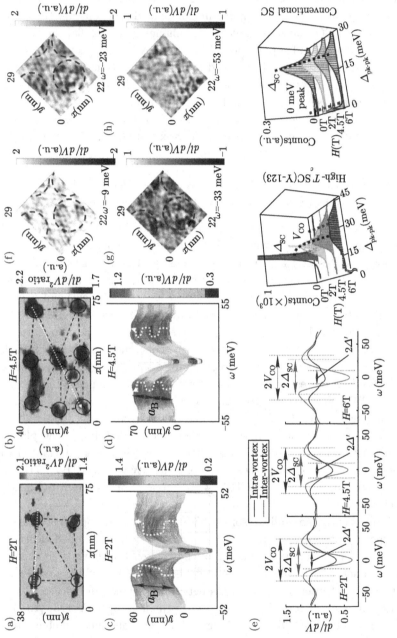

Fig. 2.12   *(Continued)*

←

Fig. 2.12   Spatially resolved STS studies of the vortex-state of Y-123 at $T = 6$ K [10, 11]: (a) Tunneling conductance power ratio r_G map over a (75×38) nm^2 area for $H = 2$ T, showing disordered vortices with an average vortex-vortex separation $a_B = (33.2 \pm 9.0)$ nm. Here the conductance power ratio at each pixel is defined by the ratio of $(dI/dV)^2$ at $V = (\Delta_{SC}/e)$ to that at $V = 0$. (b) The r_G map over a (75×40) nm^2 area for $H = 4.5$ T, showing $a_B = (23.5 \pm 8.0)$ nm. (c) Conductance spectra along the white line in (a), showing SC peaks at $\omega = \pm \Delta_{SC}$ outside vortices and PG features at $\omega = \pm V_{CO}$ inside vortices. (d) Conductance spectra along the dashed line indicated in (b). (e) Spatially averaged intra- and inter-vortex spectra for $H = 2$ T, 4.5 T and 6 T from left to right. (f) The LDOS modulations of Y-123 at $H = 5$ T over a (22×29) nm^2 area, showing patterns associated with density-wave modulations and vortices (circled objects) for $\omega = -9$ meV $\sim -\Delta'$, $\omega = -23$ meV $\sim -\Delta_{SC}$, $\omega = -33$ meV $\sim -V_{CO}$ and $\omega = -53$ meV, which is comparable to the longitudinal optical phonon frequency [196]. The vortex contrasts are the most apparent at $|\omega| \sim \Delta_{SC}$ and become nearly invisible for $|\omega| \sim V_{CO}$. (g) Energy histograms for the field-dependent spectral weight derived from the STS data for $H = 0$ T, 2 T, 4.5 T, and 6 T, showing a spectral shift from Δ_{SC} to V_{CO} and Δ' with increasing H. (h) Schematic of the histograms for a conventional type-II superconductor in the limit of $T \ll T_c$ and $H \ll H_{c2}$.

well as a "vortex halo" radius ξ_{halo} much larger than ξ_{SC}, as exemplified in Figs. 2.12(a) and 2.12(b) for $H = 2.0$ T and 4.5 T, respectively. This finding is consistent with the report from neutron scattering experiments [41, 52]. Moreover, the spatial evolution of the vortex-state spectra reveals modulating gap-like features everywhere without any zero-energy peaks, as shown in Figs. 2.12(c) and 2.12(d), and in Fig. 2.12(e) for representative spectra taken inside and outside of vortices at $H = 2.0$ T, 4.5 T and 6.0 T. For each constant field, the inter-vortex spectrum reveals a sharper set of peaks at $\omega = \pm \Delta_{SC} \sim \pm 23$ meV, whereas the intra-vortex spectrum exhibits PG features at $\omega = \pm V_{CO} \sim \pm 32$ meV and $V_{CO} > \Delta_{SC}$. Interestingly, the PG energy V_{CO} revealed inside the vortex core is in excellent agreement with the CO energy obtained from theoretical fitting to the zero-field spectra using the relation $V_{CO} = [(\Delta_{eff})^2 - (\Delta_{SC})^2]^{1/2}$, as described in Sec. 2.2.2.2. Additionally, sub-gap features at $\omega = \pm \Delta' = \pm(7 \sim 10)$ meV are found inside vortices, which become more pronounced with increasing H. The physical origin of Δ' is still unknown, although it may be associated with the energy of PDW, while V_{CO} may be associated with the CDW or disorder pinned SDW. Further, apparent LDOS modulations are visible at constant quasiparticle energies (Fig. 2.12(f)), showing patterns associated with density-wave modulations and vortices (circled objects) for $\omega = -9$ meV $\sim -\Delta'$, $\omega = -23$ meV $\sim -\Delta_{SC}$, $\omega = -33$ meV $\sim -V_{CO}$ and $\omega = -53$ meV. The vortex contrasts are the most apparent at $|\omega| \sim \Delta_{SC}$ and become nearly invisible for $|\omega| \sim V_{CO}$. The vanishing contrast at high energies may be due to the onset of Cu-O optical phonons (~ 50 meV for the cuprates [196]) so that both the collective modes and quasiparticles become scattered inelastically. Finally, energy histograms of the gap features exhibit strong spectral shifts from Δ_{SC} to V_{CO} and Δ' with increasing magnetic field, as shown in Fig. 2.12(g), which is in sharp contrast to the vortex-state spectral shifts in conventional type-II superconductors, as schematically illustrated in Fig. 2.12(h). These vortex-state spectral findings are all consistent with the co-existence of COs with SC in Y-123 [7].

In addition to the revelation of COs at characteristic energies V_{CO} and Δ', evidences for collective modes at characteristic wave-vectors may be identified from studies of the Fourier transformation (FT) of the LDOS. As exemplified in Figs. 2.13(a)–(f), the spectral intensity $|F(\mathbf{k}, \omega)|$ of the FT-LDOS reveal abundant information about the dependence of the cuprate low-energy excitations on momentum $(\mathbf{k})$, energy (ω) and magnetic field (H). Similar to the findings in zero-fields, the FT-LDOS spectra contain two types of high intensity spots. One type is associated with the strongly ω-dependent Bogoliubov QPI due to elastic scattering by impurities, as exemplified in Fig. 2.13(e) and previously shown in Figs. 2.8(b) and 2.8(d). The other type contains three sets of ω-independent spots in addition to the reciprocal lattice constants and the (π, π) resonance, including $\mathbf{Q}_{PDW}$ and $\mathbf{Q}_{CDW}$ along the $(\pi, 0)/(0, \pi)$ directions and $\mathbf{Q}_{SDW}$ along (π, π), as exemplified in Figs. 2.13(b)–(e). Further investigation of the FT-LDOS reveals interesting magnetic field dependence, showing field-enhanced spectral intensities $|F(\mathbf{k}, \omega)|$ for $\mathbf{k} = \mathbf{Q}_{PDW}$ and $\mathbf{Q}_{CDW}$, which may be compared with the significant shifts in spectral weights from Δ_{SC} to V_{CO} and Δ' with increasing H. Additionally, $|F(\mathbf{k}, \omega)|$ – vs. – ω data for both $\mathbf{k} = \mathbf{Q}_{PDW}$ and $\mathbf{Q}_{CDW}$ consistently reveal a spectral peak around Δ', as shown in Fig. 2.13(f) [197], whereas significant enhancements in $|F(\mathbf{Q}_{CDW}, \omega)|$ for $\omega > \Delta'$ only occur for $H > 0$, suggesting that the application of a finite magnetic field increases the CDW excitations that break the particle-hole symmetry. It is clear that none of these spectral dependences on $\mathbf{k}, \omega$ and H can be simply explained in terms of a pure $d_{x^2-y^2}$-wave SC ground state in the cuprates, nor can they be attributed to simple band structure effects because of the sensitive H-dependence.

In the case of STS studies of the vortex-state of electron-type cuprate superconductors, there are interesting similarities and differences when compared with the vortex-state properties of hole-type cuprates. As shown

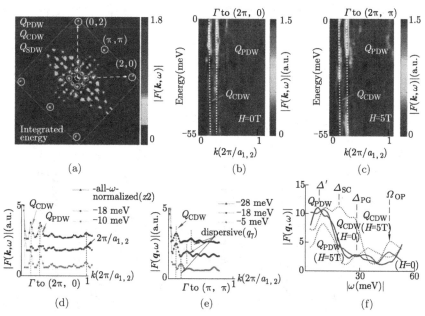

(a)   (b)   (c)

(d)   (e)   (f)

Fig. 2.13   (color online) Studies of the vortex-state FT-LDOS maps of Y-123 in the two-dimensional reciprocal space [10, 11, 197]: (a) Normalized FT-LDOS at $H = 5$ T obtained by integrating $|F(\boldsymbol{k}, \omega)|$ from $\omega = -1$ meV to -30 meV. There are three sets of ω-independent spots in addition to the reciprocal lattice constants and the (π, π) resonance, which are circled for clarity. These characteristic wave-vectors include $\boldsymbol{Q}_{\mathrm{PDW}}$ and $\boldsymbol{Q}_{\mathrm{CDW}}$ along the $(\pi, 0)/(0, \pi)$ directions and $\boldsymbol{Q}_{\mathrm{SDW}}$ along (π, π). (b) The ω-dependence of $|F(\boldsymbol{k}, \omega)|$ at $H = 0$ is plotted in the $\omega - vs. - \boldsymbol{k}$ plot against $\boldsymbol{k}(\pi, 0)$, showing ω-independent modes (bright vertical lines) at $\boldsymbol{Q}_{\mathrm{PDW}}$ and $\boldsymbol{Q}_{\mathrm{CDW}}$. (c) The ω-dependence of $|F(\boldsymbol{k}, \omega)|$ at $H = 5$ T is plotted in the ω-$vs.$-$\boldsymbol{k}$ plot against $\boldsymbol{k}\|(\pi, 0)$, showing field-enhanced spectral intensities at $\boldsymbol{Q}_{\mathrm{PDW}}$ and $\boldsymbol{Q}_{\mathrm{CDW}}$. (d) $|F(\boldsymbol{k}, \omega)|$ for different energies are plotted against $\boldsymbol{k}\|(\pi, 0)$, showing peaks at ω-independent $Q_{\mathrm{PDW}}, Q_{\mathrm{CDW}}$ and the reciprocal lattice constants at $(2\pi/a_1)$ along $(\pi, 0)$. (e) $|F(\boldsymbol{k}, \omega)|$ for different energies are plotted against $\boldsymbol{k}\|(\pi, \pi)$, showing peaks at energy-independent Q_{SDW} along (π, π). Additionally, dispersive wave-vectors due to QPI are found, as exemplified by the dispersive QPI momentum $\boldsymbol{q}_7$ specified in Figs. 2.8(b) and 2.8(d). (f) The FT-LDOS intensities $|F(\boldsymbol{q}, \omega)|$ of Y-123 for $\boldsymbol{q} = \boldsymbol{Q}_{\mathrm{PDW}}$ (red) and $\boldsymbol{Q}_{\mathrm{CDW}}$ (green) are shown as a function of ω and for $H = 0$ (solid lines) and $H = 5$ T (dashed lines) [197]. Here $\Delta_{\mathrm{PG}} = V_{\mathrm{CO}}$ and Ω_{OP} denotes the longitudinal optical phonon mode along the Cu-O bond.

in Fig. 2.14 for representative STS studies of the infinite-layer system La-112, long-range disordered vortices similar to those in the Y-123 system are also observed [17]. However, the vortex core radius appears to be comparable to ξ_{SC} [17], which is in contrast to the much larger vortex halo size ($\xi_{\mathrm{halo}} \sim 10\xi_{\mathrm{SC}}$) found in Y-123 [11]. Moreover, spatially resolved tunneling

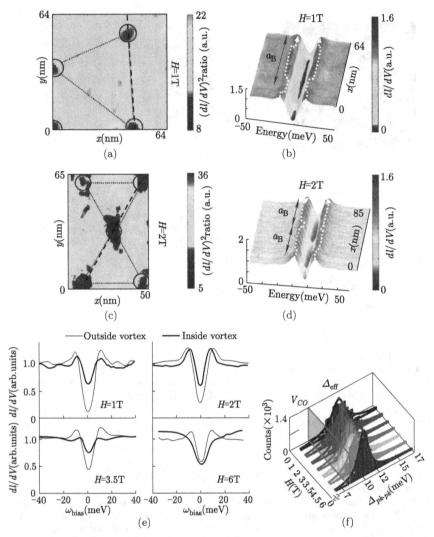

Fig. 2.14   Spatially resolved STS studies of the vortex-state of an optimally-doped electron-type cuprate La-112 ($T_c = 43\,$K) at $T = 6\,$K and for $H\|$ c-axis [10, 17]: (a) A spatial map of the conductance power ratio r_G (in log scale) taken over a (64×64) nm^2 area for $H = 1$ T, showing a zoom-in view of vortices separated by an average vortex lattice constant $a_B = 52$nm, which compares favorably with the theoretical value of 49 nm. The average radius of the vortices (indicated by the radius of the circles) is (4.7 ± 0.7) nm, comparable to the SC coherence length $\xi_{ab} = 4.9$ nm [176]. Here the conductance power ratio is defined as the ratio of $(dI/dV)^2$ at $|\omega| = \Delta_{\text{eff}}$ and that at $\omega = 0$. (b) Spatial evolution of the conductance (dI/dV) along the black dashed line

←――

Fig. 2.14   (*Continued*) cutting through two vortices in (a) for $H = 1$ T, showing significant modulations in the zero-bias conductance and slight modulations in the peak-to-peak energy gap. (c) A spatial map of the conductance power ratio r_G (in log scale) taken over a (65×50) nm^2 area with $H = 2$ T, showing a zoom-in view of vortices with an average vortex lattice constant $a_B = 35$ nm, which is consistent with the theoretical value. (d) Spatial evolution of the conductance is shown along the black dashed line cutting through three vortices in (b) for $H = 2$ T. (e) Evolution of the inter- and intra-vortex quasiparticle tunneling spectra with magnetic field in La-112 for $H = 1$ T, 2 T, 3.5 T and 6 T, where the PG spectra at the center of vortex-cores are given by the thick lines and those exterior to vortices are given by the thin lines. (f) Energy histograms of La-112 determined from our quasiparticle tunneling spectra of La-112, showing the spectral evolution with H. Note that there is no zero-bias conductance peak in the vortex-state, and that a low-energy cutoff at nearly a constant value $V_{CO} = (8.5 \pm 0.6)$ meV exists for all fields.

spectra also exhibit PG features inside the vortex-core of La-112, with a PG energy $\Delta_{PG} \sim V_{CO}$ smaller than the SC gap, $\Delta_{PG} < \Delta_{SC}$ [17], which is in contrast to the hole-type cuprates that reveal $\Delta_{PG} > \Delta_{SC}$ inside the vortex cores [10, 11]. Here we note that the gap values Δ_{SC} and V_{CO} are determined empirically by taking $1/2$ of the energy difference between the spectral peak-to-peak, Δ_{pk-pk}, for the inter- and intra-vortex spectra, respectively. This finding is again consistent with the absence of zero-field PG phenomena above T_c in the electron-type cuprate superconductors. Hence, we conclude that the rich phenomena revealed in the vortex-state STS studies of both hole- and electron-type cuprates can all be consistently understood within the two-gap scenario.

2.2.4.2   Strong quantum, thermal and disorder fluctuations

As described earlier, cuprate superconductors are doped Mott insulators with strong electronic correlation that can result in a variety of competing orders (COs) in the ground state. Therefore, significant quantum fluctuations and reduced SC stiffness are expected in the cuprates because of the existence of multiple channels of low energy excitations, which are believed to contribute to the extreme type-II nature of the cuprate superconductors [14, 15]. Moreover, external variables such as temperature (T) and applied magnetic field (H) can vary the interplay of SC and CO, such as inducing or enhancing the CO at the price of more rapid suppression of SC [41, 52], thereby leading to weakened SC stiffness and strong thermal and field-induced fluctuations [68, 77, 78, 81, 82]. These effects are likely the primary cause for the occurrence of a vortex liquid phase below the upper critical field of cuprate superconductors. Moreover, the significantly weakened SC stiffness also implies much stronger susceptibility of vortices to disorder, giving rise to various types of glassy phases at low temperatures, depending on the type and dimensionality of disorder, as exemplified in Fig. 2.15(a) for the vortex phase diagram of three-dimensional cuprate superconductors with random point disorder and for $H//c$-axis, showing the occurrence of a vortex liquid phase below the upper critical field $H_{c2}(T)$ and additional disordered vortex solid phases (i.e., the "vortex glass" [81, 82] and "Bragg glass" [86, 87]) below the vortex liquid phase. For comparison, the vortex phase diagram for conventional type-II superconductors is shown in Fig. 2.15(b), where the ordered vortex solid phase, known as the vortex lattice, extends all the way to $H_{c2}(T)$ without the occurrence of a vortex liquid phase. In the context of strong disorder fluctuations, correlated disorder such as columnar defects and twin boundaries can result in different universality classes of vortex phase transitions [83, 84, 89–97] relative to the situation of random point defects, as exemplified in Figs. 2.16(a)–(c) for the vortex glass [81, 82], Bose glass [83, 84] and splayed glass [96, 97] transitions associated with different types of defects. Details for the theory and experimental investigations of disorder-induced novel vortex dynamics in the cuprate superconductors can be found in Refs. [81–97].

In addition to the quantum criticality induced by competing orders, the quasi-two dimensional nature of the cuprates may yield a quantum criticality in the limit of decoupling of CuO_2 planes [190]. Indeed, recent studies have demonstrated experimental evidence from macroscopic magnetization measurements for field-induced quantum fluctuations among a wide variety of cuprate superconductors with different microscopic variables such as the doping level (δ) of holes or electrons, the electronic mass anisotropy (γ), and the number of CuO_2 layers per unit cell (n) [14, 15, 77, 78]. It is suggested that the manifestation of strong field-induced quantum fluctuations is consistent with a scenario that all cuprates are in close proximity to a quantum critical point (QCP) [77].

To investigate the effect of quantum fluctuations on the vortex dynamics of cuprate superconductors, vortex phase diagrams for different cuprates were studied at $T \rightarrow 0$ to minimize the effect of thermal fluctuations, and the magnetic field was applied parallel to the CuO_2 planes ($H//ab$) to minimize the effect of random point disorder [77]. The rationale for having $H//ab$ is that the intrinsic pinning effect of layered CuO_2 planes generally dominates over the pinning effects of random point disorder [85], so that

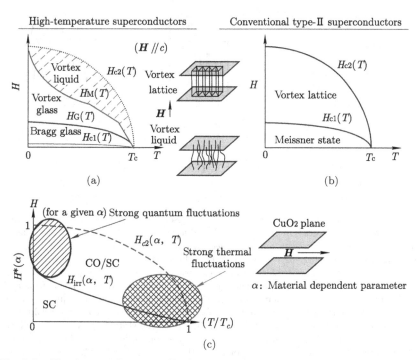

Fig. 2.15   Novel vortex dynamics of high-temperature superconducting cuprates in comparison with conventional type-II superconductors: (a) Schematic H vs. T vortex phase diagrams of cuprate superconductors for $H//c$-axis, assuming thermal fluctuations and random point disorder [81, 82, 86, 87]. The vortex phase (neglecting the lower critical field H_{c1} due to the extreme type-II nature) with increasing H and T evolves from the Bragg glass [86, 87] to the vortex glass [81, 82] through the phase boundary $H_G(T)$ and then to the vortex liquid through the phase boundary $H_M(T)$ before reaching the upper critical field $H_{c2}(T)$. The occurrence of glass and liquid phases below H_{c2} may be attributed to the strong disorder and thermal fluctuations in cuprate superconductors. (b) Schematic vortex phase diagram for conventional type-II superconductors is shown for comparison with that of the cuprate superconductors. (c) For $H//ab$-plane, assuming dominating quantum fluctuations associated with the proximity to quantum criticality and COs [6, 14, 15, 68, 77], it has been conjectured and experimentally verified that the application of high in-plane magnetic fields in the $T \rightarrow 0$ limit may suppress the phase coherent superconducting (SC) phase at a field characteristic $H^* \equiv H_{\text{irr}}(T \rightarrow 0)$ much below the upper critical field H_{c2} due to the field-induced currents along the c-axis critical currents and/or field-enhanced competing orders (COs) that result in strong quantum fluctuations [6, 14, 15, 68, 77]. The characteristic field is expected to be dependent on the material properties of the cuprates, which may be parameterized by α, where α is a function of the doping level (δ), the electronic mass anisotropy (γ) and the number of CuO_2 layers per unit cell (n) [14, 15, 77].

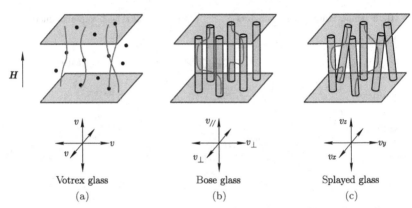

Fig. 2.16   Schematics of different universality classes of vortex phases associated with different types of disorder in cuprate superconductors [96, 97]: (a) The "vortex glass" due to random point defects and classified by an isotropic static exponent v and a dynamic exponent z [81, 82, 89–94]. (b) The "bose glass" due to correlated parallel columnar defects and classified by two static exponents $(v_{//}, v_{\perp})$ and a dynamic exponent z [83, 84, 94, 95]. (c) The "splayed glass" due to correlated canted columnar defects and classified by three static exponents (v_x, v_y, v_z) and a dynamic exponent z [96, 97]. Here the red curves represent vortices and the gray columns represent correlated columnar defects induced by heavy ion irradiation [94, 95]. Here the static exponent v is defined by the vortex correlation length ξ associated with a second-order vortex phase transition temperature T_{cr} according to the relation $\xi = \xi_0 |1 - (T/T_{cr})|^{-v}$, where ξ_0 is temperature-independent; and the dynamic exponent z is defined by the vortex relaxation time τ near T_{cr} according to the relation $\tau = \tau_0 |1 - (T/T_{cr})|^{-z}$, where τ_0 is independent of T.

the commonly observed glassy vortex phases associated with point disorder for $H//c$(*e.g.*, vortex glass and Bragg glass) [81, 82, 86, 87] can be prevented. In the absence of quantum fluctuations, random point disorder can cooperate with the intrinsic pinning effect to stabilize the low-temperature vortex smectic and vortex solid phases [85], so that the vortex phase diagram for $H//ab$ would resemble that of the vortex-glass and vortex-liquid phases observed for $H//c$ with a glass transition $H_G(T = 0)$ approaching $H_{c2}(T = 0)$. On the other hand, when field-induced quantum fluctuations are dominant [6, 68], the vortex phase diagram for $H//ab$ will deviate substantially from the predictions solely based on thermal fluctuations and intrinsic pinning, so that strong suppression of the magnetic irreversibility field $H_{\mathrm{irr}}(T)$ relative to the upper critical field H_{c2} is expected at $T \to 0$ [14, 15, 77], as schematically shown in Fig. 2.15(c), because the induced persistent current circulating along both the c-axis and the ab-plane can no longer be sustained if field-induced quantum fluctuations become too strong to maintain the c-axis superconducting phase coherence.

Indeed, experimental studies on a wide variety of cuprate super-conductors revealed consistent findings with the notion that all cuprate superconductors exhibit significant field-induced quantum fluctuations, as manifested by a characteristic field $H_{irr}(T \to 0) \equiv H^* \ll H_{c2}(T \to 0)$, and exemplified in Fig. 2.15(c) [77]. The degree of quantum fluctuations for each cuprate may be expressed in terms of a reduced field $h^* \equiv [H^*/H_{c2}(0)]$, with $h^* \to 0$ indicating strong quantum fluctuations and $h^* \to 1$ referring to the mean-field limit. Most importantly, the h^* values of all cuprates appear to follow a trend on a $h^*(\alpha)$- vs.-α plot, where α is a material parameter for a given cuprate that reflects its doping level δ, electronic mass anisotropy γ, and charge imbalance if the number of CuO_2 layers per unit cell n satisfies $n \geqslant 3$ [57, 58]. Specifically, α is defined by the following [77]:

$$\alpha \equiv \gamma^{-1}\delta(\delta_o/\delta_i)^{-(n-2)} \quad (n \geqslant 3), \qquad (2.12)$$

$$\alpha \equiv \gamma^{-1}\delta \qquad\qquad (n \leqslant 2). \qquad (2.13)$$

In Eq. (2.12) the ratio of charge imbalance in multi-layer cuprates with $n \geqslant 3$ is given by (δ_o/δ_i) [57, 58] between the doping level of the outer layers (δ_o) and that of the inner layer(s) (δ_i). Finally, in the event that $H_{c2}(0)$ exceeds the paramagnetic field $H_p \equiv \Delta_{SC}(0)/(2^{1/2}\mu_B)$ for highly anisotropic cuprates, where $\Delta_{SC}(0)$ denotes the SC gap at $T = 0$, h^* is defined by (H^*/H_p) because H_p becomes the maximum critical field for superconductivity.

Systematic studies of the in-plane irreversibility fields of various cuprate superconductors (see Fig. 2.17(a)) revealed a universal trend for $h^*(\alpha)$-vs.-α, as shown in Fig. 2.17(b) [77]. In particular, it is worth noting the h^*-vs.-α dependence in the multi-layered hole-type cuprate superconductors $HgBa_2Ca_2Cu_3O_x$ (Hg-1223, $T_c = 133$ K), $HgBa_2Ca_3Cu_4O_x$ (Hg-1234, $T_c = 125$ K) and $HgBa_2Ca_4Cu_5O_x$ (Hg-1245, $T_c = 108$ K): While these cuprate superconductors have the highest T_c and H_{c2} values, they also exhibit the smallest h^* and α values, suggesting maximum quantum fluctuations. These strong quantum fluctuations can be attributed to both their extreme two-dimensionality (i.e., $\gamma \gg 1$) [198, 199] and significant charge imbalance that leads to strong CO in the inner layers [57, 58]. This notion is corroborated by the muon spin resonance (μSR) experiments [56] that revealed increasing AFM ordering in the inner layers of the multi-layer cuprates with $n \geqslant 3$. Therefore, the investigation of the in-plane magnetic irreversibility in a wide variety of cuprate superconductors reveals strong

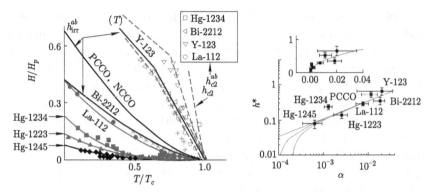

Fig. 2.17   Experimental manifestation of strong field-induced quantum fluctuations in a variety of cuprate superconductors $H//ab$. (a) Reduced in-plane fields (H_{irr}/H_p) and (H_{c2}/H_p) vs. (T/T_c) for various cuprates [77]. In the $T \to 0$ limit where $H_{\text{irr}} \to H^*$, the reduced fields $h^* \equiv (H^*/H_p) < 1$ are found for all cuprates Y-123, NCCO, Bi-2212, La-112, Hg-1234, Hg-1223, and Hg-1245 (in descending order) [77]. (b) h^*-vs.-α in logarithmic plot for different cuprates, with decreasing α representing increasing quantum fluctuations. The solid lines are power-law fitting curves given by $5(\alpha - \alpha_c)^{1/2}$, using different $\alpha_c = 0, 10^{-4}$ and 2×10^{-4}, in order from left to right [77].

field-induced quantum fluctuations [77], which is consistent with the notion that cuprate superconductors are in close proximity to quantum criticality as a result of the co-existence of competing orders and superconductivity in the ground state.

2.2.4.3   Anomalous sign-reversal Hall conductivity in the vortex state

The CO scenario also appears to be relevant to one of the outstanding issues in the cuprate superconductors, namely, the anomalous sign-reversal in the vortex-state Hall conductivity (σ_{xy}) as a function either T or H for both electron- and hole-type cuprate superconductors [106–111]. Although quantitative description for the microscopic theory of the vortex-state Hall conduction remains incomplete, several important facts have been established: First, the sign reversal is associated with the intrinsic physical properties of cuprate superconductors [200–203] and is independent of either the electronic mass anisotropy [110] or the degree of disorder in the superconductors, regardless of random [200] or correlated disorder [110]. Specifically, for a given cuprate superconductor of a mass anisotropy $\gamma \equiv (m_c/m_{ab})$ and for a magnetic field H applied at an angle θ relative to the crystalline c-axis, the Hall conductivity $\sigma_{xy}(H, T, \theta, \gamma)$ may be scaled into a universal function $\tilde{\sigma}_{xy}(\tilde{H}, T/T_c)$ through the following transformation, independent of the type of disorder or the mass anisotropy [110]:

$$\tilde{\sigma}_{xy}\left(\tilde{H}, \frac{T}{T_c}\right) \equiv \sigma_{xy}(H, T, \theta, \gamma)\sqrt{1 + \gamma^{-1}\tan^2\theta},$$

$$\tilde{H} \equiv H\sqrt{\cos^2\theta + \gamma^{-1}\sin^2\theta}. \tag{2.14}$$

Second, the occurrence of sign reversal in σ_{xy} is attributed to the non-uniform spatial distribution of carriers within and far outside the vortex core [201–203]. Third, the dc vortex-state Hall conductivity of superconducting cuprates is found to be strongly dependent on the doping level, showing anomalous sign reversal in the under-doped regime and no anomaly in the overdoped regime [204]. This important experimental finding suggests the relevance of the vortex-core electronic structures to the Hall conductivity in the SC state. Given the STS observation of PG phenomena inside the vortex cores of both hole- and electron-type cuprates [10, 11, 17] that differs fundamentally from the "normal core" approximations in conventional type-II superconductors, we may attribute the occurrence of sign reversal in the vortex-state σ_{xy} to the reduced quasiparticle LDOS inside the vortex cores as the result of COs. This conjecture is further corroborated by the increasing sign reversal effect with decreasing doping [204] because the effect of COS and therefore the suppression in the vortex-core LDOS becomes more significant in the under-doped limit.

2.2.4.4 Quantum oscillations at low temperatures

The aforementioned ubiquitous presence of strong field-induced quantum fluctuations in a large variety of cuprate superconductors implies that the low-temperature dissipative vortex-state is a strongly fluctuating vortex liquid (Fig. 2.15(c)) [77, 78], which may differ from the zero-field normal state above T_c. Indeed, recent low-temperature high-field quantum oscillations observed in under-doped hole-type cuprates $YB_2Cu_3O_{6+x}$ (Y-123) [98–102] have led to implications of excess Fermi surface structures that differ from those obtained from the zero-field ARPES experiments. Theoretical analysis of the experimental data finds that the assumption that the oscillation period is given by the underlying Fermi-surface area using the Onsager relation becomes invalid [103] in this low-temperature high-field limit. The physical origin for such differences has been attributed by some to reconstructed Fermi surfaces due to the underlying incommensurate SDW [102, 104], and by others, to the presence of excess electron pockets [105].

Generally speaking, it is not theoretically rigorous to infer the zero-field normal-state properties of cuprate superconductors directly from the studies of quantum oscillations in the low-temperature vortex-liquid state unless the effects of field-induced quantum fluctuations on the Fermi surface of the cuprates and the vortex-state quantum oscillations can be understood. For instance, a SDW state could evolve into a different magnetic order under sufficiently strong magnetic fields at low temperatures, which may give rise to a Fermi surface reconstruction such as the occurrence of four hole pockets created by a (π, π) folding [102–104]. Additionally, the observation of negative Hall effects [105] in the low-temperature high-field limit is similar to the anomalous sign-reversal Hall conductivity of both hole- and electron-type cuprates in the high-temperature and low-field vortex-liquid state [106–111]. As discussed in Sec. 2.2.4.3, the PG phenomena revealed inside the vortex cores of both hole- and electron-type cuprate superconductors [10, 11, 17] suggest that the vortex-cores contributing to the Hall conductivity in the vortex liquid state contain opposite charge carriers to regions outside the vortex-cores [7], thereby giving rise to sign-reversal Hall conductivity in the vortex liquid state. Given that the PG phenomena inside the vortex-core may be attributed to the presence of COs, the attribution of the negative Hall effects [105] together with quantum oscillations to the occurrence of electronic pockets may be naïve. Putting all empirical facts in the vortex-state of the cuprates together, it is natural to suggest that both effects of quantum fluctuations and COs must be considered to fully account for the observed mixed-state quantum oscillations.

2.2.4.5 *Anomalous Nernst effect under finite fields in the normal state*

The anomalous Nernst effect [115, 116, 205] occurring in the under-doped hole-type cuprates at temperatures well above T_c and the absence of such an effect in all electron-type cuprates remains a mystery. Generally, the Nernst effect refers to a transverse electric field generated by moving vortices in the presence of a thermal gradient. Specifically, for vortices moving with velocity v down a thermal gradient $-\nabla T // \hat{x}$, a Josephson voltage is generated and is observed as a transverse electric field $E_y = Bv_x$, where B is the mean flux density. While the vortex-Nernst effect is well explored and understood in low-T_c superconductors, the observation of the Nernst effect in various hole-type cuprates at temperatures well above T_c has generated much debate over the physical origin for such an effect. For instance, the preformed pair model [186–188] suggests that the zero-field superconducting transition temperature T_c is merely the loss of long-range phase rigidity so that pairs may in fact survive up to a much higher temperature. Therefore, vortices could exist above T_c due to local phase coherence, thus giving rise to the observed anomalous Nernst effect. However, given the inconsistency of the one-gap model with most other experimental phenomena, it seems that theoretical investigation for possible contributions from COs to the normal-state Nernst effect is necessary to settle the issue for the physical origin of the anomalous Nernst effect. For instance, the presence of SDW-like CO above T_c may give rise to a non-trivial Berry phase for carriers moving under the influence of a thermal gradient and a finite magnetic field. On the other hand, settling the physical origin for this anomalous Nernst effect may not have a major bearing on our overall understanding of the pairing mechanism of high-temperature superconductivity, because the primary issue appears to be devising means to generate a strong sign-changing pairing potential from repulsive interactions while preventing phase separations and ensuring the itinerant motion of pairs. We shall return to this point in the discussion section.

2.3   Iron-based superconductors

The discovery of a new class of iron-based superconductors in 2008 [2] with a maximum transition temperature (T_c) of $\sim$55 K to date [206–217] has rekindled intense activities in SC research. In particular, there are interesting similarities and contrasts between the cuprates and the ferrous compounds. Parallel studies of the low-energy excitations of both systems have yielded useful insights into the fundamental issue of pair formation in superconductors [218–222]. A list of similarities and differences in some of the important physical properties of these two classes of superconductors are summarized in Table 2.1.

2.3.1   *Basic structural and magnetic properties*

Similar to the cuprate superconductors, the iron-based superconductors, which include the pnictide [2, 206–213] and the iron-chalcogenide [214–217] superconductors, are correlated layered materials with magnetic instabilities. The common chemical building block of these superconductors is FeX,

Table 2.1 Comparison of various important physical properties of cuprate and ferrous superconductors.

	Cuprates (hole-type)	Cuprates (electron-type)	Iron pnictides 1111 and 122	Iron chalcogenides: 11: Fe_{1+y} (Te_{1-x} Se_x)
Record-high T_c	165 K	43 K	55 K	27 K
Parent state ($x = 0$)	Antiferromagnetic Mott insulator	Antiferromagnetic Mott insulator	Semi-metal with $(\pi, 0)$-SDW nested to the electron-hole Fermi surfaces	Semi-metal with $(\pi/2, \pi/2)$-SDW ($y < 0.11$); Semiconductors w/incommensurate SDW ($y > 0.11$)
Parent-state electronic configuration	9 d-electrons, single-band approximations	9 d-electrons, single-band approximations	6 d-electrons, five-band approximations	6 d-electrons, five-band approximations
Electronic correlation	Strong ($U \sim 8$ eV)	Strong ($U \sim 8$ eV)	Weak ($U < 2$ eV)	Intermediate
Pairing symmetry	$d_{x^2-y^2}$ (under- and optimally-doped) $d_{x^2-y^2} + s$ (over-doped)	$d_{x^2-y^2}$ (all doping levels)	Sign-changing s-wave ($s\pm$)	Sign-changing s-wave ($s\pm$) or nodeless d-wave
Ground state phases	Coexisting SC and CDW/PDW (under- and optimally-doped)	Coexisting SC and SDW (under- and optimally-doped)	Coexisting SC and $(\pi, 0)$-SDW (under-doped 122 systems)	Coexisting SC and $(\pi/2, \pi/2)$-SDW ($y < 0.11$)
	Pure SC (over-doped)	Pure SC (over-doped)	Pure SC (1111 systems)	Coexisting SC and incommensurate SDW ($y > 0.11$)

(Continued)

Table 2.1 (*Continued*)

	Cuprates (hole-type)	Cuprates (electron-type)	Iron pnictides 1111 and 122	Iron chalcogenides: 11: Fe_{1+y} ($Te_{1-x}Se_x$)
Energy gaps	SC gap at $T < T_c$ and pseudogap at $T^* > T > T_c$ (under- and optimally-doped) Pure SC gap at $T < T_c$ (over-doped)	SC gap at $T < T_c$ and pseudogap at $T < T^* < T_c$ (under- and optimally-doped) Pure SC gap at $T < T_c$ (over-doped)	Two SC gaps for hole- and electron-pockets at $T < T_c$	Two SC gaps for hole- and electron-pockets at $T < T_c$; or one SC gap for systems with vanishing holes
Intra-vortex spectra	Pseudogap > SC gap (optimally- and under-doped) Pseudogap < SC gap or bound states for over-doped?	Pseudogap < SC gap (optimally-doped)	Pseudogap < SC gap (optimally-doped) Bound states and no pseudogap for over-doped samples	Pseudogap < SC gap or bound states? Doping dependence?

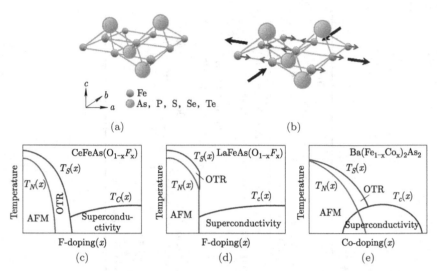

$$\begin{array}{c} c \\ \uparrow b \\ \swarrow\!\!\!\!\!\longrightarrow a \end{array}$$ ● Fe ● As, P, S, Se, Te

(a)   (b)

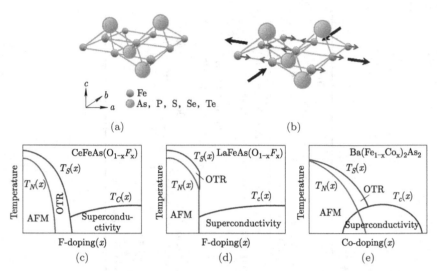

F-doping(x)   F-doping(x)   Co-doping(x)

(c)   (d)   (e)

Fig. 2.18   (a) Schematics of the basic building block of the ferrous superconductors: Top view of the FeX tri-layer, where $X =$ As, P, S, Se, Te. The triad (a, b, and c) demonstrates the three crystallographic directions. (b) The antiferromagnetic order of the stoichiometric iron-based materials. The grey arrows represent the magnetic moments, and the black arrows indicate the directions of structural distortion. (c)–(e) Schematics of three representative phase diagrams for different types of ferrous superconductors [223–225]. Here $T_S(x)$ denotes the phase boundary for a structural phase transition from a tetragonal phase at $T > T_S$ to an orthorhombic (OTR) crystalline structure at $T < T_S$; $T_N(x)$ is the Néel temperature for the onset of an AFM phase at $T < T_N$; and $T_c(x)$ represents the doping-dependent superconducting transition temperature.

where $X =$ As, P, S, Se, Te. Structurally, FeX forms a tri-layer that consists of a square array of Fe sandwiched between two checkerboard layers of X, as illustrated in Fig. 2.18(a). These tri-layers are further separated by the "bridging layers" consisting of alkali, alkaline-earth, or rare-earth atoms and oxygen/fluorine. Strong experimental and theoretical evidences have associated the origin of superconductivity in these ferrous superconductors with the d-electrons of Fe in the FeX tri-layers, with the X-layer contributing to the delocalization of the d-electrons [218–220]. Therefore, the FeX tri-layers may be considered to play the same role in ferrous superconductivity as the CuO_2 layers in the cuprates.

There are three primary structures associated with the layered rare-earth transition-metal oxypnictides. The dominant type ROT Pn ($R =$ rare-earth elements La, Nd, Sm, Pr, Ce; $T =$ transition metals Fe, Ni, Mn, Co; $Pn =$ pnictogen P, As), also denoted as the "1111" system, can be doped with either electrons or holes [2, 206–209]. In the case of LaOFeAs, the doping of fluoride ions at the oxygen sites provide electrons from the $La(O_{1-x}F_x)$ layers to the FeAs tri-layers and replacing magnetism with superconductivity (SC) for $x = 0.05$ to 0.12, leading to a maximum T_c at 26 K [2, 206, 207]. Replacing La with Pr, Nd, and Sm has been shown to further boost T_c up to 55 K [209]. On the other hand, substituting La with Sr leads to hole-doped compounds $(La_{1-x}Sr_x)OFeAs$ with T_c up to 25 K [208]. The second type of layered compounds known as the 122 system has the formula $(A'_{1-x}A_x)Fe_2As_2$ [210] or $Ba(Fe_{1-x}Co_x)_2As_2$ [211, 212], where A' = Ba or Sr, and A = K or Cs. The T_c of KFe_2As_2 and $CsFe_2As_2$ is 3.8 and 2.6 K, respectively, which rises with partial substitution of Sr for K and Cs and peaks at 37 K for 50%–60% Sr substitution [210]. Placing Fe in the 122 system by Co leads to electron-doped 122 with a maximum T_c up to 24 K [211, 212]. Moreover, SC and AFM phases are found to coexist for a range of electron-doping, similar to the CO phenomena found in the cuprates. The third type of layered compounds MFeAs (or "111") with M = Li or Na are shown to exhibit $T_c = 20$ K and 18 K [213], respectively, and the 111 system is analogous to the infinite-layer system $SrCuO_2$ in the cuprate superconductors.

Among the iron chalcogenides, the structure is known as the "11" system of $Fe(Se_{1-x}Te_x)$, which is the simplest form among the ferrous superconductors [214–217]. The first discovery of superconductivity in the 11 system was found in α-FeSe with $T_c \sim 8$ K [214]. Subsequently, dramatic pressure-enhanced T_c up to $\sim$27 K has been reported [215]. Additionally, replacing up to 50% Se by Te can further enhance T_c, and the resulting compound exhibits an even stronger pressure effect [216], although FeTe is found to not be superconducting due to structural deformation that simultaneously breaks magnetic symmetry [216]. In a very recent development, a number of intercalated FeSe compounds $A_xFe_{2-y}Se_2$ (where A = K, Cs, Tl) were made, raising T_c from 8 K for FeSe to above 30 K [217]. Overall, the Fermi surface and magnetic properties of the 11 system are very similar to those of the iron pnictides. On the other hand, there are evidences for vanished hole-pockets in the intercalated compounds $A_xFe_{2-y}Se_2$, which would be theoretically favorable for nodeless d-wave pairing.

Most of the stoichiometric parent compounds exhibit AFM at ambient pressure, and the spatial arrangement of the magnetic moments in the FeX tri-layer of the parent compounds (except the 11 system) is schematically shown in Fig. 2.18(b) [223–225]. This magnetic order couples intimately with a tetragonal-to-orthorhombic structural distortion. For the stoichiometric 122 system such as the $BaFe_2As_2$ [224], first-order structural and AFM transitions occur at the same temperature, as illustrated in Fig. 2.18(e). In the low-temperature phase, the ab-plane Fe-Fe distance elongates in the direction parallel to the magnetic moment and contracts in the direction perpendicular to it, as indicated by the blue arrows in Fig. 2.18(b). On the other hand, for the 1111 system such as LaFeAsO [223] and CeFeAsO [225], the structural transition occurs at a slightly higher temperature followed by a magnetic transition, as shown in Figs. 2.18(c) and (d). Thus, there exists a temperature window in which the stoichiometric 1111 compounds are paramagnetic with fluctuating magnetism, but the four-fold crystalline rotation symmetry is broken by the structural distortion in the OTR phase, implying that the electron-lattice coupling will be enhanced in the OTR phase [226]. This coupling could either impede or assist the electron pairing. Moreover, the AFM state is a semi-metal, which is in sharp contrast to the cuprates where the parent AFM compounds are Mott insulators.

2.3.2 Two-gap superconductivity, unconventional pairing symmetry and magnetic resonances

Calculations based on the density functional theory [212, 227–229] have shown that there are many bands near the Fermi level of these iron-based compounds and that their Fermi surfaces involve multiple disconnected Fermi pockets, as exemplified in Fig. 2.19. These electronic properties of the ferrous compounds are in contrast to the cuprates; the latter are primarily described by an effective one-band model with a large Fermi surface (Fig. 2.8(a)). The presence of multiple bands and multiple disconnected Fermi surfaces suggests that inter-Fermi surface interactions may be important to the occurrence of ferrous superconductivity. Indeed, calculations of magnetic susceptibility [229, 230] have shown that these ferrous compounds have a tendency for AFM order, and the wave-vectors associated with the AFM coupling coincide with those connecting the centers of the electron and hole Fermi pockets, as shown in Fig. 2.19. These theoretical findings have led to the conjecture of two-gap superconductivity mediated by AFM spin fluctuations, with sign-changing s-wave ($s_\pm$) order parameters for the hole and electron Fermi pockets.

The manifestation of two-gap superconductivity has been demonstrated by both ARPES [234–236] and STS [237–239] studies. As exemplified in Figs. 2.20(a)–(b) for under- and over-doped Ba(Fe$_{1-x}$Co$_x$)$_2$As$_2$ with $x =$

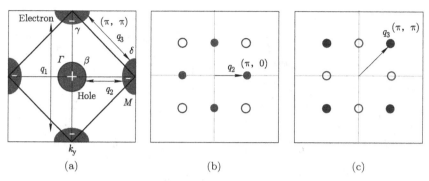

(a) (b) (c)

Fig. 2.19   (a) Schematics of the two-dimensional Fermi surfaces (FS) of ferrous super-conductors in the one-iron unit cell, showing the presence of a and b hole pockets at the Γ-point of the Brillouin zone and electron pockets g and d at the M-points. The SC order parameters are opposite in sign for the hole and electron pockets. Possible QPI wave-vectors q_1, q_2 and q_3 connecting different parts of the FS are indicated [227, 228, 231, 232]. (b) Theoretical prediction for the zero-field QPI intensities as the result of non-magnetic impurity scattering in an $s_\pm$-wave superconductor, where q_2 spots should be intense and q_3 should be suppressed if the quasiparticle energy is equal to one of the SC gap values [227, 228, 231]. Further, $q_1 = 2q_2$ may appear due to QPI induced by the CDW order, where the occurrence of CDW is associated with the AFM order [233]. (c) Theoretical prediction for the zero-field QPI intensities due to magnetic impurity scattering in an $s_\pm$-wave superconductor, where q_3 spots are intense and q_2 intensities are suppressed if the quasiparticle energy is equal to one of the SC gap values [227, 228, 231]. Alternatively, for non-magnetic impurities in the presence of magnetic fields, the intensities of q_2 spots would be reduced whereas those of q_3 spots would be enhanced [232].

0.06 and 0.12, two predominant tunneling gap features at Δ_Γ and Δ_M are apparent for both doping levels. For a given doping level both gaps decrease with increasing temperature and vanish above T_c [237]. Additionally, the tunneling gaps exhibit particle-hole symmetry, confirming that the observed gaps are associated with superconductivity [237].

The theoretical prediction for $s_\pm$-wave pairing symmetry in the ferrous superconductors has been confirmed by the inelastic neutron scattering (INS) spectroscopy [241–244], STS [237, 245] and a phase-sensitive experiment [246]. In the case of INS experiments, a neutron resonance at the AFM ordering wave-vector was theoretically expected below T_c for $s_\pm$-wave pairing [241, 242] and experimentally verified [243, 244]. Specifically, the magnetic susceptibility in the SC state of a multi-band superconductor is governed by the sign-change of the SC gaps at the "hot spots" of the Fermi surface and the following energy conservation formula for

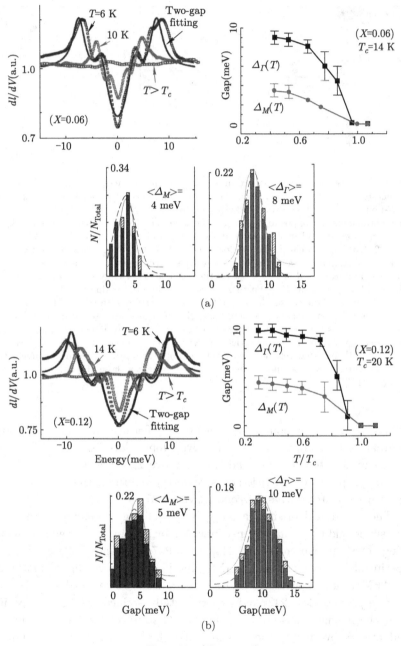

Fig. 2.20   (*Continued*)

Fig. 2.20   Spectroscopic evidence for two-gap superconductivity in $Ba(Fe_{1-x}Co_x)_2As_2$ [237]: (a) Top left panel: Normalized tunneling conductance (dI/dV) vs bias voltage (V) spectra taken at $T = 6$ K, 10 K, and 15 K for the sample with $x = 0.06$ and $T_c = 14$ K. The solid lines represent theoretical fittings to spectra using the Dynes formula [240] modified for two-gap BCS superconductors [237]. Two distinct tunneling gaps Δ_Γ and Δ_M can be identified from the spectrum at $T = 6$ K, and both gaps decrease with increasing temperature and then vanish at $T > T_c$. Top right panel: The tunneling gaps Δ_Γ and Δ_M as a function of the reduced temperature (T/T_c) are shown by the symbols and solid lines. Lower panels, histograms for the quasiparticle (solid bars) and quasihole (shaded bars) branches, showing particle-hole symmetry and the mean values of $\langle|\Delta_M|\rangle = 4$ meV and $\langle|\Delta_\Gamma|\rangle = 8$ meV. (b) Top left panel: Normalized tunneling conductance (dI/dV) vs. bias voltage (V) spectra taken at $T = 6$ K, 14 K, and 21 K for the sample with $x = 0.12$ and $T_c = 20$ K. Top right panel: The tunneling gaps Δ_Γ and Δ_M as a function of the reduced temperature (T/T_c). Lower panels, histograms for the quasiparticle (solid bars) and quasihole (shaded bars) branches, showing particle-hole symmetry and the mean values of $\langle|\Delta_M|\rangle = 5$ meV and $\langle|\Delta_\Gamma|\rangle = 10$ meV.

inelastic scattering of the Bogoliubov quasiparticles on the Fermi surface [247, 248]:

$$\Omega^{\nu\nu'}(\boldsymbol{k}_F, \boldsymbol{q}) = |\Delta_\nu(\boldsymbol{k}_F)| + |\Delta_{\nu'}(\boldsymbol{k}_F + \boldsymbol{q})|, \qquad (2.15)$$

where ν and ν' represent different energy bands, and the wave-vectors $\boldsymbol{q}$ are between various Fermi surface pieces with opposite signs in the SC pairing potential $\Delta_\nu(\boldsymbol{k}_F + \boldsymbol{q})$.

For the STS studies, the elastic scattering of quasiparticles by impurities will be dependent on whether the SC order parameter has opposite signs on electron and hole Fermi pockets [247, 248] so that

$$\Omega^{\nu\nu'}(\boldsymbol{k}_F, \boldsymbol{q}) = |\Delta_\nu(\boldsymbol{k}_F)| = |\Delta_{\nu'}(\boldsymbol{k}_F + \boldsymbol{q})|. \qquad (2.16)$$

Specifically, non-magnetic impurities will result in strong scattering of quasiparticles between the Fermi pockets of different signs in the pairing potential while suppressing the scattering between pockets of the same sign in the pairing potential [227, 228, 231], thus giving rise to the QPI patterns shown in Fig. 2.19(b). On the other hand, the presence of magnetic impurities or magnetic field would yield the QPI patterns shown in Fig. 2.19(c) [227, 228, 231, 232]. This behavior has been confirmed in $Fe_{1+x}(Se, Te)$ compounds [244] and in $Ba(Fe_{1-x}Co_x)_2As_2$ [237].

As exemplified in Figs. 2.21(a) and 2.21(b), the FT-LDOS of underdoped Ba $(Fe_{1-x}Co_x)_2As_2$ with $x = 0.06$ in the reciprocal space for the one-iron unit cell is shown for $\omega = \Delta_{\alpha,\gamma/\delta}$ and $\omega = \Delta_\beta$, respectively. Two

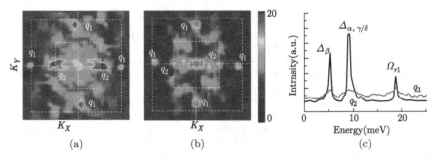

Fig. 2.21   Spectroscopic evidence for $s_{\pm}$-wave pairing symmetry in $Ba(Fe_{1-x}Co_x)_2As_2$ [237, 238]: (a) FT-LDOS of an under-doped sample $Ba(Fe_{0.94}Co_{0.06})_2As_2$ for $H = 0$ at $\omega = \Delta_{\alpha,\gamma/\delta} \sim 8$ meV is shown in the 2D reciprocal space. Strong intensities at $q = q_2$ and q_1 together with an additional nematic order are found [237]. The strong intensities at $q = q_2$ are consistent with the theoretical predictions for QPI patterns associated with $s_{\pm}$-wave pairing potential, whereas the strong intensities found at $q_1 \sim 2q_2$ are in agreement with the presence of CDW [233]. (b) $F(q,\omega)$- VS.-ω of under-doped $Ba(Fe_{0.94}Co_{0.06})_2As_2$ $(x = 0.06)$ for $H = 0$ at $q = q_2$, showing sharp peaks only at $\omega = \Delta_\beta, \Delta_{\alpha,\gamma/\delta}$ and Ω_{r1}, where $F(q,\omega)$ denotes the intensity of the FT-LDOS. The sharp QPI intensities occurring only at the SC gaps and magnetic resonance exclude the possibility of attributing these wave-vectors to Bragg diffractions of the reciprocal lattice vectors.

QPI wave-vectors q_1 and q_2 are identified [237], and the ω-dependence of $F(q_2,\omega)$ is shown in Fig. 2.21(c) [238]. The pronounced peaks of $F(q_2,\omega)$ at $\omega = \Delta_\beta, \Delta_{\alpha,\gamma/\delta}$ and an energy associated with the magnetic resonance $\Omega_{r1} \sim (\Delta_\beta + \Delta_{\gamma/\delta})$ [238] following Eqs. (2.15) and (2.16) support the notion that q_2 is associated with the QPI wave-vector between the electron and hole pockets rather than due to Bragg diffraction because the latter would have been ω-independent. The absence of q_3 in Figs. 2.21(a)–(b) further corroborates the $s_{\pm}$-wave pairing [237]. Moreover, the strong q_1 intensity is consistent with the presence of CDW [233].

In addition to the information obtained from QPI due to elastic impurity scattering, the quasiparticle tunneling spectra also contain inelastic scattering information at higher quasiparticle energies. Indeed, STS studies of $Ba(Fe_{1-x}Co_x)_2As_2$ [237, 238] single crystals have revealed spectral features that are consistent with the magnetic resonances. The characteristic energies identified from the quasiparticle tunneling spectra of $Ba(Fe_{1-x}Co_x)_2As_2$ and attributed to the magnetic resonances in these samples are found to satisfy the relationship $\Omega_{r1} \sim (\Delta_\beta + \Delta_{\gamma/\delta}) \sim 1.5\Delta_{\alpha,\gamma/\delta} \sim 3\Delta_\beta$ and $\Omega_{r2} \sim (\Delta_\alpha + \Delta_{\gamma/\delta}) \sim 2\Delta_{\alpha,\gamma/\delta}$ [238]. Therefore, only one magnetic

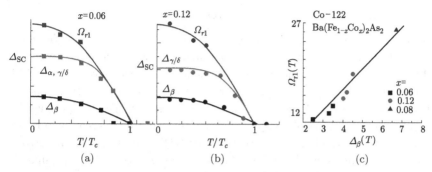

Fig. 2.22   Correlation between the SC gaps $\Delta_{\alpha,\gamma/\delta}$, Δ_β and the magnetic resonant mode Ω_{r1} of the Ba(Fe$_{1-x}$Co$_x$)$_2$As$_2$ superconductors [238]: (a) T-dependence of $\Delta_{\alpha,\gamma/\delta}$, Δ_β and Ω_{r1} for the under-doped sample ($x = 0.06$). (b) T-dependence of $\Delta_{\gamma/\delta}$, Δ_β and Ω_{r1} for the over-doped sample ($x = 0.12$). (c) Correlation of the magnetic resonant mode $\Omega_{r1}(T)$ with the SC gap $\Delta_\beta(T)$ for Co-122 samples with three different doping levels of $x = 0.06, 0.08$ and 0.12. The slope agrees with the relation $\Omega_{r1} \sim 3\Delta_\beta \sim 1.5\Delta_{\gamma/\delta}$. This universal relationship among samples of Ba(Fe$_{1-x}$Co$_x$)$_2$As$_2$ with doping levels and at different temperatures below T_c suggests the relevance of AFM spin fluctuations to the Cooper pairing in these superconductors [238, 248].

resonance Ω_{r1} is observed for optimally- and over-doped Ba(Fe$_{1-x}$Co$_x$)$_2$As$_2$ due to vanished α-pocket, whereas $\Omega_{r2} \sim (16 \pm 1)$ meV for under-doped Ba(Fe$_{1-x}$Co$_x$)$_2$As$_2$. The correlation between the SC gaps and magnetic resonances is manifested in Figs. 2.22(a)–(c) for three different doping levels. These findings from the STS studies for the magnetic resonances are consistent with the observation in INS experiments [244, 247], which again confirm the presence of sign-changes in the pairing potential associated with different disconnected Fermi surfaces [248].

In general, the magnetic resonance behavior directly probed by the INS spectroscopy provides valuable information about the pairing mechanism of unconventional superconductors. In the cuprate superconductors, INS exhibits a clear signature of a resonance mode that becomes strongly enhanced below T_c in addition to its characteristic dispersion known as the "hourglass" behavior, and the resonant energy scales universally with the SC gap amplitude [249]. The generalization of the observed magnetic resonances from the single-band cuprate superconductors to the multi-band multi-gap ferrous superconductors [238, 248] points to the importance of spin fluctuations in the SC state of these unconventional superconductors.

2.3.3 Vortex-state characteristics

As mentioned in Sec. 2.3.1 and exemplified in Figs. 2.18(c)–(e), there are a number of non-universal properties among different ferrous superconductors. For instance, the SC state appears to be exclusive to the AFM phase in the 1111 system, whereas the electron-type 122 system appears to have coexisting SC and AFM phases for a finite range of doping levels, similar to the finding of competing orders in cuprate superconductors. In this context, spatially resolved vortex-state STS studies of different ferrous superconductors should exhibit different types of intra-vortex quasiparticle spectra, depending on whether AFM coexists with SC in the ground state. Such results can also provide useful comparison of the iron-based superconductors with the cuprates.

To date there have only been two published reports with varying findings from the vortex-state STS studies of the ferrous superconductors [239, 250]. In one report, asymmetric vortex bound states appearing as sub-gap peaks inside the vortex cores were observed in a hole-type 122 system $(Ba_{0.6}K_{0.4})Fe_2As_2$ [239], which implies pure SC in this compound. In contrast, STS studies of an electron-type 122 system $Ba(Fe_{0.9}Co_{0.1})_2As_2$ found complete suppression of SC coherence peaks but no apparent sub-gap peaks inside the vortex-cores [250]. However, in the latter case of investigation, the zero-field tunneling spectra of the specific sample only revealed one spatially varying SC gap for all scanned areas [250], which differed from the expected two-gap SC characteristics [237–239] that have been exemplified in Fig. 2.20. The missing two-gap phenomena may be the result of surface reconstructions due to the reactive nature of the sample surface. Therefore, it is not conclusive whether vortex bound states or pseudogap features exist inside the vortex-core of the electron-type $Ba(Fe_{1-x}Co_x)_2As_2$ system. Moreover, whether the vortex-core states of both the electron-type and hole-type 122 systems may exhibit doping-dependence remains an open issue because no doping-dependence of the vortex-core states has been investigated. A comprehensive STS experimental survey of the doping-dependent vortex-state spectral characteristics in various families of iron-based superconductors can provide useful information about possible correlations between the vortex-core states and the existence of AFM spin fluctuations. These studies are nonetheless quite challenging, because STS experiments are extremely time-consuming, whereas the surface layers of most ferrous superconductors are very reactive and are prone to surface reconstructions as well as surface degradation with time.

2.4   Implications for the high-temperature superconducting mechanism

In this section we discuss the physical implications on the pairing mechanism of high-temperature superconductivity based on the aforementioned comparative studies of the cuprate and ferrous superconductors.

The quest for the pairing mechanism of high-temperature superconductivity [2–4, 251, 252] and the debate over the role of electron-phonon coupling [196, 253] has never ceased since the discovery of cuprate superconductors. Based on comprehensive experimental surveys in both the cuprate and ferrous superconductors, there are strong evidences that the relevant pairing interaction must be repulsive in these two classes of high-temperature superconductors. In fact, the same situation applies to the heavy-fermion superconductors that also exhibit unconventional pairing symmetry, spin resonances, layered structures, competing orders, and quantum criticality [254–263]. Hence, the pairing mechanism for all these unconventional superconductors must involve sign-changing pairing potentials in different parts of the Fermi surfaces [123, 124, 237, 238, 245]. The apparent link between the Fermi surface topology and the sign-reversal SC pairing potential [222, 248] as well as the proportionality between the SC transition temperature T_c and the spin fluctuation temperature T_0 [249, 264] for all these unconventional superconductors strongly suggests that the presence of a magnetic mode mediates the Cooper pairing via a repulsive interaction. In this context, the attractive and small-magnitude pairing energy mediated by the electron-phonon coupling is unlikely to be the pairing mechanism for high-temperature superconductivity. Rather, the pairing mechanism would favor repulsive electronic interactions with Coulomb-like correlations, and it is tempting to suggest that stronger correlated cuprates with an insulating parent state would yield larger pairing potentials and therefore higher T_c values than the iron-based compounds that have a semi-metallic parent state.

However, the formation of Cooper pairing in unconventional superconductors must involve a subtle balance between the large repulsive interaction and the tendency to localize charge carriers [220]. In particular, the repulsive interaction must be gingerly arranged among pairs of carriers in different parts of the Fermi surfaces with opposite signs in the pairing potential, which may be achieved by either pairing the carriers in the same band with an orbital angular momentum larger than 0 [118], or pairing the carriers with zero orbital angular momentum in different bands of opposite signs in potential [229]. The latter pairing arrangement is consistent with most iron-based superconductors other than a special class of iron chalcogenides $A_x Fe_{2-y} Se_2$, whereas the former pairing scenario is consistent with the situation in cuprate superconductors and also in $A_x Fe_{2-y} Se_2$ where the hole pockets are found to completely vanish and so nodeless d-wave pairing has been predicted theoretically [265, 266]. In addition to the orbital degree of freedom for pairing, AFM coupling is most favorable for the spin degree of freedom to achieve singlet Cooper pairs. Thus, a parent system with AFM order appears to be an important common feature among the cuprate, ferrous and heavy-fermion superconductors. While long-range AFM must be suppressed before SC can appear, dynamic AFM fluctuations [252] may be favorable for high-temperature superconductivity. Moreover, the cuprate, ferrous and heavy-fermion superconductors all exhibit very similar doping-dependent phase diagrams as illustrated in Fig. 2.1(a) and Figs. 2.18(c)–(e). The proximity of SC to AFM instabilities implies the relevance of quantum criticality and quantum fluctuations to the occurrence of SC. Finally, all these three types of unconventional superconductors exhibit quasi-two dimensionalities, with effective layered structures playing the key role in superconductivity.

On the other hand, the aforementioned clever arrangements for pairing the carriers in different parts of the Fermi surfaces under a repulsive interaction are not sufficient to yield superconductivity, because the complexity of many-body interactions in a correlated electronic system could favor different instabilities from superconductivity upon lowering the temperature. It is therefore not difficult to understand the complications of competing orders in the under-doped cuprate superconductors where the electronic correlations are most significant. Similarly, in the case of heavy-fermion superconductors where the T_c values are much lower than those of the cuprate and ferrous superconductors, long-range magnetic orders are generally prevailing at higher temperatures and superconductivity cannot appear until the temperature lowers below the Kondo temperature so that localized magnetic moments become completely shielded by conduction electrons. Thus, the occurrence of competing orders in these unconventional superconductors seems to be a natural consequence of strong electronic correlations rather than a necessary condition for high-temperature superconductivity.

Although the electron-phonon interaction is unlikely to be the "glue" for Cooper pairs in the cuprate and ferrous superconductors, the presence of electron-phonon coupling under special circumstances may assist stronger

charge transfer and thus enhance the electronic density of states at the
Fermi level and result in an increase of T_c under favorable conditions [196,
253]. For instance, in the under-doped hole-type cuprate superconductors,
isotope effects on T_c are found to decrease with increasing doping and then
completely vanish at the optimal doping [267–271]. In contrast, no isotope
effects have been reported in any of the electron-type cuprate superconductors. Such findings may be attributed to the slower carrier mobility
in under-doped cuprate superconductors so that coupling between typically faster-moving carriers in the dynamic mixture of $3d^9 2p^5$ and $3d^9 2p^6$
electronic configurations to the slower-moving longitudinal optical phonon
modes along the Cu-O bond becomes possible. This stronger electron-
phonon coupling could assist better charge transfer along the anti-nodal
direction of the pairing potential [196]. In contrast, the electronic configurations of electron-doped cuprate superconductors consist of a dynamic
mixture of $3d^{10} 2p^6$ and $3d^9 2p^6$, which is less favorable for optical phonon-
assisted charge transfer along the Cu-O bond [7] and therefore is consistent
with the empirical finding of insignificant isotope effect. Overall, it appears
that electron-phonon coupling is unlikely to be solely responsible for the
occurrence of high-temperature superconductivity, at least not in the case
of the cuprate, ferrous and heavy-fermion superconductors, even though
in some special cases the coupling may help enhance the superconducting
transition.

The realization of repulsive pairing potentials and competing orders
appears to have settled many puzzling issues and debates in the cuprate and
ferrous superconductivity. However, a conclusive experiment to unambigu-
ously point to the AFM spin fluctuations as the mediator for Cooper pairing
in these high-temperature superconductors has yet to be devised. Moreover,
in the search for a good "recipe" for high-temperature superconductivity,
it is important to ask whether other pairing mechanisms different from the
AFM spin fluctuations may be viable candidates. These profound and yet
unsettled issues will certainly keep the research field of superconductivity
intellectually challenging and exciting.

2.5   Conclusion

We have reviewed the experimental findings and the corresponding theoretical understandings of various unconventional low-energy excitation phenomena in two types of high-temperature superconductors, the cuprate and

iron-based superconductors. In the cuprate superconductors that are known as doped Mott insulators with strong electronic correlations and are in close proximity to an AFM instability, sign-changing unconventional $d_{x^2-y^2}$-wave and $(d_{x^2-y^2} + s)$-wave pairing symmetries are established among different cuprates with varying doping levels, and the s-component appears to increase with increasing doping. The unconventional $d_{x^2-y^2}$-wave and $(d_{x^2-y^2} + s)$-wave pairing symmetries minimize the on-site Coulomb repulsion, and have significant consequences on the low-energy excitations and impurity-induced quasiparticle scattering in the cuprates. The strong correlation in the cuprates and their proximity to an AFM instability result in coexisting SC with various COs in the ground state, yielding non-universal phenomena (such as the pseudogap and Fermi arc phenomena) among different cuprates as well as the occurrence of quantum criticality, strong quantum fluctuations, and weakened superconducting stiffness. A phenomenology based on coexisting COs and SC in the cuprates appears to provide consistent account for a wide range of experimental findings, and is also compatible with the possibility of pre-formed Cooper pairs and significant phase fluctuations in cuprate superconductors.

In the case of iron-based superconductors whose parent states are AFM semi-metals, studies of the low-energy quasiparticle and spin excitations reveal unconventional sign-changing $s_\pm$-wave or nodeless d-wave pairing symmetries with two SC gaps and two magnetic resonant modes that scale with the SC gaps. Our comparative studies therefore suggest that the commonalities among the cuprate and the ferrous superconductors include the proximity to AFM instabilities, the existence of AFM spin fluctuations and magnetic resonances in the SC state, the unconventional pairing symmetries with sign-changing order parameters on different parts of the Fermi surfaces, the layered structures, and the appearance of multi-channel low-energy excitations in the SC state either due to COs as in the cuprates or due to multi-band pairing as in the iron-based superconductors. These common features imply that the pairing potential in these high-temperature superconductors is repulsive and therefore is predominantly electronic in nature. Moreover, the apparent link between the Fermi surface topology and the sign-reversal SC pairing potential as well as the proportionality between the SC transition temperature T_c and the spin fluctuation temperature T_0 strongly suggests that the Cooper pairing in these high-temperature superconductors is mediated by a magnetic mode through repulsive interactions. Although under special circumstances the electron-phonon interaction may help enhance the T_c value, it seems that the attractive and relatively small electron-phonon interaction is unlikely to be the sole pairing mechanism for high-T_c superconductivity.

In the context of singlet pairing through repulsive electronic interaction, it is tempting to suggest that the strongly correlated cuprates with an insulating parent state are likely to acquire larger pairing potentials and higher SC transition temperatures, whereas the semi-metallic parent state of the iron-based superconductors may result in overall lower T_c values relative to the cuprates. On the other hand, strong electronic correlations have the tendency to localize carriers and/or to induce other instabilities than superconductivity. Hence, proper balance between the electronic correlation and the itinerancy of pairs is essential to the occurrence of high-temperature superconductivity. Finally, whether pairing mechanisms other than spin fluctuations may be feasible for high-temperature superconductivity remains an open issue for exploration.

Acknowledgement

Much of the research results presented here are in collaboration with Dr. Marcus L. Teague, Dr. Andrew D. Beyer, Dr. C. T. Chen, Dr. M. S. Grinolds, Professor Setsuko Tajima, Professor Hai-Hu Wen, Professor Jochen Mannhart and late Professor Sung-Ik Lee. The author also acknowledges research support from the National Science Foundation and the Kavli Foundation through the facilities at the Kavli Nanoscience Institute at Caltech.

References

[1] J. G. Bednorz and K. A. Muller, *Z. Phys. B* **64**, 189 (1986).

[2] Y. Kamihara, T. Watanabe, M. Hirano and H. Hosono, *J. Am. Chem. Soc.* **130**, 3296 (2008).

[3] P. A. Lee, N. Nagaosa and X. G. Wen, *Rev. Mod. Phys.* **78**, 17 (2006); and references therein.

[4] S. A. Kivelson, I. P. Bindloss, E. Fradkin, V. Oganesyan, J. M. Tranquada, A. Kapitulnik, and C. Howald, *Rev. Mod. Phys.* **75**, 1201 (2003).

[5] S. Sachdev, *Rev. Mod. Phys.* **75**, 913 (2003).

[6] E. Demler, W. Hanke and S. C. Zhang, *Rev. Mod. Phys.* **76**, 909 (2004).

[7] N. C. Yeh and A. D. Beyer, *Int. J. Mod. Phys. B* **23**, 4543 (2009).

[8] M. A. Kastner, R. J. Birgeneau, G. Shirane and Y. Endoh, *Rev. Mod. Phys.* **70**, 897 (1998).

[9] M. B. Maple, *MRS Bulletin* **15** (6), 60 (1990).

[10] N. C. Yeh, A. D. Beyer, M. L. Teague, S. P. Lee, S. Tajima and S. I. Lee, *J. Supercond. Nov. Magn.* **23**, 757 (2010).

[11] A. D. Beyer, M. S. Grinolds, M. L. Teague, S. Tajima and N. C. Yeh, *Europhys. Lett.* **87**, 37005 (2009).

[12] C. T. Chen, A. D. Beyer and N. C. Yeh, *Solid State Commun.* **143**, 447 (2007).

[13] A. D. Beyer, C. T. Chen and N. C. Yeh, *Physica C* **468**, 471 (2008).

[14] N. C. Yeh, A. D. Beyer, M. L. Teague, S. P. Lee, S. Tajima, and S. I. Lee, *Int. J. Mod. Phys. B* **19**, 285 (2005).

[15] N. C. Yeh, C. T. Chen, A. D. Beyer and S. I. Lee, *Chinese J. Phys.* **45**, 263 (2007).

[16] B. L. Yu, J. Wang, A. D. Beyer, M. L. Teague, J. S. A. Horng, S. P. Lee and N. C. Yeh, *Solid State Commun.* **149**, 261 (2009).

[17] M. L. Teague, A. D. Beyer, M. S. Grinolds, S. I. Lee and N. C. Yeh, *Europhys. Lett.* **85**, 17004 (2009).

[18] T. Timusk and B. Statt, Rep. Prog. Phys. **62**, 61 (1999); and references therein.

[19] A. Damascelli, Z. Hussain and Z. X. Shen, *Rev. Mod. Phys.* **75**, 473 (2003).

[20] Ø. Fischer, Ø. Fischer, M. Kugler, I. Maggio-Aprile, C. Berthod and Ch. Renner, *Rev. Mod. Phys.* **79**, 353 (2007); and references therein.

[21] W. S. Lee, I. M. Vishik, K. Tanaka, D. H. Lu, T. Sasagawa, N. Nagaosa, T. P. Devereaux, Z. Hussain and Z. X. Shen, *Nature* **450**, 81 (2007).

[22] Ch. Renner, B. Revaz, J. Y. Genoud, K. Kadowaki, and Ø. Fischer, *Phys. Rev. Lett.* **80**, 149 (1998).

[23] M. Vershinin, S. Misra, S. Ono, Y. Abe, Y. Ando and A. Yazdani, *Science* **303**, 1995 (2004).

[24] M. C. Boyer, W. D. Wise, K. Chatterjee, M. Yi, T. Kondo, T. Takeuchi, H. Ikuta, Y. Wang, and E. W. Hudson, *Nature Phys.* **3**, 802 (2007).

[25] K. K. Gomes, A. N. Pasupathy, A. Pushp, S. Ono, Y. Ando and A. Yazdani, *Nature* **447**, 569 (2007).

[26] R. E. Walstedt, R. F. Bell and D. B. Mitzi, *Phys. Rev. B* **44**, 7760 (1991).

[27] R. Stern, M. Mali, I. Mangelschots, J. Roos and D. Brinkmann, *Phys. Rev. B* **50**, 426 (1994).

[28] R. Stern, M. Mali, J. Roos, D. Brinkmann, J. Y. Genoud, T. Graf, and J. Muller, *Phys. Rev. B* **52**, R15734 (1995).

[29] R. L. Corey, N. J. Curro, K. OHara, T. Imai, C. P. Slichter, K. Yoshimura, M. Katoh and K. Kosuge, *Phys. Rev. B* **53**, 5907 (1996).

[30] M. H. Julien, P. Carretta, M. Horvatic, C. Berthier, Y. Berthier, P. Segransan, A. Carrington and D. Colson, *Phys. Rev. Lett.* **76**, 4238 (1996).

[31] Y. W. Hsueh, B. W. Statt, M. Reedyk, J. S. Xue and J. E. Greedan, *Phys. Rev. B* **56**, R8511 (1997).

[32] A. V. Puchkov, P. Fournier, D. N. Basov, T. Timusk, A. Kapitulnik and N. N. Kolesnikov, *Phys. Rev. Lett.* **77**, 3212 (1996).

[33] M. Opel, R. Nemetschek, C. Hoffmann, R. Philipp, P. F. Müler, R. Hackl, I. Tüttö A. Erb, B. Revaz, E. Walker, H. Berger and L. Forró, *Phys. Rev.* B **61**, 9752 (2000).

[34] M. LeTacon, A. Georges, G. Kotliar, Y. Gallais, D. Colson, and A. Forget, *Nature Phys.* **2**, 537 (2006).

[35] C. T. Chen, P. Seneor, N. C. Yeh, R. P. Vasquez, L. D. Bell, C. U. Jung, J. Y. Kim, Min-Seok Park, Heon-Jung Kim, and Sung-Ik Lee, *Phys. Rev. Lett.* **88**, 227002 (2002).

[36] H. Matsui, K. Terashima, T. Sato, T. Takahashi, M. Fujita, and K. Yamada, *Phys. Rev. Lett.* **95**, 017003 (2005).

[37] S. Kleefisch, B. Welter, A. Marx, L. Alff, R. Gross and M. Naito, *Phys. Rev.* B **63**, 100507 (2001).

[38] L. Alff, Y. Krockenberger, B. Welter, M. Schonecke, R. Gross, D. Manske, and M. Naito, *Nature* **422**, 698 (2003).

[39] G. Blumberg, A. Koitzsch, A. Gozar, B. S. Dennis, C. A. Kendziora, P. Fournier and R. L. Greene, *Phys. Rev. Lett.* **88**, 107002 (2002).

[40] K. Yamada, S. Wakimoto, G. Shirane, C. H. Lee, M. A. Kastner, S. Hosoya, M. Greven, Y. Endoh and R. J. Birgeneau, *Phys. Rev. Lett.* **75**, 1626 (1995).

[41] B. Lake, H. M. Ronnow, N. B. Christensen, G. Aeppli, K. Lefmann, D. F. McMorrow, P. Vorderwisch, P. Smeibidl, N. Mangkorntong, T. Sasagawa, M. Nohara, H. Takagi and T. E. Mason, *Nature* **415**, 299 (2002).

[42] C. H. Lee, K. Yamada, Y. Endoh, G. Shirane, R. J. Birgeneau, M. A. Kastner, M. Greven and Y. J. Kim, *J. Phys. Soc. Japan* **69**, 1170 (2000).

[43] K. Yamada, K. Kurahashi, T. Uefuji, M. Fujita, S. Park, S. H. Lee, and Y. Endoh, *Phys. Rev. Lett.* **90**, 137004 (2003).

[44] M. Fujita, M. Matsuda, S. Katano and K. Yamada, *Phys. Rev. Lett.* **93**, 147003 (2004).

[45] H. J. Kang *et al.*, *Phys. Rev.* B **71**, 214512 (2005).

[46] Y. Onos, Y. Taguchi, K. Ishizaka and Y. Tokura, *Phys. Rev. Lett.* **87**, 217001 (2001).

[47] Y. Gallais, A. Sacuto, T. P. Devereaux and D. Colson, *Phys. Rev.* B **71**, 012506 (2005).

[48] K. Yamada, C. H. Lee, K. Kurahashi, J. Wada, S. Wakimoto, S. Ueki, H. Kimura, Y. Endoh, S. Hosoya, G. Shirane, R. J. Birgeneau, M. Greven, M. A. Kastner and Y. J. Kim, *Phys. Rev.* B **57**, 6165 (1998).

[49] J. M. Tranquada, B. J. Sternlieb, J. D. Axe, Y. Nakamura, and S. Uchida, *Nature* **375**, 561 (1995).

[50] J. M. Tranquada, J. D. Axe, N. Ichikawa, A. R. Moodenbaugh, Y. Nakamura and S. Uchida, *Phys. Rev. Lett.* **78**, 338 (1997).

[51] B. O. Wells, Y. S. Lee, M. A. Kastner, R. J. Christianson, R. J. Birgeneau, K. Yamada, Y. Endoh, and G. Shirane, *Science* **277**, 1067 (1997).

[52] B. Lake, G. Aeppli, K. N. Clausen, D. F. McMorrow, K. Lefmann, N. E. Hussey, N. Mangkorntong, M. Nohara, H. Takagi, T. E. Mason, and A. Schroder, *Science* **291**, 1759 (2001).

[53] H. A. Mook, P. C. Dai, and F. Dogan, *Phys. Rev. Lett.* **88**, 097004 (2002).

[54] M. Fujita, H. Goka, K. Yamada and M. Matsuda, *Phys. Rev. Lett.* **88**, 167008 (2002).

[55] M. Matsuda, S. Katano, T. Uefuji, M. Fujita, and K. Yamada, *Phys. Rev. B* **66**, 172509 (2002).

[56] K. Tokiwa, H. Okumoto, T. Imamura, S. Mikusu, K. Yuasa, W. Higemoto, K. Nishiyama, A. Iyo, Y. Tanaka, and T. Watanbe, *Int. J. Mod. Phys. B* **17**, 3540 (2003).

[57] H. Kotegawa, Y. Tokunaga, K. Ishida, G. Q. Zheng, Y. Kitaoka, K. Asayama, H. Kito, A. Iyo, H. Ihara, K. Tanaka, K. Tokiwa, T. Watanabe, *J. Phys. Chem. Solids* **62**, 171 (2001).

[58] H. Kotegawa Y. Tokunaga, K. Ishida, G. Q. Zheng, Y. Kitaoka, H. Kito, A. Iyo, K. Tokiwa, T. Watanabe, *Phys. Rev. B* **64**, 064515 (2001).

[59] T. Kondo, T. Takeuchi, A. Kaminski, S. Tsuda, and S. Shin, *Phys. Rev. Lett.* **98**, 267004 (2007).

[60] M. Hashimoto, R. H. He, K. Tanaka, J. P. Testaud, W. Meevasana, R. G. Moore, D. Lu, Ho. Yao, Y. Yoshida, H. Eisaki, T. P. Devereaux, Z. Hussain and Z. X. Shen, *Nat. Phys.* **6**, 414 (2010).

[61] J. E. Hoffman, E. W. Hudson, K. M. Lang, V. Madhavan, H. Eisaki, S. Uchida and J. C. Davis, *Science* **295**, 466 (2002).

[62] C. Howald, H. Eisaki, N. Kaneko, M. Greven and A. Kapitulnik, *Phys. Rev. B* **67**, 014533 (2003).

[63] T. Hanaguri, C. Lupien, Y. Kohsaka, D. H. Lee, M. Azuma, M. Takano, H. Takagi, and J. C. Davis, *Nature* **430**, 1001 (2004).

[64] W. D. Wise, M. C. Boyer, K. Chatterjee, T. Kondo, T. Takeuchi, H. Ikuta, Y. Wang and E. W. Hudson, *Nature Phys.* **4**, 696 (2008).

[65] S. C. Zhang, *Science* **275**, 1089 (1997).

[66] M. Vojta, Y. Zhang and S. Sachdev, *Phys. Rev. B* **62**, 6721 (2000).

[67] C. M. Varma, *Phys. Rev. B* **55**, 14554 (1997).

[68] E. Demler, S. Sachdev and Y. Zhang, *Phys. Rev. Lett.* **87**, 067202 (2001).

[69] A. Polkovnikov, M. Vojta and S. Sachdev, *Phys. Rev. B* **65**, 220509(R) (2002).

[70] Y. Chen, H. Y. Chen and C. S. Ting, *Phys. Rev. B* **66**, 104501 (2002).

[71] H. D. Chen, J. P. Hu, S. Capponi, E. Arrigoni and S. C. Zhang, *Phys. Rev. Lett.* **89**, 137004 (2002).

[72] H. D. Chen *et al.*, *Phys. Rev. Lett.* **93**, 187002 (2004).

[73] S. Chakravarty, R. B. Laughlin, D. K. Morr and C. Nayak, *Phys. Rev. B* **63**, 094503 (2001).

[74] U. Schollwöck, S. Chakravarty, J. O. Fjærestad, J. B. Marston and M. Troyer, *Phys. Rev. Lett.* **90**, 186401 (2003).

[75] J. X. Li, C. Q. Wu and D. H. Lee, *Phys. Rev. B* **74**, 184515 (2006).

[76] S. Chakravarty, H. Y. Kee and K. Volker, *Nature (London)* **428**, 53 (2004).

[77] A. D. Beyer, V. S. Zapf, H. Yang, F. Fabris, M. S. Park, K. H. Kim, S. I. Lee and N. C. Yeh, *Phys. Rev. B* **76**, 140506(R) (2007).

[78] V. S. Zapf, N. C. Yeh, A. D. Beyer, C. R. Hughes, C. Mielke, N. Harrison, M. S. Park, K. H. Kim and S. I. Lee, *Phys. Rev. B* **71**, 134526 (2005).

[79] V. J. Emery and S. A. Kivelson, *Nature* **374**, 434 (1995).

[80] J. Corson, R. Mallozzi, J. Orenstein and J. N. Eckstein, *Nature* **398**, 221 (1999).

[81] G. Blatter, M. V. Feigel'man, V. B. Geshkenbein, A. I. Larkin and V. M. Vinokur, *Rev. Mod. Phys.* **66**, 1125 (1994).

[82] D. S. Fisher, M. P. A. Fisher and D. Huse, *Phys. Rev. B* **47**, 130 (1991).

[83] D. R. Nelson and V. M. Vinokur, *Phys. Rev. Lett.* **68**, 2398 (1992).

[84] D. R. Nelson and V. M. Vinokur, *Phys. Rev. B* **48**, 13060 (1993).

[85] L. Balents and D. R. Nelson, *Phys. Rev. Lett.* **73**, 2618 (1994).

[86] T. Giamarchi and P. Le Doussal, *Phys. Rev. Lett.* **72**, 1530 (1994).

[87] T. Giamarchi and P. Le Doussal, *Phys. Rev. B* **52**, 1242 (1995).

[88] N. C. Yeh, D. S. Reed, W. Jiang, U. Kriplani, C. C. Tsuei, C. C. Chi and F. Holtzberg, *Phys. Rev. Lett.* **71**, 4043 (1993).

[89] N. C. Yeh, D. S. Reed, W. Jiang, U. Kriplani, M. Konczykowski, F. Holtzberg, C. C. Tsuei and C. C. Chi, *Physica A* **200**, 374 (1993).

[90] N. C. Yeh, D. S. Reed, W. Jiang, U. Kriplani, and F. Holtzberg, *Physica C* **235–240**, 2659 (1994).

[91] D. S. Reed, N. C. Yeh, W. Jiang, U. Kriplani, and F. Holtzberg, *Phys. Rev. B* **47**, 6150 (1993).

[92] W. Jiang, N. C. Yeh, D. S. Reed, U. Kriplani, T. A. Tombrello, A. P. Rice and F. Holtzberg, *Phys. Rev. B* **47**, 8308 (1993).

[93] D. S. Reed, N. C. Yeh, W. Jiang, U. Kriplani, D. A. Beam and F. Holtzberg, *Phys. Rev. B* **49**, 4384 (1994).

[94] W. Jiang, N. C. Yeh, D. S. Reed, D. A. Beam, U. Kriplani, M. Konczykowski and F. Holtzberg, *Phys. Rev. Lett.* **72**, 550 (1994).

[95] D. S. Reed, N. C. Yeh, M. Konczykowski, A. V. Samoilov and F. Holtzberg, *Phys. Rev. B* **51**, 16448 (1995).

[96] N. C. Yeh, D. S. Reed, W. Jiang, U. Kriplani, D. A. Beam, M. Konczykowski, F. Holtzberg and C. C. Tsuei, in "Advances in Superconductivity — VII", Vol. 1, 455–461, Spinger-Verlag, Tokyo (1995).

[97] N. C. Yeh, W. Jiang, D. S. Reed, U. Kriplani, M. Konczykowski, F. Holtzberg, and C. C. Tsuei, *Ferroelectrics* **177**, 143–159 (1996).

[98] N. Doiron-Leyraud, D. LaBoeuf, J. Levallois, J. B. Bonnemaison, R. Liang, D. A. Bonn, W. N. Hardy, C. Proust and L. Taillefer, *Nature (London)* **447**, 565 (2007).

[99] A. Bangura, J. D. Fletcher, A. Carrington, J. Levallois, M. Nardone, B. Vignolle, D. J. Heard, N. Doiron-Leyraud, D. LaBoeuf, L. Taillefer, S. Adachi, C. Proust and N. E. Hussey, *Phys. Rev. Lett.* **100**, 047004 (2008).

[100] E. A. Yelland, J. Singleton, C. H. Mielke, N. Harrison, F. F. Balakirev, B. Dabrowski, and J. R. Cooper, *Phys. Rev. Lett.* **100**, 047003 (2008).

[101] C. Jaudet, D. Vignolles, A. Audouard, J. Levallois, D. LaBoeuf, N. Doiron-Leyraud, B. Vignolle, M. Nardone, A. Zitouni, R. Liang, D. A. Bonn, W. N. Hardy, L. Taillefer and C. Proust, *Phys. Rev. Lett.* **100**, 187005 (2008).

[102] S. E. Sebastian, N. Harrison, E. Palm, T. P. Murphy, C. H. Mielke, R. Liang, D. A. Bonn, W. N. Hardy, and G. G. Lonzarich, *Nature (London)* **454**, 200 (2008).

[103] K. T. Chen and P. A. Lee, *Phys. Rev. B* **79**, 180510 (2009).

[104] A. J. Millis and M. R. Norman, *Phys. Rev. B* **76**, 220503(R) (2007).

[105] D. LeBoeuf, N. Doiron-Leyraud, J. Levallois, R. Daou, J. B. Bonnemaison, N. E. Hussey, L. Balicas, B. J. Ramshaw, R. Liang, D. A. Bonn, W. N. Hardy, S. Adachi, C. Proust and L. Taillefer, *Nature (London)* **450**, 533 (2007).

[106] Y. Iye, S. Nakamura and T. Tamegai, *Physica C* **159**, 616 (1989).

[107] S. J. Hagen, A. W. Smith, M. Rajeswari, J. L. Peng, Z. Y. Li, R. L. Greene, S. N. Mao, X. X. Xi, S. Bhattacharya, Q. Li and C. J. Lobb, *Phys. Rev. B* **47**, 1064 (1993).

[108] J. M. Harris, N. P. Ong and Y. F. Yan, *Phys. Rev. Lett.* **71**, 1455 (1993).

[109] T. W. Clinton, A. W. Smith, Q. Li, J. L. Peng, R. L. Greene, C. J. Lobb, M. Eddy and C. C. Tsuei, *Phys. Rev. B* **52**, R7046 (1995).

[110] D. A. Beam, N. C. Yeh and F. Holtzberg, *J. Phys.: Condens. Matter* **10**, 5955 (1998).

[111] D. A. Beam, N. C. Yeh and R. P. Vasquez, *Phys. Rev. B* **60**, 601 (1999).

[112] K. McElroy, D. H. Lee, J. E. Hoffman, K. M. Lang, J. Lee, E. W. Hudson, H. Eisaki, S. Uchida and J. C. Davis, *Phys. Rev. Lett.* **94**, 197005 (2005).

[113] X. J. Zhou, T. Yoshida, D. H. Lee, W. L. Yang, V. Brouet, F. Zhou, W. X. Ti, J. W. Xiong, Z. X. Zhou, T. Sasagawa, T. Kakeshita, H. Eisaki, S. Uchida, A. Fujimori, H. Hussain and Z. X. Shen, *Phys. Rev. Lett.* **92**, 187001 (2004).

[114] S. H. Pan, E. W. Hudson, A. K. Gupta, K. W. Ng, H. Eisaki, S. Uchida and J. C. Davis, *Phys. Rev. Lett.* **85**, 1536 (2000).

[115] Y. Wang, N. P. Ong, Z. A. Xu, T. Kakeshita, S. Uchida, D. A. Bonn, R. Liang and W. N. Hardy, *Phys. Rev. Lett.* **88**, 257003 (2002).

[116] Y. Wang, L. Li and N. P. Ong, *Phys. Rev. B* **73**, 024510 (2006).

[117] M. R. Norman, A. Kanigel, M. Randeria, U. Chatterjee and J. C. Campuzano, *Phys. Rev. B* **76**, 174501 (2007).

[118] F. C. Zhang and T. M. Rice, *Phys. Rev. B* **37**, 3759 (1988).

[119] D. J. van Harlingen, *Rev. Mod. Phys.* **67**, 515 (1995); and references therein.

[120] C. C. Tsuei and J. R. Kirtley, *Rev. Mod. Phys.* **72**, 969 (2000); and references therein.

[121] C. C. Tsuei and J. R. Kirtley, *Phys. Rev. Lett.* **85**, 182 (2000).

[122] J. R. Kirtley, C. C. Tsuei and K. A. Moler, *Science* **285**, 1373 (1999).

[123] J. Y. T. Wei, N. C. Yeh, D. F. Garrigus and M. Strasik, *Phys. Rev. Lett.* **81**, 2542 (1998).

[124] N. C. Yeh, C. T. Chen, G. Hammerl, J. Mannhart, A. Schmehl, C. W. Schneider, R. R. Schulz, S. Tajima, K. Yoshida, D. Garrigus and M. Strasik,, *Phys. Rev. Lett.* **87**, 087003 (2001).

[125] N. C. Yeh, C. T. Chen, G. Hammerl, J. Mannhart, S. Tajima, K. Yoshida, A. Schmehl, C. W. Schneider and R. R. Schulz, *Physica C* **364–365**, 450 (2001).

[126] J. Y. T. Wei, N. C. Yeh, W. D. Si and X. X. Xi, *Physica B* **284**, 973 (2000).

[127] N. C. Yeh, C. T. Chen, R. P. Vasquez, C. U. Jung, S. I. Lee, K. Yoshida and S. Tajima, *J. Low Temp. Phys.* **131**, 435 (2003).

[128] C. R. Hu, *Phys. Rev. Lett.* **72**, 1526 (1994).

[129] Y. Tanaka and S. Kashiwaya, *Phys. Rev. Lett.* **74**, 3451 (1995).

[130] S. Kashiwaya and Y. Tanaka, *Phys. Rev. B* **53**, 2667 (1996).

[131] G. E. Blonder, M. Tinkham and T. M. Klapwijk, *Phys. Rev. B* **25**, 4515 (1982).

[132] N. C. Yeh, Bulletin of Assoc. Asia Pacific Phys. Soc. **12**, 2 (2002).

[133] A. G. Sun, D. A. Gajewski, M. B. Maple and R. C. Dynes, *Phys. Rev. Lett.* **72**, 2267 (1994).

[134] A. G. Sun, A. Truscott, A. S. Katz, R. C. Dynes, B. W. Veal and C. Gu, *Phys. Rev. B* **54**, 6734 (1996).

[135] Q. Li, Y. N. Tsay, M. Suenaga, R. A. Klemm, G. D. Gu and N. Koshizuka, *Phys. Rev. Lett.* **83**, 4160 (1999).

[136] R. A. Klemm, *Philos. Mag.* **85**, 801 (2005).

[137] R. Kleiner, A. S. Katz, A. G. Sun, R. Summer, D. A. Gajewski, S. H. Han, S. I. Woods, E. Dantsker, B. Chen, K. Char, M. B. Maple, R. C. Dynes and J. Clarke, *Phys. Rev. Lett.* **76**, 2161 (1996).

[138] T. Masui, M. Limonov, H. Uchiyama, S. Lee, S. Tajima and A. Yamanaka, *Phys. Rev. B* **68**, 060506(R) (2003).

[139] R. Khasanov, A. Shengelaya, A. Maisuradze, F. La Mattina, A. Bussmann-Holder, H. Keller and K. A. Mueller, *Phys. Rev. Lett.* **98**, 057007 (2007).

[140] R. Khasanov, S. Straessle, D. Di Castro, T. Masui, S. Miyasaka, S. Tajima, A. Bussmann-Holder and H. Keller, *Phys. Rev. Lett.* **99**, 237601 (2007).

[141] A. A. Abrikosov and L. P. Gor'kov, *Soviet Phys. JETP* **12**, 1243 (1961).

[142] H. Shiba, *Prog. Theor. Phys.* **40**, 435 (1968).

[143] P. Fulde and R. A. Ferrell, *Phys. Rev.* **135**, A550 (1964).

[144] M. A. Wolf and F. Reif, *Phys. Rev.* **137**, A557 (1965).

[145] P. Schlottmann, *Phys. Rev. B* **13**, 1 (1976).

[146] A. Yazdani, B. A. Jones, C. P. Lutz, M. F. Crommie and D. A. Eigler, *Science* **275**, 1767 (1997).

[147] P. W. Anderson, *J. Phys. Chem. Solids* **11**, 26 (1959).

[148] G. Q. Zheng, T. Odaguchi, T. Mito, Y. Kitaoka, K. Asayama and Y. Kodama, *J. Phys. Soc. Japan* **62**, 2591 (1989).

[149] H. Alloul, P. Mendels, H. Casalta, J. F. Marucco and J. Arabski, *Phys. Rev. Lett.* **67**, 3140 (1991).

[150] T. Miyatake, K. Yamaguchi, T. Takata, N. Koshizuka and S. Tanaka, *Phys. Rev. B* **44**, 10139 (1991).

[151] G. Q. Zheng, T. Odaguchi, Y. Kitaoka, K. Asayama, Y. Kodama, K. Mizuhashi and S. Uchida, *Physica C* **263**, 367 (1996).

[152] N. L. Wang, S. Tajima, A. I. Rykov and K. Tomimoto, *Phys. Rev. B* **57**, R11081 (1999).

[153] K. Tomimoto, I. Terasaki, A. I. Rykov, T. Mimura and S. Tajima, *Phys. Rev. B* **60**, (1999).

[154] Y. Sidis *et al.*, *Phys. Rev. Lett.* **84**, 5900 (2000).

[155] H. F. Fong *et al.*, *Phys. Rev. Lett.* **82**, 1939 (1999).

[156] J. Figueras, T. Puig, A. E. Carrillo and X. Obradors, *Supercond. Sci. Technol.* **13**, 1067 (2000).

[157] K. Ishida, Y. Kitaoka, K. Yamazoe, K. Asayama and Y. Yamada, *Phys. Rev. Lett.* **76**, 531 (1996).

[158] S. H. Pan, E. W. Hudson, K. M. Lang, H. Eisaki, S. Uchida and J. C. Davis, *Nature* **403**, 746 (2000).

[159] E. W. Hudson, K. M. Lang, V. Madhavan, S. H. Pan, H. Eisaki, S. Uchida and J. C. Davis, *Nature* **411**, 920 (2001).

[160] A. V. Balatsky, M. I. Salkola and A. Rosengren, *Phys. Rev. B* **51**, 15547 (1995).

[161] M. I. Salkola, A. V. Balatsky and D. J. Scalapino, *Phys. Rev. Lett.* **77**, 1841 (1996).

[162] M. E. Flatte and J. M. Byers, *Phys. Rev. B* **56**, 11213 (1997); M. E. Flatte and J. M. Byers, *Phys. Rev. Lett.* **80**, 4546 (1998).

[163] M. I. Salkola, A. V. Balatsky and J. R. Schrieffer, *Phys. Rev. B* **55**, 12648 (1997).

[164] M. E. Flatte, *Phys. Rev. B* **61**, 14920 (2000).

[165] J. Bobroff, W. A. MacFarlane, H. Alloul, P. Mendels, N. Blanchard, G. Collin and J. F. Marucco, *Phys. Rev. Lett.* **83**, 4381 (1999).

[166] A. Polkovnikov, S. Sachdev and M. Vojta *Phys. Rev. Lett.* **86**, 296 (2001).

[167] M. Vojta and R. Bulla, *Phys. Rev. B* **65**, 014511 (2001).

[168] R. Kilian, S. Krivenko, G. Khaliullin and P. Fulde, *Phys. Rev. B* **59**, 14432 (1999).

[169] A. V. Chubukov and N. Gemelke, *Phys. Rev. B* **61**, R6467 (2000).

[170] C. L. Wu, C. Y. Mou and D. Chang, *Phys. Rev. B* **63**, 172503 (2001).

[171] A. Lanzara, P. V. Bogdanov, X. J. Zhou, S. A. Kellar, D. L. Feng, E. D. Lu, T. Yoshida, H. Eisaki, A. Fujimori, K. Kishio, J. I. Shimoyama, T. Noda, S. Uchida, Z. Hussain, Z. X. Shen, *Nature* **412**, 510 (2001).

[172] J. Bobroff, H. Alloul, S. Ouazi, P. Mendels, A. Mahajan, N. Blanchard, G. Collin, V. Guillen and J. F. Marucco, *Phys. Rev. Lett.* **89**, 157002 (2002).

[173] K. M. Lang, V. Madhavan, J. E. Hoffman, E. W. Hudson, H. Eisaki, S. Uchida and J. C. Davis, *Nature* **415**, 412 (2002).

[174] S. H. Pan, J. P. O'Neal, R. L. Badzey, C. Chamon, H. Ding, J. R. Engelbrecht, Z. Wang, H. Eisaki, S. Uchida, A. K. Gupta, K. W. Ng, E. W. Hudson, K. M. Lang and J. C. Davis, *Nature* **413**, 282 (2001).

[175] C. U. Jung, J. Y. Kim, M. S. Park, M. S. Kim, H. J. Kim, S. Y. Lee and S. I. Lee, *Phys. Rev. B* **65,** 172501 (2002).

[176] C. U. Jung, J. Y. Kim, Mun-Seog Kim, Min-Seok Park, Heon-Jung Kim, Yushu Yao, S. Y. Lee and S. I. Lee, *Physica C* **366**, 299 (2002).

[177] G. Blumberg, A. Koitzsch, A. Gozar, B. S. Dennis, C. A. Kendziora, P. Fournier and R. L. Greene, *Phys. Rev. Lett.* **88**, 107002 (2002).

[178] J. R. Schrieffer, X. G. Wen and S. C. Zhang, *Phys. Rev. B* **39**, 11663 (1989).

[179] A. Polkovnikov, S. Sachdev and M. Vojta, *Physica C* **388**, 19 (2003).

[180] P. Dai, H. A. Mook, R. D. Hunt and F. Dogan, *Phys. Rev. B* **63**, 054525 (2001).

[181] E. M. Motoyama, P. K. Mang, D. Petitgrand, G. Yu, O. P. Vajk, I. M. Vishik and M. Greven, *Phys. Rev. Lett.* **96**, 137002 (2006).

[182] T. Kondo, R. Khasanov, T. Takeuchi, J. Schmalian and A. Kaminski, *Nature* **457**, 296 (2009).

[183] C. T. Chen and N. C. Yeh, *Phys. Rev. B* **68**, 220505(R) (2003).

[184] A. Kanigel, M. R. Norman, M. Randeria, U. Chatterjee, S. Souma, A. Kaminski, H. M. Fretwell, S. Rosenkranz, M. Shi, T. Sato, T. Takahashi, Z. Z. Li, H. Raffy, K. Kadowaki, D. Hinks, L. Ozyuzer and J. C. Campuzano, *Nature Phys.* **2**, 447 (2006).

[185] P. W. Anderson, *Science* **235**, 1196 (1987).

[186] M. R. Norman, A. Kanigel, M. Randeria, U. Chatterjee and J. C. Campuzano, *Phys. Rev. B* **76**, 174501 (2007).

[187] C. Gros, B. Edegger, V. N. Muthukumar and P. W. Anderson, *Proc. Natl. Acad. Sci.* **103**, 14298 (2006).

[188] A. V. Chubukov, M. R. Norman, A. J. Millis and E. Abrahams, *Phys. Rev. B* **76**, 180506(R) (2007).

[189] F. Onufrieva and P. Pfeuty, *Phys. Rev. Lett.* **92**, 247003 (2004).

[190] G. Kotliar and C. M. Varma, *Phys. Rev. Lett.* **77**, 2296 (1996).

[191] A. A. Abrikosov, *Sov. Phys. JETP* **5**, 1174 (1957).

[192] C. Caroli, P. G. deGennes and J. Matricon, *J. Phys. Lett.* **9**, 307 (1964).

[193] F. Gygi and M. Schluter, *Phys. Rev. B* **43**, 7609 (1991).

[194] H. Hess, R. B. Rubinson and J. V. Waszczak, *Phys. Rev. Lett.* **64**, 2711 (1990).

[195] S. H. Pan, E. W. Hudson, A. K. Gupta, K. W. Ng, H. Eisaki, S. Uchida and J. C. Davis, *Phys. Rev. Lett.* **85**, 1536 (2000).

[196] M. Tachiki, M. Machida and T. Egami, *Phys. Rev. B* **67**, 174506 (2003).

[197] N. C. Yeh, M. L. Teague, A. D. Beyer, B. Shen and H. H. Wen, conference proceedings for the 26th International Low-Temperature Physics Conference (LT26), accepted for publication in *J. Phys. Conf. Series* (2011); [arXiv:1107.0697].

[198] M. S. Kim, S. I. Lee, S. C. Yu, I. Kuzemskaya, E. S. Itskevich and K. A. Lokshin, *Phys. Rev. B* **57**, 6121 (1998).

[199] M. S. Kim, C. U. Jung, S. I. Lee and A. Iyo, *Phys. Rev. B* **63**, 134513 (2001).

[200] V. M. Vinokur, V. B. Geshkenbein, M. V. Feigelman and G. Blatter, *Phys. Rev. Lett.* **71**, 1242 (1993).

[201] M. V. Feigelman, V. B. Geshkenbein, A. I. Larkin and V. M. Vinokur, *Pisma Zh. Eksp. Teor. Fiz.* **62**, 811 (1995).

[202] M. V. Feigelman, V. B. Geshkenbein, A. I. Larkin and V. M. Vinokur, *JETP Lett.* **62**, 834 (1995).

[203] A. van Otterlo, M. V. Feigelman, V. B. Geshkenbein and G. Blatter, *Phys. Rev. Lett.* **75**, 3736 (1995).

[204] T. Nagaoka *et al.*, *Phys. Rev. Lett.* **80**, 3594 (1998).

[205] L. Li, Y. Wang, M. J. Naughton, S. Komiya, S. Ono, Y. Ando and N. P. Ong, *J. Magn. Magn. Mater.* **310**, 460 (2007).

[206] H. Takahashi, K. Igawa, K. Arii, Y. Kamihara, M. Hirano and H. Hosono, *Nature* **453**, 376 (2008).

[207] X. H. Chen, T. Wu, G. Wu, R. H. Liu, H. Chen and D. F. Fang, *Nature (London)* **453**, 761 (2008).

[208] H. H. Wen, C. Mu, L. Fang, H. Yang and X. Zhu, *Europhys. Lett.* **82**, 17009 (2008).

[209] Z. A. Ren, J. Yang, W. Lu, X. L. Shen, Z. C. Li, G. C. Che, X. L. Dong, L. L. Sun, F. Zhou and Z. X. Zhao, *Europhys. Lett.* **82**, 57002 (2008).

[210] K. Sasmal, B. Lv, B. Lorenz, A. M. Guloy, F. Chen, Y. Y. Xue and C. W. Chu, *Phys. Rev. Lett.* **101**, 107007 (2008).

[211] M. Rotter, M. Tegel and D. Johrendt, *Phys. Rev. Lett.* **101**, 107006 (2008).

[212] A. S. Sefat, Rongying Jin, Michael A. McGuire, Brian C. Sales, David J. Singh and David Mandrus, *Phys. Rev. Lett.* **101**, 117004 (2008).

[213] J. H. Tapp, Z. Tang, B. Lv, K. Sasmal, B. Lorenz, C. W. Chu and A. M. Guloy, *Phys. Rev. B* **78**, 060505(R) (2008).

[214] K. W. Yeh, T. W. Huang, Y. L. Huang, T. K. Chen, F. C. Hsu, P. M. Wu, Y. C. Lee, Y. Y. Chu, C. L. Chen, J. Y. Luo, D. C. Yan and M. K. Wu, *Europhys. Lett.* **84**, 37002 (2008).

[215] F. C. Hsu, J. Y. Luo, K. W. Yeh, T. K. Chen, T. W. Huang, P. M. Wu, Y. C. Lee, Y. L. Huang, Y. Y. Chu, D. C. Yan and M. K. Wu, *Proc. Nat. Acad. Sci.* **105**, 14262 (2008).

[216] Y. Mizuguchi, F. Tomioka, S. Tsuda, T. Yamaguchi and Y. Takano, *Appl. Phys. Lett.* **93**, 152505 (2008).

[217] J. Guo, S. Jin, G. Wang, S. Wang, K. Zhu, T. Zhou, M. He and X. Chen, *Phys. Rev. B* **82**, 180520(R) (2010).

[218] V. Cvetkovic and Z. Tesanovic, *Europhys. Lett.* **85**, 37002 (2009).

[219] M. R. Norman, *Physics* **1**, 21 (2008).

[220] Z. Tesanovic, *Physics* **2**, 60 (2009).

[221] A. V. Balatsky and D. Parker, *Physics* **2**, 59 (2009).

[222] F. Wang and D. H. Lee, *Science* **332**, 200 (2011).

[223] C. de la Cruz, Q. Huang, J. Li, W. Ratcliff, II, J. L. Zarestky, H. A. Mook, G. F. Chen, J. L. Luo, N. L. Wang and P. C. Dai, *Nature* **453**, 899 (2008).

[224] Q. Huang, Y. Qiu, W. Bao, M. A. Green, J. W. Lynn, Y. C. Gasparovic, T. Wu, G. Wu and X. H. Chen, *Phys. Rev. Lett.* **101**, 257003 (2008).

[225] J. Zhao, Q. Huang, C. de la Cruz, S. L. Li, J. W. Lynn, Y. Chen, M. A. Green, G. F. Chen, G. Li, Z. Li, J. L. Luo, N. L. Wang and P. C. Dai, *Nat. Mater.* **7**, 953 (2008).

[226] T. Yildirim, *Physica C* **469**, 425 (2009).

[227] F. Wang, H. Zhai and D. H. Lee, *Europhys. Lett.* **85**, 37005 (2009).

[228] F. Wang, H. Zhai, Y. Ran, A. Vishwanath and D. H. Lee, *Phys. Rev. Lett.* **102**, 047005 (2009).

[229] I. I. Mazin, D. J. Singh, M. D. Johannes and M. H. Du, *Phys. Rev. Lett.* **101**, 057003 (2008).

[230] K. Kuroki *et al.*, *Phys. Rev. Lett.* **101**, 087004 (2008).

[231] Y. Y. Zhang *et al.*, *Phys. Rev. B* **80**, 094528 (2009).

[232] E. Plamadeala, T. Pereg-Barnea and G. Refael, *Phys. Rev. B* **81**, 134513 (2010).

[233] A. V. Balatsky, D. N. Basov and J.X. Zhu, *Phys. Rev. B* **82**, 144522 (2010).

[234] D. H. Lu, M. Yi, S. K. Mo, A. S. Erickson, J. Analytis, J. H. Chu, D. J. Singh, Z. Hussain, T. H. Geballe, I. R. Fisher and Z. X. Shen, *Nature (London)* **455**, 81 (2008).

[235] H. Ding, P. Richard, K. Nakayama, K. Sugawara, T. Arakane, Y. Sekiba, A. Takayama, S. Souma, T. Sato, T. Takahashi, Z. Wang, X. Dai, Z. Fang, G. F. Chen, J. L. Luo and N. L. Wang, *Europhys. Lett.* **83**, 47001 (2008).

[236] K. Terashima, Y. Sekiba, J. H. Bowen, K. Nakayama, T. Kawahara, T. Sato, P. Richard, Y. M. Xu, L. J. Li, G. H. Cao, Z. A. Xu, H. Ding and T. Takahashi, *Proc. Natl. Acad. Sci. U.S.A.* **106**, 7330 (2009).

[237] M. L. Teague, G. K. Drayna, G. P. Lockhart, P. Cheng, B. Shen, H. H. Wen and N. C. Yeh, *Phys. Rev. Lett.* **106**, 087004 (2011).

[238] N. C. Yeh, M. L. Teague, A. D. Beyer, B. Shen and H. H. Wen, Conference Proceedings for the 26th International Low-Temperature Physics Conference (LT26), accepted for publication in *J. Phys. Conf. Series* (2012); [arXiv:1107.0697].

[239] L. Shan, Y. L. Wang, B. Shen, B. Zeng, Y. Huang, A. Li, D. Wang, H. Yang, C. Ren, Q. H. Wang, S. H. Pan and H. H. Wen, *Nat. Phys.* **7**, 325 (2011).

[240] R. C. Dynes, V. Narayanamurti and J. P. Garno, *Phys. Rev. Lett.* **41**, 1509 (1978).

[241] M. M. Korshunov and I. Eremin, *Phys. Rev. B* **78**, 140509(R) (2008).

[242] T. A. Maier and D. J. Scalapino, *Phys. Rev. B* **78**, 020514(R) (2008).

[243] M. D. Lumsden *et al.*, *Phys. Rev. Lett.* **102**, 107005 (2009).

[244] A. D. Christianson, E. A. Goremychkin, R. Osborn, S. Rosenkranz, M. D. Lumsden, C. D. Malliakas, I. S. Todorov, H. Claus, D. Y. Chung, M. G. Kanatzidis, R. I. Bewley and T. Guidi, *Nature* **456**, 930 (2008).

[245] T. Hanaguri, S. Niitaka, K. Kuroki, H. Takagi, *Science* **328**, 474 (2010).

[246] C. T. Chen, C. C. Tsuei1, M. B. Ketchen, Z. A. Ren and Z. X. Zhao, *Nature Phys.* **6**, 260 (2010).

[247] I. Eremin *et al.*, *Phys. Rev. Lett.* **94**, 147001 (2005).

[248] T. Das and A. V. Balatsky, *Phys. Rev. Lett.* **106**, 157004 (2011).

[249] G. Yu, Y. Li, E. M. Motoyama and M. Greven, *Nature Phys.* **5**, 873 (2009).

[250] Y. Yin, M. Zech, T. L. Williams, X. F. Wang, G. Wu, X. H. Chen and J. E. Hoffman, *Phys. Rev. Lett.* **102**, 097002 (2009).

[251] P. W. Anderson, *The Theory of Superconductivity in the High-T_c Cuprates*, Princeton: Princeton University Press (1997).

[252] D. J. Scalapino, *Phys. Rep.* **250**, 329 (1995); and many references therein.

[253] X. J. Chen, V. V. Struzhkin, Z. Wu, H. Q. Lin, R. J. Hemley and H. K. Mao, *PNAS* **104**, 3732 (2007).

[254] J. L. Sarrao, L. A. Morales, J. D. Thompson, B. L. Scott, G. R. Stewart, F. Wastin, J. Rebizant, P. Boulet, E. Colineau and G. H. Lander, *Nature* **420**, 297 (2002).

[255] E. D. Bauer, J. D. Thompson, J. L. Sarrao, L. A. Morales, F. Wastin, J. Rebizant, J. C. Griveau, P. Javorsky, P. Boulet, E. Colineau, G. H. Lander and G. R. Stewart, *Phys. Rev. Lett.* **93**, 147005 (2004).

[256] J. Custers, P. Gegenwart, H. Wilhelm, K. Neumaier, Y. Tokiwa, O. Trovarelli, C. Geibel, F. Steglich, C. Pepin and P. Coleman, *Nature* **424**, 524 (2003).

[257] J. R. Jeffries, N. A. Frederick, E. D. Bauer, H. Kimura, V. S. Zapf, K. D. Hof, T. A. Sayles and M. B. Maple, *Phys. Rev. B* **72**, 024551 (2005).

[258] P. Gegenwart, Q. Si, and F. Steglich, *Nature Physics* **4**, 186 (2008).

[259] T. Park, F. Ronning, H. Q. Yuan, M. B. Salamon, R. Movshovich, J. L. Sarrao and J. D. Thompson, *Nature* **440**, 65 (2006).

[260] N. K. Sato *et al.*, *Nature* **410**, 340 (2001).

[261] C. Stock, C. Broholm, J. Hudis, H. J. Kang and C. Petrovic, *Phys. Rev. Lett.* **100**, 087001 (2008).

[262] F. Steglich, J. Arndt, S. Friedemann, C. Krellner, Y. Tokiwa, T. Westerkamp, M. Brando, P. Gegenwart, C. Geibel, S. Wirth and O. Stockert, *J. Phys.: Condens. Matter* **22**, 164202 (2010).

[263] E. D. Bauer, R. P. Dickey, V. S. Zapf and M. B. Maple, *J. Phys.: Cond. Matter* **13**, L759 (2001).

[264] N. J. Curro, T. Caldwell, E. D. Bauer, L. A. Morales, M. J. Graf, Y. Bang, A. V. Balatsky, J. D. Thompson, J. L. Sarrao, *Nature* **434**, 622 (2005).

[265] T. Das and A. V. Balatsky, *Phys. Rev. B* **84**, 014521 (2011).

[266] T. Das and A. V. Balatsky, *Phys. Rev. B* **84**, 115117 (2011).

[267] T. Egami, J. H. Chung, R. J. McQueeney, M. Yethiraj, H. A. Mook, C. Frost, Y. Petrov, F. Dogan, Y. Inamura, M. Arai, S. Tajima and Y. Endoh, *Physica B* **316-317**, 62 (2002).

[268] D. Zech, K. Conder, H. Keller, E. Kaldis and K. A. Muller, *Physica B* **219&220**, 136 (1996).

[269] M. K. Crawford, M. N. Kunchur, W. E. Farneth, E. M. McCarron III and S. J. Poon, *Phys. Rev. B* **41**, 282 (1990).

[270] D. J. Pringle, G. V. M. Williams and J. L. Tallon, *Phys. Rev. B* **62**, 12527 (2000).

[271] G. M. Zhao, V. Kirtikar and D. E. Morris, *Phys. Rev. B* **63**, 220506 (2001).

3

Transport Studies on Metal (M) Doped $PrBa_2Cu_3O_7$

T. P. Chen[1], K. Wu[2], Q. Li[3], B. Chen[4], Z. X. Wang[2], J. C. Chen[5], M. Mohammed[1] and A. Al-Hilo[1]

Abstract

We report here the fabrication of our metal (M)-doped $PrBa_2Cu_3O_7$ (PBCO) or $PrBa[Cu_{1-x}M_x]O_7$ (PBCMO) bulk materials and epitaxial thin films in which M = Al, Co, Fe, Ga, Ni, and Zn, and x = 0.00, 0.05, 0.10, 0.15, and 0.20. We will also present the results of our transport and Raman studies on these PBCMO samples. From the results these studies we are able to identify the doping sites for the dopants and their effects on the electrical resistivity. The transport property studies on the YBCO/PBCMO multilayer allow us to deduce the superconducting coupling length, the finite size effect and the two-dimensional (2D) phase transition of YBCO.

3.1 Introduction

Among the rare-earth cuprates the most thoroughly studied is $YBa_2Cu_3O_{7-\delta}$ (YBCO). In these 123 compounds, almost all of the rare-earth compounds can be made superconducting except for Tb, Ce and Pr. This is understandable for the mixed valence elements Tb and Ce, because they can be in the valence of 3^+ and the extra electron in each atom may

---

[1]Department of Physics, University of Arkansas, Little Rock, AR USA.
[2]Department of Physics, Peking University, China.
[3]Department of Physics, Penn. State University, PA USA.
[4]Department of Physics, University of Buffalo, USA.
[5]Academic Affairs & Institutional Studies, University of Arkansas for Medical Sciences, USA.

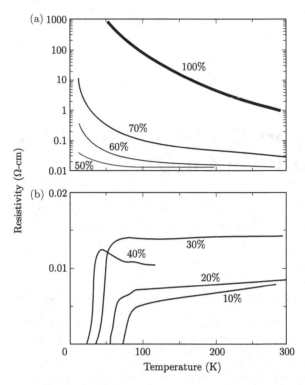

Fig. 3.1   Resistivity vs. temperature plot for Pr replacing Y in YBCO.

fill the oxygen 2 hole. Thus, few empty orbits are available for the electrons to hop, preventing electron transport. Pr in general is a 2^+ element which does not have the extra electron to fill the oxygen 2 hole. This makes the study of the non-superconducting $PrBa_2Cu_3O_{7-\delta}$ (PBCO) interesting. PBCO was first fabricated in 1989 by Peng et $al.$ [1] through replacing Y by Pr in YBCO. Their experimental data is shown in Fig. 3.1. This figure indicates that the compound makes a transition from a superconducting state into a semiconductor/insulator state when the doping level x in the $Y_{1-x}Pr_xCu_3O_{7-\delta}$ reaches 50%. The resistivity for $x = 1.00$, i.e. PBCO is about $200\,\Omega$-cm at $77\,K$, which falls into the semiconductor range. However, at liquid helium temperature, the PBCO does behave as an insulator.

Table 3.1 shows the crystal structure, the lattice parameters a, b, and c, the thermal expansion coefficient α, the processing temperature for YBCO and a group of materials which are commonly used as substrates for YBCO.

Table 3.1   Structures and lattice parameters for PBCO and selected substrates.

Material	Structure	a (Å)	b (Å)	c (Å)	α ($10^{-6}/°C$)	T_{pr} (°C)
YBCO	Orthorhombic	3.83	3.89	11.68	8.5	600–850
PBCO	Orthorhombic	3.88	3.93	11.74	8.5	600–850
$SrTiO_3$	Cubic	3.91	3.91	3.91	9.4	735–800
MgO	Cubic	4.21	4.21	4.21	14	300–700
YSZ	Cubic	5.15	5.15	5.15	11	
$LaAlO_3$	Fcc	7.58	7.58	7.58	9.8	700–760
$LaGaO_3$	Pseudocubic	3.89	3.89	3.89		
$NdGaO_3$	Orthorhombic	5.33	5.49	7.73	a:12,b:7,c:6	675–765

The data in Tab. 3.1 show that the structure of PBCO is the same as YBCO and their lattice parameters are very close. This guarantees that YBCO films grown on PBCO retain their high T_c. Their thermal expansion coefficients α are identical and thus the YBCO-superconducting electronics and devices, using PBCO as an insulator, will have a much longer temperature recycling life. Their processing temperatures T_{pr} are also the same, making the growth of thin films on top of one another much simpler. Based on these data, PBCO has been widely considered as the best substrate and insulating material for making YBCO-superconducting electronics. The drawback is that even for polycrystals the resistivity of PBCO at 77 K is only about 200 Ω-cm [1, 2] and thus it has a resistance of 2×10^{-5} Ω if it is fabricated into a 1 cm^2-wide and 1 nm-thick thin films. This is too low to be used as an insulating layer for devices.

Experiments on YBCO thin films sandwiched in PBCO layers were first carried out by Li et $al.$ and Lowndes et $al.$ in 1990 [2, 3]. Their experimental data indicated that for a fixed thickness of YBCO in a PBCO/YBCO/PBCO multi-layer, T_c decreases with increasing thickness of the PBCO layer. This indicates that the superconducting electrons have migrated from the YBCO superconducting layer into the PBCO layers, a proximity effect. The proximity effect further reduces the insulation of PBCO to YBCO. For this reason, a 2000Å-thick PBCO layer is needed to provide a good potential barrier for a Josephson YBCO/PBCO/YBCO superconductor-insulator-superconductor (SIS) junction. However, the Josephson current density J_c for an SIS junction is given by

$$J_c = [e\hbar(n_1 n_2)^{1/2}]/[4\pi m\{\sinh(2a/\xi)\}], \quad (3.1)$$

in which e is the electron charge, $\hbar$ is the Planck constant, n_1 and n_2 are the superconducting charge densities of the S-layer on the opposite side of the I-layer, m is the mass of an electron, a is the thickness of the I-layer, and ξ is the coherence length of the superconductor. If $a \gg \xi$, then $2a/\xi \gg 1$, and $\sinh(2a/\xi) \approx e^{-2a/\xi}/2$, i.e. the Josphson current density decreases exponentially.

The coherence length of YBCO along its ab-plane $\sim$20 Å while along its c-axis is only about 5 Å. For a 2000Å-thick PBCO layer and such a short coherence length, J_c will be too small for electronic devices. Therefore, to make SIS Josephson junctions, the thickness for the PBCO layer should be comparable to the coherence length, i.e. $\sim$5 to 20 Å. On the other hand, the temperature-dependence of the Josephson critical current is given by

$$J_c(T) \sim (1 - T/T_c)^2. \tag{3.2}$$

Therefore, if a Josephson junction is made to be operable at nitrogen temperature, the thickness of the I-layer in the junction should not only be made very close to the coherence length, without proximity effect, but can also provide sufficient potential barrier for electron-tunneling, otherwise the junction can only be operated at a temperature lower than 77 K, and in this situation the advantage of using the high-T_c cuperates is lost.

It is well understood that the SIS Josephson junction is the heart of most superconducting electronics. For this reason, improving the insulation properties for PBCO is needed in order to make YBCO/PBCO/YBCO SIS junctions, and as an insulator for other electronic devices. We present here our methods and results of improving PBCO through partial replacement of Cu by other metal elements.

3.2   Experiments

3.2.1   *Fabrication and characterization on ceramic*
$PrBa_2[Cu_{1-x}M_x]_3O_7$ (PBCMO)

To improve the insulation properties of PBCO, we have carried out a series of experiments substituting Cu in PBCO by other metal elements including M = Al, Co, Fe, Ga, Ni, and Zn at doping levels of $x = 0.05$, 0.10, 0.15, and 0.20. High purity (99.999%) metal oxide/carbide powders were properly weighted and mixed in according to the chemical formula $PrBa_2[Cu_{1-x}M_x]O_7$. Then the uniformly mixed powder was placed in the furnace and heated up to 950°C for 48 hours for chemical reaction. After the chemical reaction has concluded, the sample was ground into fine powder again and XRD spectra were taken. The processes were repeated until no significant impurity peak could be observed. The X-ray data of these samples are shown in Figs. 3.2 and 3.3. From these figures, we find that the positions of the X-ray peaks for the metal-doped and undoped samples are the same, indicating that except the lattice parameters, no crystal structural change occurred when PBCO is doped by M elements. From the X-ray

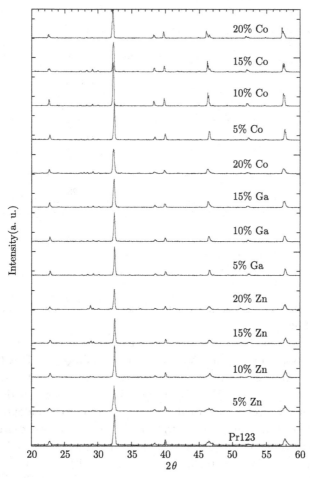

Fig. 3.2   X-ray data for $PrBa_2[Cu_{1-x}M_x]_3O_7$ with M = Co, Ga, and Zn and doping level $x = 0.0, 0.05, 0.10, 0.15$, and 0.20.

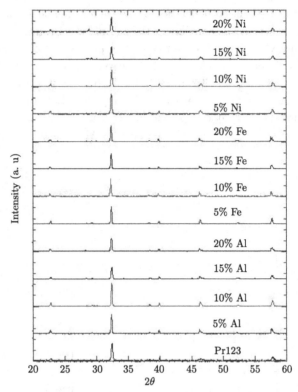

Fig. 3.3   X-ray data for $PrBa_2[Cu_{1-x}M_x]_3O_7$ with M = Al, Fe, and Ni and doping level x = 0.0, 0.05, 0.10, 0.15, and 0.20.

peaks, the lattice parameters for the crystal were calculated, and they are listed in Table 3.2.

Electrical resistivity measurements have also been carried out. The data are plotted in Fig. 3.4 for Co-, Ga- and Zn-doped PBCO, and in Fig. 3.5 for Al-, Fe-, and Ni-doped PBCO. From Figs. 3.4 and 3.5, it is obvious that except for 5% Ni- and Zn-doped samples, all doped samples have much higher electrical resistivities than the undoped, PBCO. Al- and Ga-doped samples have the highest electrical resistivities which, at 77 K, are order of magnitudes higher than the undoped PBCO and therefore can be used to replace the insulating PBCO layer for YBCO superconducting electronic devices or in the YBCO/PBCO/YBCO SIS junction.

Table 3.2   Crystallographic data for $PrBa_2(Cu_{1-x}T_x)_3O_{7-\delta}$.

Composition	a (Å)	b (Å)	c (Å)	V (Å^3)
$YBa_2Cu_3O_{7-\delta}$	3.82030(8)	3.88548(10)	11.68349(23)	
$PrBa_2Cu_3O_{7-\delta}$	3.90608(46)	3.92504(42)	11.63817(129)	178.433(49)
$PrBa_2(Cu_{1-x}Al_x)_3O_{7-\delta}$				
$x = 0.05$	3.92023(102)	3.92160(94)	11.68440(139)	179.631(47)
$x = 0.10$	3.91198(60)	3.91476(53)	11.70080(146)	179.187(45)
$x = 0.15$	3.92081(59)	3.92519(54)	11.67906(150)	179.740(53)
$x = 0.20$	3.91683(60)	3.92462(55)	11.67391(165)	179.455(62)
$PrBa_2(Cu_{1-x}Fe_x)_3O_{7-\delta}$				
$x = 0.05$	3.91108(87)	3.91348(93)	11.73872(290)	179.672(43)
$x = 0.10$	3.91731(95)	3.91903(85)	11.73948(161)	180.225(39)
$x = 0.15$	3.92383(61)	3.92728(55)	11.72063(144)	180.615(47)
$x = 0.20$	3.92372(86)	3.92496(85)	11.70641(136)	180.284(45)
$PrBa_2(Cu_{1-x}Ga_x)_3O_{7-\delta}$				
$x = 0.05$	3.91092(59)	3.91634(51)	11.70261(165)	179.243(53)
$x = 0.10$	3.91738(48)	3.92439(41)	11.70455(130)	179.938(45)
$x = 0.15$	3.92052(90)	3.93069(82)	11.69938(246)	180.291(93)
$x = 0.20$	3.92032(60)	3.93380(55)	11.69724(166)	180.392(64)
$PrBa_2(Cu_{1-x}Ni_x)_3O_{7-\delta}$				
$x = 0.05$	3.90433(49)	3.91607(43)	11.66708(137)	178.385(51)
$x = 0.10$	3.90247(57)	3.91207(47)	11.67310(157)	178.210(55)
$x = 0.15$	3.90374(75)	3.91515(63)	11.67351(206)	178.415(75)
$x = 0.20$	3.90379(123)	3.91523(101)	11.67141(331)	178.388(122)
$PrBa_2(Cu_{1-x}Zn_x)_3O_{7-\delta}$				
$x = 0.05$	3.90503(47)	3.92735(43)	11.64535(138)	178.598(51)
$x = 0.10$	3.90509(48)	3.92233(42)	11.67745(143)	178.865(52)
$x = 0.15$	3.90876(68)	3.92132(59)	11.68568(191)	179.112(71)
$x = 0.20$	3.91032(89)	3.92262(77)	11.68899(248)	179.294(92)
$PrBa_2(Cu_{1-x}Co_x)_3O_{7-\delta}$				
$x = 0.05$	3.9143(6)	3.9193(6)	11.7267(19)	179.90(5)
$x = 0.10$	3.9180(10)	3.9210(11)	11.7294(19)	180.19(7)
$x = 0.15$	3.9251(12)	3.9307(11)	11.7193(19)	180.81(8)
$x = 0.20$	3.9348(11)	3.9367(12)	11.7167(18)	181.49(7)

3.2.2   Fabrication and studies on YBCO/PBCMO multi-layers

Among all doped samples, the 20% Ga- and Al-doped have the highest electrical resistivity. For this reason we used Ga- and Al-doped samples to make YBCO/PBCMO (M = Ga, or Al) multi-layers. The multi-layers

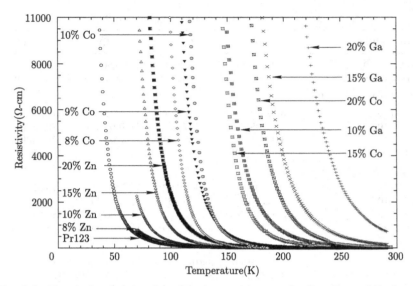

Fig. 3.4   Electrical resistivity plotted against temperature for Co-, Ga- and Zn-doped PBCO.

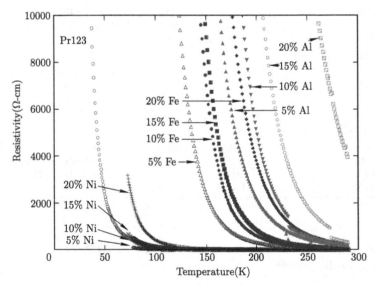

Fig. 3.5   Electrical resistivity plotted against temperature for Al-, Fe- and Ni-doped PBCO.

were made by graining the ceramic sample into powder which was then compressed into a $1''$-diameter and $1/4''$-thick disk-shaped target. A high-power UV pulse laser or a 3 AC gun magnetron sputtering system may be used to make thin films. The outgoing molecules from the target were collected by a $LaAlO_3$ (LAO) single crystal substrate heated at $\sim 800°C$ and at a $45°$-orientation with respect to the target. Because there are 3 sputtering guns in the magnetron sputtering system and an 8-target holder in the laser ablation system, the multi-layers were made *in situ* with no need to open the chamber to air, when changing targets. X-ray data for a (110)-oriented PBCAO thin film and a (110)-oriented PBCGO film are shown in Figs. 3.6 and 3.7. These figures showed that the films are high quality epitaxy films. To prove that the Al- and Ga- doped samples are free from proximity effects, we fabricated a 200 Å sandwiched in a variety of thicknesses of PBCAM and PBCGO films. The thicknesses of PBCAO or PBCGO of the tri-layers are 12.5 Å, 25 Å, 37.5 Å, 50 Å, 75 Å, 100 Å, and 200 Å respectively. Resistivities measured and plotted as a function of temperature of these tri-layers are shown in Figs. 3.8 and 3.9 for PBCAO and PBCGO respectively. As shown in these figures, regardless of the thicknesses of the PBCMO films, the fact that the onset T_c occurs at the same temperature indicates that no proximity effect exists in any of these tri-layers. This shows that the metal-doped PBCO materials not only have a much higher electrical resistivity but are also free from the proximity effect

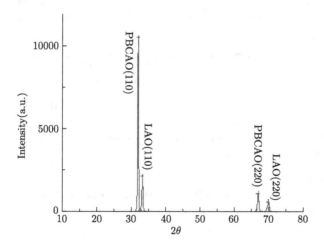

Fig. 3.6   (110)-oriented PBCAO thin film grown at $750°C$.

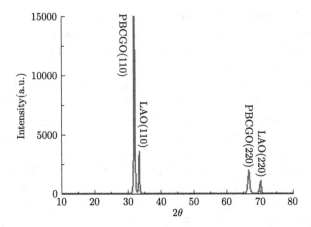

Fig. 3.7   (110)-oriented PBCGO thin film grown at 750°C.

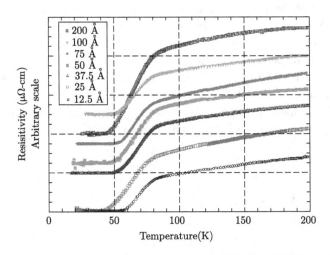

Fig. 3.8   Resistivity plotted against temperature for YBCO/PBCAO multi-layer.

and are thus much better materials to be used in YBCO electronics or in making SIS Josephson junctions.

3.2.3   Two-dimension (2D) to three-dimesion (3D) transition and finite size studies

We have showed in the last section that YBCO thin films fabricated on PBCGO and PBCAO are free from the proximity effect.

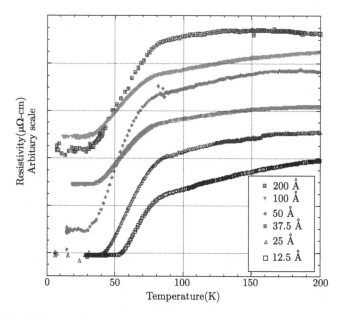

Fig. 3.9   Resistivity plotted against temperature for YBCO/PBCAO multi-layer.

Thus, PBCGO and PBCAO can be used to cut off the interaction between YBCO layers in YBCO/PBCAO and YBCO/PBCAO multi-layers. We may use the multi-layer structure to study the superconducting coupling length, the finite size effect, and 2D to 3D transition for YBCO. Based on this motivation, we fabricated 12.5Å-, 25Å-, 37.5Å-, 50Å-, 75Å-, 100Å-, and 200Å- thick YBCO films sandwiched in 20% Al- or Ga- doped PBCO. Electrical resistivity for these multi-layers were measured as a function of temperature. Three types of multi-layers were fabricated: the tri-layer (PBCMO/YBCO/PBCMO), 5-layer (PBCMO/YBCO/PBCMO/YBCO/PBCMO), and 7-layer (PBCMO/YBCO/PBCMO/YBCO/PBCMO/PBCO/PBCMO). Experimental data for YBCO films of varying thicknesses from 25–200 Å sandwiched in 2 PBCAO layers were plotted in Fig. 3.10. The same YBCO films sandwiched in 2 PBCGO layers were plotted in Fig. 3.11. On the other hand, data for 2 YBCO films sandwiched in 3 PBCAO layers and in 3 PBCGO layers were plotted in Figs. 3.12 and 3.13, respectively.

We found that the 12.5 Å YBCO film (i.e. one unit cell thick) in all multi-layer structures does not have a drop and thus, does not have a superconducting transition. YBCO films thicker than 37.5 Å (i.e. 3 unit cells)

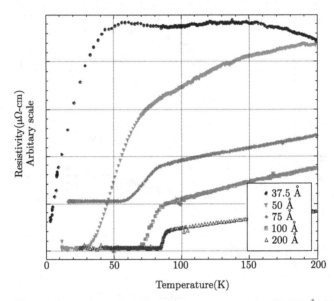

Fig. 3.10   Electrical resistivity plotted against temperature for 25–200 Å-thick YBCO films in Al doped tri-layers.

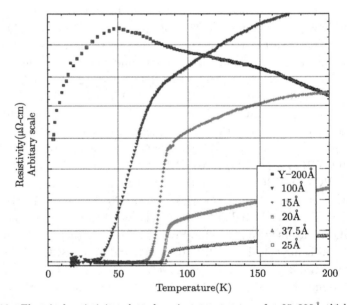

Fig. 3.11   Electrical resistivity plotted against temperature for 25–200Å-thick YBCO films in Ga-doped 3-layers.

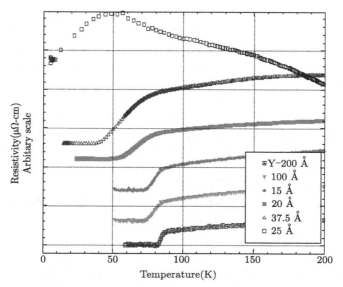

Fig. 3.12 Electrical resistivity plotted against temperature for 25–200Å-thick YBCO films in Al-doped 5-layers.

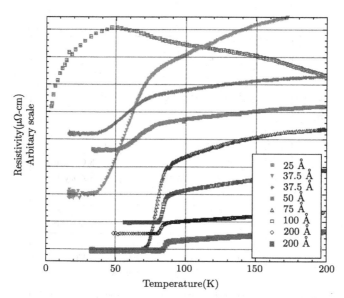

Fig. 3.13 Electrical resistivity plotted against temperature for 25–200Å-thick YBCO films in Ga-doped 5-layers.

have the same behavior. Their resistivity curves decrease with temperature, then drop sharply when the transition temperature is reached. However, the range for superconducting onset is broader when the film thickness is lower. The 25Å-thick YBCO films (2 unit cells) behave differently. They increase with decreasing temperature and then have a very broad turn before dropping. Our explanation is that the films undergo a 3D to 2D transition and become 2D superconductors at the film thickness of 25 Å. This is consistent with the fact that the coherence lengths of the YBCO ~5 Å along its c-axis and ~40 Å along its ab-plane. The onset of the 25 Å film is about 50 K, which may be considered as the 2D superconducting transition temperature for YBCO.

The superconducting onset temperatures for all plots (including that for the 7-layer multi-layers, which was not shown) were estimated and plotted in Fig. 3.14. The vertical bars on the data points in Fig. 3.14 are the uncertainties estimated from the broad onset temperature. It is believed that superconducting coupling between the YBCO unit cell help in increasing transition temperature. For this reason, the T_c for a 3D superconductor is higher than for a 2D superconductor. However, the coupling will be saturated at a length which we call the superconducting coupling length. The onset T_c does increase with the YBCO film thickness and reaches a

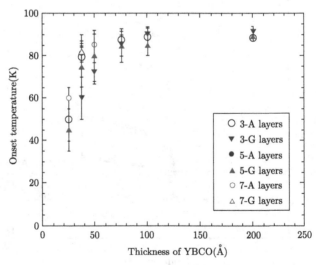

Fig. 3.14   The onset superconducting temperature plotted against the YBCO film thickness in a multi-layer.

plateau at about 150Å ± 50Å. This value is considered the superconducting coupling length for YBCO. When the film reaches this thickness it becomes a 3D superconductor.

3.3 Conclusion

We fabricate Al-, Co-, Fe-, Ga-, Ni-, and Zn-doped PBCO and performed transport studies on these doped materials. We find that the electrical resistivities, except that for 5% Ni- and Zn-doped samples, increase by orders of magnitude. We have also fabricated epitaxy films and multi-layers. Resistivity measurements on the multi-layers show that no proximity effect occurs when YBCO is fabricated on a 20% Al- or Ga-doped layer. This makes the 20% Al- and Ga-doped PBCO a better insulator for YBCO superconducting electronic devices or for Josephson SIS junctions. The transport studies also provide us the 2D to 3D superconducting transition. From the data, we find that the 2D superconducting transition for YBCO is about 50 K, which is very different from that found in Ref. [3]. Plotting the onset temperatures versus the YBCO film thicknesses provides us information on the superconducting coupling length, which is estimated to be (150 ± 50) Å.

References

[1] Peng, et al., PRB **40**, 45 (1989).
[2] Q. Li, X. X. Xi, X. D. Wu, A. Inam, S. Vadlamaniiati and W. L. Mclean, Phys. Rev. Lett. **64**, 3086 (1990).
[3] Douglas H. Lowndes, David P. Norton and J. D. Budai, Phys. Rev. Lett. **65**, 1160 (1990).
[4] T. Usagawa, et al., Jap. J. Appl. Phys. **36**, L1583 (1997).

4

Tweaking Superconductivity at Nanoscales

J. Wu[1]

4.1 Introduction

The year 2011 marks the 100th anniversary of the first discovery of super-conductivity by Kamerlingh Onnes, in mercury. This work was awarded the 1913 Nobel Prize in Physics and triggered exciting discoveries of many other superconductors during the past century, including low-temperature superconductors (LTS) that are mostly metals and metal alloys, high-temperature superconductor (HTS) cuprates that captured another Nobel Prize in 1986 [1], MgB_2 [2] and related materials, and most recently, iron-based pnictides [3, 4]. A superconductor is a material that is characterized by the absence of electrical resistance below a transitional temperature (T_c), which also defines the upper limit of the temperature range in which applications that are based on the superconductivity may be considered. T_c has been raised continuously from 4.2 K in mercury to 138 K in HTS mercury-based cuprates, [5, 6] the highest so far achieved in ambient, and halfway between absolute zero to room temperature. Higher T_c implies higher operating temperatures, and demonstrations followed shortly in applications in both superconducting electronic and electric devices at temperatures above 100 K [5, 7–11]. In addition, a significant trait of a superconductor in the superconducting state is the exclusion of magnetic flux. Up to a critical magnetic field H_c, the flux exclusion may be considered as the result of a complete diamagnetic counteraction of the applied field, and is commonly referred to as the Meissner effect, as it was first discovered by Meissner and

---

[1]Department of Physics and Astronomy, University of Kansas, USA.

Ochsenfeld [12]. From the point of view of energy, as shown by H. London, the Meissner effect was a consequence of the minimization of the electromagnetic free energy carried by superconducting current [13].

In parallel to the discovery of superconducting materials, the theory of superconductivity has been developed. Central to the superconductivity mechanism is the pairing of the electrons, or so-called Cooper pairs. For metal-based superconductors, phonon-mediated attraction was found to play a primary role based on the BCS theory, which was awarded another Nobel Prize in Physics in 1972. While the BCS theory could be applied to explain superconductivity in many superconductors discovered so far, the mechanism of high-temperature superconductivity remains unresolved and presents a grand challenge as well as an active topic to researchers in the field of superconductivity; in particular, to those in search of higher-temperature superconductors.

Applications of superconductivity were initiated in large scale by two important events in 1962. The first commercial superconducting NbTi wire made by Westinghouse initiated the electrical applications of superconductors in magnets, which was followed by many other kinds of electrical applications. On the other hand, the theoretical work by Josephson on the tunneling of the superconducting Cooper pairs through an insulating layer sandwiched by superconductors, the so-called Josephson junction [14], laid the groundwork for superconducting electronics and won him a Nobel Prize in 1973. Various superconducting electronic devices were later constructed based on the concept of the Josephson junctions, in particular, the superconducting quantum interference devices, or SQUIDs, which are the most sensitive magnetometers and have been widely used in NMR/MRI (for nuclear magnetic resonance/magnetic resonance imaging used widely in medicine), mass spectrometry, quantum computing, etc.

Upon entering the next century of superconductivity, a question arises regarding the future of superconductivity. In particular, will superconductivity play an important role in the technological advances of both electronics and electrical devices and systems after a century of its material research and development? What would be the next step to take superconductivity forward to the next level of innovation?

The last decade or so has witnessed explosive advances made in nanoscience and nanotechnology. These include the development of nanostructured materials with at least one dimension in the nano-regime (the definition is a bit blurred, although the common understanding is that it should be 100 nm or smaller), such as nanowires, nanotubes, nanobelts, etc. (one-dimensional, 1D); and extremely thin films (two-dimensional, 2D). This has prompted the invention of various new tools and approaches for characterization and manipulation of nanostructures, particularly in high resolution electron microscopy and scanning probe microscopy with multiple functionalities that enable correlation of various physics properties at the nanoscale. The availability of this whole new class of nanostructured materials and approaches to manipulate them has enabled the exploration of new physics as well as novel devices. The distinctive difference of nanostructured materials when compared to their bulk counterparts is the greatly enhanced surface effect, as the surface-to-volume ratio increases monotonically with decreasing dimension of materials. In addition, many relevant length scales in classical models of materials are in the nano-regime and below where the behaviors of particles, such as electrons and holes, photons, and phonons, can no longer be described by the classical models. Instead, quantum physics will have to be employed and it is where the true excitement of nanoscience and nanotechnology stems from. While its impact on science and technology has been dramatic, more significant impacts are certain to affect almost all aspects of everyday life.

The advances in nanoscience and nanotechnology have generated fresh opportunities for the exploration of the new materials and physics of superconductivity at nanoscales. In fact, extensive researches have been conducted in the last several years, focusing on two major areas: generating nanostructures in existing superconductors, mostly for enhancing specific properties such as the critical current density J_c via nanostructure-enabled magnetic vortex pinning landscape, and exploring superconductivity in nanostructures of either superconductors or hybrids of superconductors with other materials. The work in the former area has been driven by large-scale applications of superconductor wires for electrical devices and systems, including in zero-loss power transmission cables, generators and motors of light weight and high power for applications such as wind turbines, high-field magnets for applications in NMR/MRI (for nuclear magnetic resonance/magnetic resonance imaging used widely in medicine) and higher-energy particle beam accelerators beyond Tevatron in the Fermi Lab and the Large Hadron Collider in CERN, etc. Research exploration in the latter focuses more on understanding superconductivity in small systems, in which quantum confinement is important. Novel electronic devices have been reported based on such an understanding and this trend will certainly continue when hybrids of superconductors with other nanostructured materials, such as carbon nanotubes and graphene, can be generated in a more controllable fashion.

This book chapter intends to provide an overview of recent research on superconductivity using nanotechnology as a unique approach to either the improvement of superconducting properties by generating nanostructures in existing superconductors, or to the probing of behaviors of Cooper pairs in nanostructures such as nanowires and nanotubes, graphene and ultrathin films and interfaces. This chapter is divided into two major parts. The first part focuses on the nanostructures generated inside superconductors, in particular those in the form of thin or thick films, primarily in HTSs for 2nd generation coated conductor applications. The second part will cover superconductivity in nanostructures, either in the form of superconducting nanowires, nanobelts, etc. or in the hybrid nanostructures of superconductors and other materials, or at the interface of non-superconductors. Instead of trying to be thorough, which seems impossible with the vast advances achieved in recent years and limited time and space to put this chapter together, the author will only highlight some works she is familiar with, and sincerely regrets that she cannot include other important works she has ignored in writing this book chapter.

4.2   Nanostructures in superconductors

Superconductors are classified into two types. Type I superconductors, which are mostly elemental metals, have one H_c, and the superconducting state can be reverted to the normal state by applied magnetic fields above H_c. In contrast, Type II superconductors have a lower (H_{c1}) and an upper (H_{c2}) critical field. It should be noted that upon increasing beyond H_{c1}, magnetic field lines are admitted, contributing to the magnetic flux in quantized increments (flux quanta or vortices) until the field intensity exceeds H_{c2}, and the normal state is again restored. For non-zero flux through the superconductor, in the field range $H_{c1} < H < H_{c2}$, the material is in a vortex state [15, 16]. A vortex is defined by two parameters: the superconducting coherence length ξ, which defines the dimension of the normal core, and the London penetration depth λ, which determines the extent of magnetic field around the normal core. The vortices can arrange themselves in a regular structure known as the vortex lattice, also named the Abrikosov vortex, after Alexei Abrikosov, who was awarded the 2003

Nobel Prize in Physics for his pioneering contributions [17]. The movement of the flux lines leads to a resistive effect and limits the current-carrying capacity. This maximum current set by the resistive effect adds the vortex-motion contribution to the critical current, and thus motivates the search for a means to counter the interaction by inducing a state in which the vortices are stabilized, or pinned, in an energetically preferred region.

Most LTS alloys and all HTSs are type II superconductors and are promising for practical applications involving magnetic fields. Since the discovery of the HTS by Bednorz and Müller [1] in 1986, numerous prototypes of electronic as well as electric power devices based on HTS materials have been demonstrated. A great impact on reducing energy loss and environment impact is anticipated from adopting superconducting devices [18]. Significant market penetration, however, requires performance-cost-balanced HTS devices and systems. This puts an urgent demand for material research that can lead to capability in manipulating superconducting properties at nanoscales based on basic physics.

The electrical current-carrying capability of HTS materials is usually described by critical current density J_c and critical current I_c, and has been regarded as one of the most important parameters for practical applications [19, 20]. For example, HTS coated conductors (HTScc), so called 2nd generation HTS wires that employ a strategy of epitaxy of HTS films on metal tapes of textured surface templates, are expected to carry large I_c on the order of a few hundreds to thousands of Amperes per centimeter width at the liquid nitrogen temperature on many kilometers of lengths, to meet the requirement of power applications [18, 21, 22]. In the Josephson junctions, on the other hand, highly controllable I_c is required to achieve uniform device performance [23]. In passive microwave devices employed for telecommunications, J_c was found to correlate well with the power handling capability and nonlinearity of the device at elevated temperatures $\geqslant 77$ K [10, 24].

J_c is a difficult parameter since it is affected by both intrinsic and extrinsic properties of HTS materials. Among others, grain boundaries (GBs) — decisive barriers to current flow — pose a particularly interesting and difficult problem whose solution is vital to HTS applications. First of all, J_c is highly anisotropic in HTSs due to their layered structures and higher J_c flows along the ab-planes. This requires HTS films to have c-axis orientation to eliminate the J_c obstructing c-axis tilt GBs. In addition, the grains must be aligned also in the plane of the film. When crossing the GBs, J_c decreases exponentially as a function of the in-plane misalignment angle above $\sim 2°$–$3°$ for YBCO [25]. This "Dimos" curve sets a stringent upper limit for the in-plane grain alignment. Although the texture of epitaxial HTS films on single crystal substrates is not a concern, reducing the GB misalignment towards the Dimos curve in HTScc has been the research focus in the past 1.5 decades in the development of HTScc. Understanding and being able to engineer the GB properties is clearly essential and challenging [26–28].

4.2.1   *Ideal pinning landscape*

In an ideal epitaxial HTS film, in which GB is not an issue, the ultimate limit of the J_c is dictated by depairing of the Cooper pair electrons, and the depairing current density J_d can be estimated from the Ginzburg–Landau equation: $J_d = \frac{c\phi_0}{12\sqrt{3}\pi^2\lambda_a^2\xi_a}$, with ϕ_0 being the flux quantum and λ_a and ξ_a, the transverse London penetration depth and Pippard coherence length, respectively [29]. The J_c observed in practical superconductors is usually much lower than the theoretically predicted J_d. For YBCO films, $J_d \sim 40$–50 MA/cm^2 at 77 K and self field (SF) is about an order of magnitude higher than the J_c reported on optimized epitaxial YBCO films on single crystal substrates. One of the major hurdles in achieving J_d is in minimizing the dissipation associated with the entry and motion of magnetic vortices in a HTS film. Strong nanoscale vortex pins with lateral dimension approaching ξ_a on the order of few nanometers for HTSs must be generated to suppress the dissipation of vortex motion.

In type-II superconductors such as HTSs, magnetic vortices penetrate the superconductor above the lower critical field H_{c1} and form a lattice of quantized line vortices that must be pinned to maintain a finite J_c. To pin the vortices at an applied magnetic field H, the superconductor matrix must contain defects of desired shape, dimension and density. This has prompted an extensive effort in impurity doping in HTS, and many exciting results have been obtained in generating nanoscale defects in HTSs [19, 26, 30–40]. Regardless of the particular means of adding secondary phases as pinning centers in HTS, the pinning enhancement is largely dependent on the geometry of the particular pinning centers, such as in $YBa_2Cu_3O_{7-x}$ (YBCO) thin films. There have been both nanoparticulate dispersions of non-superconducting additions [41–43] and nanorod pinning centers using

$BaZrO_3$ (BZO), $BaSnO_3$ (BSO) or the more recent Ba_2YNbO_6 (BYNO) or YBa_2NbO (YBNO) [31, 44–48]. Both improved J_c in YBCO, providing either an isotropic pinning enhancement or preferential enhancement in the c-axis direction, respectively. Either way, each can provide an overall angular improvement.

Although these pinning additions do increase the J_c of the YBCO superconductor based on the particular geometry of the inclusions, this may not be a complete perspective of how the additions affect J_c. For example, it is known that chemical interactions between the doping additions and the YBCO can degrade T_c of the sample with an attended reduction in the J_c performance. This can limit the amount of some additions such as BZO, where excessive amounts present in the sample heavily degrade the J_c. Other additive effects can be more forgiving such as BSO. A chemical interaction can be beneficial such as the minute doping of YBCO with deleterious rare-earths [39]. In the particular case of Y_2BaCuO_5(Y211) additions, Ca-doping of this secondary phase provides an added increase to the $J_c(H)$ of YBCO [27, 28].

The pinning strength of a defect is quantitatively determined by the interface quality between the defect and the superconductor matrix [49]. Among many other defects, aligned linear defects have received the most attention since they provide correlated pinning to the vortex lattice [49]. Generating linear defects along the c-axis of HTS is particularly important for the layer-structured HTS materials since the c-axis is the weakest pinning direction. When measured as function of the orientation of H, a strong peak of J_c at $H//ab$-plane was observed, in contrast to the much lower J_c values at $H//c$-axis. To reduce the J_c anisotropy, generating linear defects aligned in the c-axis has been the focus of many groups using either high energy ion beam irradiation [50–52] or aligned nanoparticle (NP) growth [19, 32–36]. Significantly improved correlated pinning has been demonstrated as reflected in enhanced J_c at $H//c$-axis, if H is comparable to or less than the accommodation field $H^* = n_{col}\phi_0$, where n_{col} is the density of the linear defects.

Another benefit of aligned linear defects in c-axis is the constant pinning force (F_p) through HTS film thickness. HTS cables must carry I_c values on the order of hundreds to thousands of Amperes per centimeter width to be used for power applications [18–20]. Given the "optimized" J_c of typically 4–5 MA/cm^2 at 77K and SF on the standard YBCO films, this requires HTS coating in HTScc to be several μms thick. Surprisingly, a monotonic decrease of J_c at 77 K and SF was observed in HTS films with film thickness (t) [19, 53]. This mysterious J_c-t behavior, fitted well by $J_c \sim 1/\sqrt{t}$, persists despite the improvement made in achieving uniform chemical composition and crystalline structure across the film thickness [54, 55], suggesting the intrinsic nature of the issue related to the defect structure of HTS films. In undoped YBCO films, the dominant pinning centers are point defects. According to the weak collective pinning (CP) model [49], the F_p exerted on a vortex of length $L < L_c$ ($L \sim t$ when H is along the film normal) scales as $L^{1/2}$ while the Lorentz force scales as L. This leads to the $J_c \sim 1/\sqrt{t}$ behavior because the vortex lines are rigid within the Larkin length L_c, which is on the order of a few μm for a strong pinning system like YBCO [56], and cannot bend to accommodate different pinning centers nearby. Although thermally activated flux motion adds further complication on the $J_c - t$ issue, the CP behavior has been confirmed in YBCO films in a wide range of temperature (T) and H [57]. One resolution is to chop vortices shorter [58] by making thick YBCO films into multi-layers with insulating spacers. For an n-layer sample, F_p can be increased by a factor of $\sqrt{n}$ at more overhead cost in processing [59]. Aligned linear defects in c-axis can provide constant F_p/vortex length and can overpower the point defects and hence surpass the CP mechanism as we have recently demonstrated in YBCO films with BZO-NRs [60] and nanotube (hollow tube-like) pores (NTPs) [61]. An issue associated with the nearly perfect alignment of the 1D pinning centers, such as BZO-NRs along the c-axis, is that the in-field J_c is not much improved or even reduced at field orientations, defined by an angle θ between H and c-axis, deviating from the c-axis. In particular, the J_c at $H//ab$-plane ($\theta = 90°$), is significantly reduced. This is unfortunate for many applications of HTScc, such as motors and generators, wherein the generated strong magnetic field may bend or twist causing the vortices to stray from the correlated pins. Splaying linear defects in YBCO films may extend the benefit of correlated pinning of the linear defects to a larger θ range. A theoretical work by Hwa et al. predicted that splayed linear defects force vortex entanglement and enhance J_c. An optimal splaying angle of $\theta_c \sim 10°$ was suggested [51]. In experiment, splayed linear defects produced by high-energy ion irradiation in HTS samples were found to carry higher J_c in H-fields than those with the same density of uniformly oriented linear tracks [50, 52]. The challenge is hence whether splayed linear defects could be generated in HTS matrix via a practical way, such as growth.

A further challenge in achieving a three-dimensional (3D) pinning land-scape stems from the need to add additional defects to optimize pinning when H is applied at $\theta > \theta_c$. In fact, the effort towards 3D pinning just began recently [62, 63]. Achieving such a goal requires an understanding of the microscopic growth mechanism of nanostructures in HTS films and ulti-mately, developing control parameters to manipulate nanostructure growth at nanoscales.

4.2.2   Edge pinning on narrow superconducting bridges

Raising J_c has been the focus of world-wide efforts in the field of applied superconductivity in recent years. In particular, a long-standing question is whether the theoretical depairing limit J_d can ever be reached in HTS, especially in practical HTS conductors through self-assembly of nanostruc-tures precisely designed to pin the magnetic vortices with optimal efficiency. Answering this question demands capability in engineering the microstruc-ture of HTS materials at nanoscales, based on a thorough understanding of fundamental physics relevant to both superconducting charge transport and microscopic growth mechanisms of superconducting nanostructures. Such an approach of nanoscale manipulation does not correspond well to and represents a transition from the essentially empirical way in which the HTS materials have been developed so far.

J_d is the thermodynamic depairing limit — it cannot be reached in the presence of dissipation of magnetic vortex motion. In conventional super-conductors, J_d was observed in narrow bridges of superconductors [64–66], and the edge pinning effect on J_c is theoretically predicted to increase mono-tonically with decreasing bridge width as shown in Fig. 4.1 (solid line) [67]. In a recent study of edge pinning in YBCO bridges, we have found that J_c indeed increases with decreasing bridge width in the range of few to ~ 100 μm, shown in Fig. 4.1 [68]. A question arises on whether this trend will continue if the bridge width is further reduced, and if it does, can the benefit of the edge pinning on a narrow bridge be extended to a regular HTS film?

Synthesizing and characterizing HTS narrower bridges below several μm's is necessary to complete the comparison with theory [67] (solid line in Fig. 4.1). However, the fabrication of HTS narrow bridges has pre-sented a major experimental challenge because material degradation, typ-ically showing in decreased T_c and broadened transition width, occurs

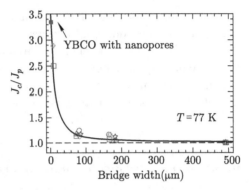

Fig. 4.1   The dotted line represents J_p (bulk J_c) and the solid line, the calculated J_c assuming a perfect edge. Data points (except the highest one) were taken on YBCO bridges of different width.

for HTS narrow bridges made with lithographic methods [69–72]. To generate narrow HTS bridges, we have explored several different strategies including direct patterning of YBCO films using photolithography (PL) and e-beam lithography (EBL), patterning a step on substrates for YBCO nanowire growth, and growing a $PrBa_2Cu_3O_7$ (PBCO) cap layer on YBCO or PBCO/YBCO/PBCO tri-layers followed by lithography. On PBCO/YBCO/PBCO tri-layer bridges of width as small as 240 nm, only slight degradation of $\sim$1–2 K in T_c was observed, suggesting that this approach is promising, while minor damages on the edge of the bridge still occurred due to the high sensitivity of oxides HTS materials to the local oxygen defects. To improve the quality of the lithographically generated edge, thermal annealing in O_2 will be applied to narrow bridges to repair the possible edge damage in fabrication process. An over-layer coating of polymers, such as PMMA and photoresist, will be used to passivate the bridge edge. It remains to be seen whether transport $J_c(H)$ close to the theoretical predicted J_d could be eventually achieved on narrow HTS bridges.

4.2.3   Growth of 1D nanostructures in HTS superconductor films

For large-scale applications, generating linear defects via growth is preferred. It is particularly worth mentioning that $BaZrO_3$ nanorods (BZO-NRs) self-assembled into aligned linear arrays along the c-axis were shown to provide strong correlated pinning, resulting in a prominent peak of J_c at $H//c$-axis at several Teslas [32–36]. Consequently, the J_c anisotropy between the $H//c$ and $H//ab$ directions was decreased.

The benefit of BZO-NRs comes, however, at a cost of the considerably reduced J_c values in SF and low H fields. In addition, reduced T_c is common in doped YBCO and serious T_c degradation occurs beyond few percent Vol.% of BZO-doping [32, 33]. These problems have been attributed to the YBCO lattice strain caused by insertion of the BZO-NRs of large lattice mismatch [44, 73–75]. Typically, chemical contamination occurs at the interface between the YBCO matrix and defects of secondary phases. Although BZO is reportedly stable on the YBCO matrix, a substantial lattice strain can affect the electronic structure of the YBCO matrix.

An alternative strategy is to generate nanotubes of pores in YBCO, in which both chemical contamination and strain across the interface between YBCO matrix and defects can be eliminated, or at least minimized. In our prior research, nanopores have been obtained via nanoparticle-facilitated strain manipulation on YBCO lattice [76–78]. The density of the nanopores in the range of 5 ± 3 pores/μm^2 corresponds to $H^* \sim 4.1$–16.6 mT. Significantly enhanced J_c up to 8.3 MA/cm^2 at 77 K and SF has been obtained on these samples. A close correlation between J_c and the magnetic pinning potential U_p of the nanopores has been demonstrated below H^*, confirming that nanopores are strong pins on the magnetic vortices. It should be noted that in LTS films such as Pb films, nano-voids were also found recently to be strong vortex pins [79, 80]. Nevertheless, the best J_c in YBCO films with nanopores is still far below the J_d. It should be realized that the reported nanopores in YBCO film have lateral dimensions at least an order of magnitude larger than the coherence length. Although the pore surface provides excellent pinning, the benefit may be significantly reduced by the reduction of current cross-sectional area. Reducing the pore lateral dimension, increasing the pore density, and controlling the pore array geometry, requires understanding of the pore nucleation and formation mechanism.

These superconducting "sponges" as shown in Fig. 4.2(a) may provide a practical way to approach J_d through edge pinning provided on the pore surface. In particular, the quality of the edge formed on a pore surface may be much better than that in a lithographically defined one since edge damage caused by ion beam or chemical etching can be avoided. The bridge width (YBCO between pores) is on the order of 200–800 nm as shown in Fig. 4.2(b). The trend showing in Fig. 4.1 suggests that the dimension

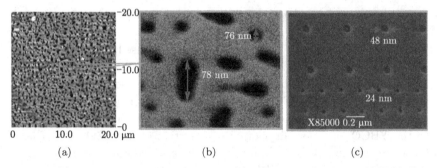

Fig. 4.2 (a) AFM image of a YBCO sponge; (b) SEM images of a porous YBCO film; and (c) an EBL generated mask of array of pores of diameters of 24 nm and 48 nm, respectively, for definition of an array of nanoparticle catalyst for nucleation of nanopores in YBCO film with pore dimension, density and array geometry precisely defined.

(density) of the pores must be further reduced (increased) to optimize the efficiency of pore-surface edge pining.

This raises a question on the mechanism of pore nucleation. To probe this, atomic force microscopy (AFM) images were taken on YBCO films after consecutive growth of a few nm-thick YBCO films on vicinal $SrTiO_3$ (STO) substrates [78]. It was found that a large number of NPs form at a few nanometers from the film/substrate interface and serve as nucleation centers of the pores forming atop. These NPs have been confirmed to be Y_2O_3 NPs using high resolution transmission electron microscopy (HRTEM) and ion milling with energy dispersive X-ray spectroscopy. Reproduction of the pores using Y_2O_3 NPs seed layer on non-vicinal substrates confirms the critical role of the NPs on nucleation [81]. These observations suggest that the generation of smaller size and denser catalytic NP array is a key to obtaining smaller and denser pores or NTPs for better pinning and hence, J_c. Figure 4.3(c) shows an EBL generated mask for deposition of NPs of designed dimension and density, as a part of our ongoing research.

The strain near the film/substrate interface seems to act as a driving force to the formation of Y_2O_3 NPs. This argument is supported by the much higher density NPs observed on the more stressful interface between YBCO film and vicinal STO as compared to the non-vicinal STO. In addition, the strain of the vicinal YBCO films plays a critical role in evolution of pores through a larger film thickness of several μm into NTPs. When such a strain is not present, pores annihilate within a small thickness of a few hundreds of nm [81]. Understanding the role of the vicinal strain on the

YBCO film

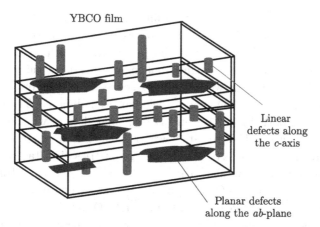

Linear
defects along
the c-axis

Planar defects
along the ab-plane

Fig. 4.3 An example of 3D pinning landscape in YBCO film with linear defects along c-axis and planar defects in ab-plane.

formation of NTPs is certainly important towards achieving controllable growth and optimization of these strong vortex cages.

4.2.4 Towards 3D pinning landscape

To obtain an overall enhanced in-field J_c, 3D pinning landscape is necessary. In practical applications such as electromagnets, the lowest value of $J_c(H, \theta)$ will limit the device performance. Thus, an isotropic $J_c(H, \theta)$ with optimized value is desired in the complete range of magnetic field orientation. This requires controllable incorporation of defects of desired geometry, dimension, density and distribution. One scheme to achieve 3D pinning is to combine several different types of defects, for example, BZO-NRs in composite YBCO thin films doped with 5 mol % Y_2O_3 [63]. The excess Y_2O_3 is less mobile due to its high melting temperature, and the smaller lattice misfit to YBCO reduces the likelihood for strain-driven self-assembly, thus remaining in the form of nanoparticle inclusions. An interaction between the multiple dopant phases (e.g. Y_2O_3 and $BaZrO_3$) during film growth is expected, which leads to variation of BZO-NRs' linearity, length, and splay of BZO nanorods as compared to the case of single dopant of BZO [82]. Consequently, significant microstructural and vortex pinning effects have been observed in BZO-NR self-assembly with and without the second dopant.

The geometry of the BZO dopant may be altered directly via strain manipulation to obtain defects aligned along both the c-axis and ab-plane

directions in order to provide 3D pinning landscape for optimization of the
current-carrying capability. Figure 4.3 shows an example of the combination
of splayed linear defects in c-axis and planar one in ab-plane obtained in
BZO-doped YBCO films through strain engineering using vicinal substrate
with the vicinal angle between 5–20 degrees [62]. Much improved pinning
over all orientations of the magnetic field has been observed. The formation
mechanism of both BZO-NRS and BZO planar defects will be discussed
in the next section. This result, however, demonstrates that it is indeed
possible to generate defects of different geometries simultaneously using
strain manipulation. As shown in Fig. 4.4, at smaller vicinal angles up
to 5 degrees, splayed BZO-NRs are the dominant defect while at larger
angles approaching 20 degrees, only planar BZO defects can be observed.
The J_c-θ measurement at $H = 1$ T and 77 K on these samples (Fig. 4.4)
in comparison with their non-vicinal counterpart (red) shows significantly
enhanced J_c on the vicinal samples. Besides $\sim$100% of enhancement of
J_c value at $H//c$-axis ($\theta \sim 0$) when BZO-NRs are splayed ($\phi = 5°$) as
compared to aligned, the J_c enhancement extends to all H orientations with
400% enhancement at $H//ab$-plane ($\theta \sim 90°$). The benefit of splayed BZO-
NRs on 5° vicinal STO also extends to thicker films of $\sim 1\mu$m thickness
[83]. When more BZO-NRs are switched from c-axis to ab-aligned (from
$\phi = 5°$, to 10° to 20°), the J_c enhancement is shifted also from $H//c$-
axis to $H//ab$-plane, confirming the critical role of BZO-NRs in vortex
pinning.

It should be noted that the BZO-NRs may not need to be perfectly
connected into a line along the c-axis. They can be small segments. Considering the case where the density of the segments (or NPs) is high enough,
vortices along any orientations may encounter enough segments and are
therefore pinned well, leading to 3D pinning landscape as shown in YBCO-
BZO nanocomposites [34].

4.2.5   Modeling of the nanostructure self-assembly
for 3D pinning landscape

As many experimental results have been obtained on the growth of nanostructures in YBCO films, most of them through empirical ways, there
have been strong desires in this research community to develop theoretical
modeling to provide understanding of the microscopic mechanism of the
nanostructure self-assembly. Such an understanding will not only help in

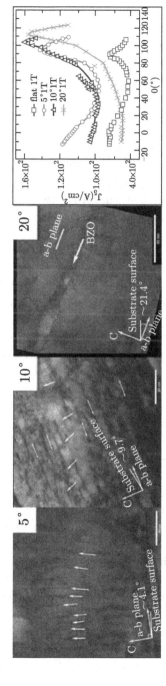

Fig. 4.4   HRTEM images of the cross sections and J_c-θ curves at $H = 1$ T and 77 K of YBCO/BZO-NR films on 5°, 10° and 20° vicinal STO substrates respectively [56]. Scale bars for HRTEM: 50 nm. The J_c-θ curve for non-vicinal YBCO-BZO-NR counterpart is also included for comparison. The BZO doping level for all four samples is 2 Vol.%.

explaining experimental observations but also provide guidance for design nanostructures and hence, for physical properties of nanostructured materials based on fundamental physics.

A question arises immediately on the microscopic mechanism of BZO-NR alignment, or in general, the formation of linear defects in YBCO. At low BZO doping level such that a coherent or semi-coherent interface between BZO dopant and YBCO matrix is maintained, one may speculate a mechanism similar to that responsible for the vertical assembly of quantum dots in heteroepitaxy of semiconductors of large lattice mismatch [84]. The modulated strain generated, which is determined by the elasticity of the embedding material and its lattice mismatch with the dopant, serve as the driving force for the aligned nanostructures.

We have also made attempts to probe the microscopic growth mechanism of the BZO-NRs using elastic strain model [85–87]. Some interesting insights to the growth mechanism of the nanorods in YBCO have been obtained. First, the energetically preferred orientation of nanorods of dopants with a large positive lattice mismatch with YBCO matrix, such as BZO [33, 44], BSO [35] and many rare-earth tantalates and niobates [36, 88], is always along the normal of the c-axis oriented YBCO film grown on substrates with negligible lattice mismatch, irrespective of the nanorod doping level as long as a coherent interface (epitaxial relation between YBCO matrix and nanorod) is maintained [89]. This explains the only c-axis oriented alignment of nanorods observed in experiment [32–40, 90]. On the other hand, if the YBCO lattice is under in-plane tensile strain through using a lattice-mismatched substrate, such as small angle vicinal ones, there will be a threshold of the tensile strain across which the alignment of the nanorods will switch from c-axis aligned on lower strain side to ab-aligned on the higher side. On the vicinal substrate, the tensile strain is controlled quantitatively by the vicinal angle ϕ. The theoretically calculated threshold $\phi_c \sim 9°$ together with the predicted change in a-/b- (purple line) and c-axis (blue line) lattice constants are depicted in Fig. 4.5. An excellent agreement between theory and experiment was obtained in observation of BZO-NR alignment switch in the range of $\phi \sim 5° - 10°$ shown in Fig. 4.4 [62]. The measured change of the c-axis lattice constant (black squares in Fig. 4.5) follows the trend of the theory (purple line) as a consequence of this orientation transition.

The elastic energy model can be applied to analyze and simulate the geometry of a large spectrum of oxide dopants in YBCO. For example,

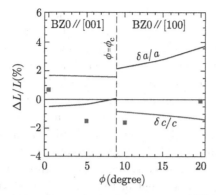

Fig. 4.5   Predicted variation of a- and c-lattice constants of YBCO after doping BZO-NRs. ϕ_c defines the transition vicinal angle at which BZO-NRs.

Y_2O_3 has a much smaller lattice mismatch with YBCO as compared to the BZO case and the modeling shows aligned Y_2O_3 nanorod geometry in YBCO is not energetically favorable. This agrees well with experimentally observed Y_2O_3NPs in YBCO and suggests Y_2O_3 may be combined with BZO-NRs for 3D pinning landscape, as shown recently in experiment [34].

4.3   Superconducting nanostructures

Superconducting properties of nanostructures in 1D (nanowires, nanotubes, nanobelts, etc) and 2D forms can be entirely different from those of the bulk superconductors. The effect of thermal and quantum fluctuations is significantly enhanced at nanoscales. Investigation of the size effect on the superconducting properties, in particular the electrical transport properties including electrical and thermal conductance and J_c, has been a highly active topic in many groups [91–93].

4.3.1   One-dimensional superconductor nanostructures

1D superconductors typically have their thickness and width smaller than the Ginzburg-Landau coherence length and magnetic penetration depth. These materials have a uniform current distribution over their cross-sections [94], and the onset of resistance in the current I-biasing condition is characterized by a sequence of regular voltage V steps, corresponding to the nucleation of phase-slip centers [95–97]. Recent advances in the synthesis of nanostructured materials have yielded a variety of 1D superconductors, including electrodeposited Pb and Sn nanowires [98, 99], nanotube-templated amorphous MoGe [100] and Nb [101], nanowires, laser-ablated $YBa_2Cu_3O_7$ nanowires [102], and MgB_2 nanowires [103], NbN [104] and NbS_2 [105] nanoribbons. The low dimensionality of these superconductor nanostructures presents a strong effect on the electron transport. Current-driven electrical measurements on the Nb nanobelts show voltage steps, indicating the nucleation of phase-slip structures. At temperatures much below T_c, the position of the voltage steps exhibits a sharp, periodic further dependence as a function of the magnetic field $\boldsymbol{H}$. Two mechanisms were proposed to explain this observation: the interference of the order parameter and the periodic rearrangement of the vortex lattice within the nanorib-bon [105]. In addition, the investigation of the confinement effects on the resistive anisotropy of a superconducting Nb nanobelt with a rectangular cross-section, of transverse dimensions comparable to the superconducting coherence length, showed that the angle-dependent magnetoresistances at a fixed temperature can be scaled as $R(\theta, H) = R(H/H_{c\theta}).\gamma$ is the width to thickness ratio and H_{c_0} is the critical field in thickness direction at $\theta = 0°$. The results can be understood in terms of the anisotropic diamagnetic energy for a given field in a one-dimensional superconductor.

4.3.2   Two-dimensional superconductor nanostructures

An unusual, non-bulk-like charge state could be formed at a polar hetero-interface such as that between insulating $LaAlO_3$ (LAO) and $SrTiO_3$ (STO) [106]. In particular, the properties of this interface are dictated sensitively by the atomic termination of the two materials, and the hole-doped inter-face of $(SrO)^0$-LaO^+ was found to be insulating while that of the $(TiO_2)^0$-$(AlO_2)^-$ is electron-doped and highly conducting with electron mobility up to 10000 $cm^2V^{-1}s^{-1}$. Moreover, superconductivity was also observed at the electron-doped interface with $T_c \sim 200$ mK within the interface thickness, up to 10 nm (Reyren et al., 2007). The characteristics of the transition are consistent with those of a 2D electron system undergoing a Berezinskii–Kosterlitz–Thouless (BKT) transition [30, 32, 42].

Superconductivity has also been observed in other two-dimensional (2D) systems. Graphene is one example, and the superconductivity observed is induced by the proximity effect from either superconductor electrodes [107, 108] or superconducting nanoparticles [109]. It should be noted that in proximity devices, the T_c of the graphene channel is typically

Table 4.1   List of superconductors obtained through chemical intercalation in graphite.

Chemical formula	$T_c(K)$	Structure	References
CaC_6	11.5	R^-3m	[110, 112]
C_8K	0.39		[111]
C_6Yb	6.5	$P6_3/mmc$	[113]
C_6Ca	11.5		[113]
C_2Li	1.9		[115]
NaC_2	5		[116]
CsC_8		$P6_222$	[117]

lower than that of the superconductor electrodes (or nanoparticles). For example, when Al electrodes ($T_c \sim 1$ K) were employed, the graphene channel's T_c is in the range 100 mK, and electrical-field-induced charge doping was required. Higher $T_c > 1$ K in the graphene channel in Sn nanoparticle-decorated graphene was attributed to both higher original $T_c \sim 3.7$ K of Sn and also the smaller channel length as defined by the distance between the Sn nanoparticles. Employment of high T_c superconductors in this kind of proximity device, assuming a clean interface between graphene and superconductor could be maintained, is certainly a pathway for achieving higher T_c superconductivity in 2D graphene sheet.

Chemical intercalation in between graphene sheets may provide an alternative scheme to obtaining superconductivity in 2D nanostructures. T_c in exceeding 10 K has been predicted theoretically in Ca intercalated bilayer graphene [110]. In fact, intercalation of various chemicals into graphite was initiated several decades ago and superconductivity in these intercalated systems has been demonstrated in the bulk materials using many intercalated chemicals including Ca, Yb, K, Rb, Cs, Li, Na, etc [111–114] (see Table 4.1). It remains to be seen whether superconductivity can be sustained in the 2D form.

4.3.3   Proximity effect in hybrid superconductor nanostructures

Superconductivity may occur at the interface between superconductor and normal material (nonsuperconducting, mostly metals) through the proximity effect, which occurs when the Cooper pairs diffuse into the normal material from the superconductor while the two are in good contact [118]. When the normal material is sandwiched between two superconductors and has a small enough thickness, it may become superconducting at a temperature below the original T_c. While the proximity effect has been well-known in SNS (N for metals) structures and S/N superlattices [119, 120], nanostructures with superconductors and other materials have attracted considerable interest recently, and novel physics and device concepts are anticipated [121].

In the 1D case, superconductivity in carbon nanotubes has received extensive study. While carbon nanotubes are not superconducting intrinsically, superconductivity has been reported when they are connected to superconductor electrodes [122–126]. The supercurrent in the carbon nanotubes may be described by the Andreev bound states, which are discrete states of entangled electron-holes confined in the nanotube. Confirmation of the Andreev bound states, a key concept in mesoscopic superconductivity in quantum-coherent nanostructures, has been recently demonstrated using tunneling spectroscopy in carbon nanotubes [127]. The understanding of universal Josephson-like behaviors and exploration of various applications in 1D nanostructures represent the current interests of many groups. In the 2D case, proximity effect in graphene has received much attention recently. The first report of superconducting proximity effect is on an exfoliated graphene flake contacted by two closely spaced (≈ 0.5 μm) superconducting electrodes (Ti/Al bi-layer: 10/70 nm) [107]. With $T_c \approx 1.3$ K for the electrodes, supercurrent was clearly demonstrated in graphene at 30 mK with the value depending on the gate voltage or charge carrier doping. Shortly afterwards, Andreev reflections were also reported in S-G-S devices using a bi-layer graphene flake with similar superconducting electrodes [128]. The employment of higher T_c superconductors, either for electrodes or as nanoparticles on the graphene to reduce the proximity channel length, have been reported to raise the proximity superconducting T_c in graphene [129–131]. There is no doubt that this will remain an active field in the near future.

4.4   Conclusion

After a century of persistent studies, superconductivity remains an intriguing research topic. The interplay between superconductivity with nanoscience and nanotechnology adds further complexity and richness, making superconductivity a fascinating subject to both fundamental researches and practical applications. Many outstanding issues will continue to draw the attention of this community due to their importance to and impact on both the understanding of the basic science underlying superconductivity far beyond the context of the BCS theory, and practical applications driven by various applications such as high-energy beam accelerators, NMR/MRI systems, etc. One such issue is the unresolved mechanism of high temperature superconductivity. Understanding the HTS mechanism is not only important to the current research of HTS materials, but also to the search for new superconductors that can surpass them. In particular, the search for room-temperature superconductors has never been exhausted, while understanding the HTS mechanism will certainly provide guidelines for this search. Another outstanding issue relates to achieving the theoretical limit of J_c — the depairing critical current, which is particularly important in superconducting wires and cables. Nanoscale engineering of magnetic pinning landscape in superconductors has provided a viable approach towards this goal. The last decade has witnessed much exciting progress, leading to continuously improving current-carrying capability. Reading the depairing limit of J_c may likely occur in the near future through implementation of nanoscale artificial pinning centers to provide an optimal pinning landscape in superconductors.

While steady advances have been maintained in improving properties of superconductors, investigation of superconductivity in hybrid superconductor nanostructures is taking off and emerging as a hot field in the last few years. This is partly because of the recent discovery of various 1D and 2D nanostructures, which provide ideal systems examining the effect of quantum confinement on superconductivity. In addition, the maturity of nanofabrication techniques in recent years makes it possible to generate superconductor-other material hybrid nanostructures in a controllable fashion, enabling the exploration of superconductivity in individual 1D and 2D devices. This trend will continue growing in the many years to come, since the availability of new nanostructures and the capability to manipulate them with nanometer precision will continue to grow.

Acknowledgements

The authors acknowledge support in part by ARO contract No. ARO–W911NF-09-1-0295, NSF contract No. NSF–DMR–0803149, 1105986, 1337737, 1508494 and NSF contract No. EPSCoR–0903806, and matching support from the State of Kansas through the Kansas Technology

Enterprise Corporation. They would also like to thank Dr. Jianwei Liu and Jack Shi for technical assistance in putting this chapter together.

References

[1] J. G. Bednorz and K. A. Muller, *Zeitschrift Fur Physik B-Condensed Matter* **64**, 189 (1986).

[2] J. Nagamatsu, N. Nakagawa, T. Muranaka, Y. Zenitani and J. Akimitsu, *Nature* **410**, 63–64 (2001).

[3] Y. Kamihara, H. Hiramatsu, M. Hirano, R. Kawamura, H. Yanagi, T. Kamiya and H. Hosono, *J. Am. Chem. Soc.* **128**, 31, 10012–10013 (2006).

[4] Y. Kamihara, T. Watanabe, M. Hirano and H. Hosono, *J. Am. Chem. Soc.* **130**, 3296 (2008).

[5] A. Gupta, J. Z. Sun and C. C. Tsuei, *Science* **265**, 1075 (1994).

[6] G. F. Sun, K. W. Wong, B. R. Xu, Y. Xin and D. F. Lu, *Phys. Lett. A* **192**, 122 (1994).

[7] R. S. Aga, X. Wang, J. Dizon, J. Noffsinger and J. Z. Wu, *Appl. Phys. Lett.* **86**, 23 (2005).

[8] J. R. Dizon, H. Zhao, J. Baca, S. Mishra, R. L. Emergo, R. S. Aga and J. Z. Wu, *Appl. Phys. Lett.* **88**, 9 (2006).

[9] Y. Y. Xie, T. Aytug, J. Z. Wu, D. T. Verebelyi, M. Paranthaman, A. Goyal and D. K. Christen, *Appl. Phys. Lett.* **77**, 4193 (2000).

[10] H. Zhao, J. R. Dizon and J. Z. Wu, *Appl. Phys. Lett.* **91**, 4 (2007).

[11] L. Ji, S. L. Yan and J. Z. Wu, *Supercond. Sci. Technol.* **27**, 1 (2014).

[12] W. Meissner and R. Ochsenfeld, *Naturwissenschaften* **21**, 787 (1933).

[13] F. London and H. London, *Proceedings of the Royal Society of London Series a-Mathematical and Physical Sciences* **149**, 0071 (1935).

[14] B. D. Josephson, *Phys. Lett.* **1**, 251 (1962).

[15] W. Buckel and R. Kleiner, *Superconductivity: Fundamentals and Applications*, Weinheim: Wiley-VCH, (2004).

[16] R. P. Huebener, *Magnetic Flux Structures in Superconductors*, Berlin: Springer (1979).

[17] A. A. Abrikosov, *Rev. Mod. Phys.* **76**, 975 (2004).

[18] B. A. Ruzicka, S. Wang, J. W. Liu, K. P. Loh, J. Z. Wu and H. Zhao, *Opt. Mat. Express* **2**, 708 (2012).

[19] S. R. Foltyn, L. Civale, J. L. Macmanus-Driscoll, Q. X. Jia, B. Maiorov, H. Wang and M. Maley, *Nat. Mat.* **6**, 631 (2007).

[20] A. P. Malozemoff, *Nat. Mat.* **6**, 617 (2007).

[21] D. Larbalestier, A. Gurevich, D. M. Feldmann and A. Polyanskii, *Nature* **414**, 368 (2001).

[22] M. P. Paranthaman and T. Izumi, *Mrs Bull.* **29**, 533 (2004).

[23] K. A. Delin and A. W. Kleinsasser, *Supercond. Sci. Technol.* **9**, 227 (1996).

[24] D. E. Oates, *J. Supercond. Novel Magn.* **20**, 3 (2007).

[25] D. Dimos, P. Chaudhari and J. Mannhart, *Phys. Rev. B* **41**, 4038 (1990).

[26] G. Hammerl, A. Schmehl, R. R. Schulz, B. Goetz, H. Bielefeldt, C. W. Schneider, H. Hilgenkamp and J. Mannhart, *Nature* **407**, 162 (2000).

[27] A. Schmehl, B. Goetz, R. R. Schulz, C. W. Schneider, H. Bielefeldt, H. Hilgenkamp and J. Mannhart, *EPL* **47**, 110 (1999).

[28] X. Y. Song, G. Daniels, D. M. Feldmann, A. Gurevich and D. Larbalestier, *Nat. Mat.* **4**, 470 (2005).

[29] M. Tinkham, *Introduction to Superconductivity*, New York: McGraw-Hill, (1996) pp. 124–146.

[30] T. Aytug, M. Paranthaman, A. A. Gapud, S. Kang, H. M. Christen, K. J. Leonard, P. M. Martin, J. R. Thompson, D. K. Christen, R. Meng, I. Rusakova, C. W. Chu and T. H. Johansen, *J. Appr. Phys.* **98**, 5 (2005).

[31] J. L. Macmanus-Driscoll, S. R. Foltyn, Q. X. Jia, H. Wang, A. Serquis, L. Civale, B. Maiorov, M. E. Hawley, M. P. Maley and D. E. Peterson, *Nat. Mat.* **3**, 439 (2004).

[32] K. Matsumoto, T. Horide, K. Osamura, M. Mukaida, Y. Yoshida, A. Ichinose and S. Horii, *Physica C-Superconductivity and Its Applications* **412**, 1267 (2004).

[33] S. Kang, A. Goyal, J. Li, A. A. Gapud, P. M. Martin, L. Heatherly, J. R. Thompson, D. K. Christen, F. A. List, M. Paranthaman and D. F. Lee, *Science* **311**, 1911 (2006).

[34] J. Gutierrez, A. Llordes, J. Gazquez, M. Gibert, N. Roma, S. Ricart, A. Pomar, F. Sandiumenge, N. Mestres, T. Puig and X. Obradors, *Nat. Mat.* **6**, 367 (2007).

[35] C. V. Varanasi, J. Burke, H. Wang, J. H. Lee and P. N. Barnes, *Appl. Phys. Lett.* **93** (9) (2008).

[36] S. A. Harrington, J. H. Durrell, B. Maiorov, H. Wang, S. C. Wimbush, A. Kursumovic, J. H. Lee and J. L. MacManus-Driscoll, *Supercond. Sci. Technol.* **22**(2) (2009).

[37] T. J. Haugan, P. N. Barnes, T. A. Campbell, N. A. Pierce, F. J. Baca and I. Maartense, *IEEE Transactions on Applied Superconductivity* **17**, 3724 (2007).

[38] Y. Yoshida, K. Matsumoto, Y. Ichino, M. Itoh, A. Ichinose, S. Horii, M. Mukaida and Y. Takai, *Japan. J. Appl. Phys. Part 2-Letters & Express Letters* **44**, L129 (2005).

[39] P. N. Barnes, J. W. Kell, B. C. Harrison, T. J. Haugan, C. V. Varanasi, M. Rane and F. Ramos, *Appl. Phys. Lett.* **89**, 3 (2006).

[40] Y. Yoshida, K. Matsumoto, M. Miura, Y. Ichino, Y. Takai, A. Ichinose, M. Mukaida and S. Horii, *Physica C-Supercond. Its Applications* **445**, 637 (2006).

[41] J. Hanisch, C. Cai, R. Huhne, L. Schultz and B. Holzapfel, *Appl. Phys. Lett.* **86**, 3 (2005).

[42] T. Haugan, P. N. Barnes, R. Wheeler, F. Meisenkothen and M. Sumption, *Nature* **430**, 867 (2004).

[43] N. Long, N. Strickland, B. Chapman, N. Ross, J. Xia, X. Li, W. Zhang, T. Kodenkandath, Y. Huang and M. Rupich, *Supercond. Sci. Technol.* **18**, S405 (2005).

[44] A. Goyal, S. Kang, K. J. Leonard, P. M. Martin, A. A. Gapud, M. Varela, M. Paranthaman, A. O. Ijaduola, E. D. Specht, J. R. Thompson, D. K. Christen, S. J. Pennycook and F. A. List, *Supercond. Sci. & Technol.* **18**, 1533 (2005).

[45] P. Mele, K. Matsumoto, T. Horide, A. Ichinose, M. Mukaida, Y. Yoshida and S. Horii, *Supercond. Sci. & Technol.* **20**, 244 (2007).

[46] C. V. Varanasi, P. N. Barnes, J. Burke, L. Brunke, I. Maartense, T. J. Haugan, E. A. Stinzianni, K. A. Dunn and P. Haldar, *Supercond. Sci. Technol.* **19**, L37 (2006).

[47] C. V. Varanasi, J. Burke, L. Brunke, H. Wang, J. H. Lee and P. N. Barnes, *J. Mater. Res.* **23**, 3363 (2008).

[48] S. H. Wee, A. Goyal, Y. L. Zuev, C. Cantoni, V. Selvamanickam and E. D. Specht, *Appl. Phys. Express.* **3**, 023101 (2010).

[49] G. Blatter, M. V. Feigelman, V. B. Geshkenbein, A. I. Larkin and V. M. Vinokur, *Rev. Mod. Phys.* **66**, 1125 (1994).

[50] L. Civale, L. Krusinelbaum, J. R. Thompson, R. Wheeler, A. D. Marwick, M. A. Kirk, Y. R. Sun, F. Holtzberg and C. Feild, *Phys. Rev. B* **50**, 4102 (1994).

[51] T. Hwa, P. Ledoussal, D. R. Nelson and V. M. Vinokur, *Phys. Rev. Lett.* **71**, 3545 (1993).

[52] L. Krusinelbaum, J. R. Thompson, R. Wheeler, A. D. Marwick, C. Li, S. Patel and D. T. Shaw, *Appl. Phys. Lett.* **64**, 3331 (1994).

[53] S. R. Foltyn, Q. X. Jia, P. N. Arendt, L. Kinder, Y. Fan and J. F. Smith, *Appl. Phys. Lett.* **75**, 3692 (1999).

[54] B. W. Kang, A. Goyal, D. R. Lee, J. E. Mathis, E. D. Specht, P. M. Martin, D. M. Kroeger, M. Paranthaman and S. Sathyamurthy, *J. Mater. Res.* **17**, 1750 (2002).

[55] K. J. Leonard, S. Kang, A. Goyal, K. A. Yarborough and D. M. Kroeger, *J. Mater. Res.* **18**, 1723 (2003).

[56] A. Gurevich, *Presented at the Annual DOE Peer Review*, Washington DC (2004).

[57] X. Wang and J. Z. Wu, *Phys. Rev. B* **76**, 5 (2007).

[58] X. Wang and J. Z. Wu, *Appl. Phys. Lett.* **88**, 3 (2006).

[59] Q. X. Jia, S. R. Foltyn, P. N. Arendt and J. F. Smith, *Appl. Phys. Lett.* **80**, 1601 (2002).

[60] X. Wang, F. J. Baca, R. L. S. Emergo, J. Z. Wu, T. J. Haugan and P. N. Barnes, *J. Appl. Phys.* **108**, 5 (2010).

[61] X. Wang, A. Dibos and J. Z. Wu, *Phys. Rev. B* **77**, 5 (2008).

[62] F. J. Baca, P. N. Barnes, R. L. S. Emergo, T. J. Haugan, J. N. Reichart and J. Z. Wu, *Appl. Phys. Lett.* **94**, 3 (2009).

[63] B. Maiorov, S. A. Baily, H. Zhou, O. Ugurlu, J. A. Kennison, P. C. Dowden, T. G. Holesinger, S. R. Foltyn and L. Civale, *Nat. Mat.* **8**, 398 (2009).

[64] R. E. Glover and H. T. Coffey, *Rev. Mod. Phys.* **36**, 299 (1964).

[65] T. K. Hunt, *Phys. Rev.* **151**, 325 (1966).

[66] W. J. Skocpol, *Phys. Rev. B* **14**, 1045 (1976).

[67] M. Benkraouda and J. R. Clem, *Phys. Rev. B* **58**, 15103 (1998).

[68] W. A. Jones, P. N. Barnes, M. J. Mullins, F. J. Baca, R. L. S. Emergo, J. Wu, T. J. Haugan and J. R. Clem, *Appl. Phys. Lett.* **97**, 3 (2010).

[69] C. A. J. Damen, H. J. H. Smilde, D. H. A. Blank and H. Rogalla, *Supercond. Sci. Technol.* **11**, 437 (1998).

[70] F. M. Kamm, A. Plettl and P. Ziemann, *Supercond. Sci. Technol.* **11**, 1397 (1998).

[71] C. Peroz, J. C. Villegier, A. F. Degardin, B. Guillet and A. J. Kreisler, *Appl. Phys. Lett.* **89**, 3 (2006).

[72] S. Tahara, S. M. Anlage, J. Halbritter, C. B. Eom, D. K. Fork, T. H. Geballe and M. R. Beasley, *Phys. Rev. B* **41**, 11203 (1990).

[73] P. Mele, K. Matsumoto, T. Horide, A. Ichinose, M. Mukaida, Y. Yoshida, S. Horii and R. Kita, *Supercond. Sci. Technol.* **21**, 5 (2008).

[74] M. Peurla, H. Huhtinen, M. A. Shakhov, K. Traito, Y. P. Stepanov, M. Safonchik, P. Paturi, Y. Y. Tse, R. Palai and R. Laiho, *Phys. Rev. B* **75**, 6 (2007).

[75] M. Peurla, P. Paturi, Y. P. Stepanov, H. Huhtinen, Y. Y. Tse, A. C. Boodi, J. Raittila and R. Laiho, *Supercond. Sci. & Technol.* **19**, 767 (2006).

[76] R. L. S. Emergo, J. Z. Wu, T. J. Haugan and P. N. Barnes, *Appl. Phys. Lett.* **87**, 3 (2005).

[77] R. L. S. Emergo, J. Z. Wu, T. J. Haugan and P. N. Barnes, *Supercond. Sci. & Technol.* **21**, 7 (2008).

[78] J. Z. Wu, R. L. S. Emergo, X. Wang, G. Xu, T. J. Haugan and P. N. Barnes, *Appl. Phys. Lett.* **93**, 3 (2008).

[79] M. M. Ozer, J. R. Thompson and H. H. Weitering, *Nat. Phys.* **2**, 173 (2006).

[80] A. Yazdani, *Nat. Phys.* **2**, 151 (2006).

[81] F. J. Baca, D. Fisher, R. L. S. Emergo and J. Z. Wu, *Supercond. Sci. & Technol.* **20**, 554 (2007).

[82] F. Baca, T. Haugan, P. Barnes, T. Holesinger, B. Maiorov, R. Lu, X. Wang, J. Reichart and J. Wu, *Appl. Phys. Lett.* **83**, 19 (2003).

[83] R. L. S. Emergo, F. J. Baca, J. Z. Wu, T. J. Haugan and P. N. Barnes, *Supercond. Sci. & Technol.* **23**, 11 (2010).

[84] Y. F. Gao, J. Y. Meng, A. Goyal and G. M. Stocks, *JOM* **60**, 54 (2008).

[85] L. D. Landau and E. M. Lifshitz, *Theory of Elasticity*, Third ed. Oxford: Pergamon Press (1986).

[86] A. L. Roitburd, "Martensitic transformation as a typical phase transformation in solid", in *Solid State Physics, Advances in Research and Applications*, edited by H. Ehrenreich, F. Seitz and D. Turnbull (Academic, New York), Vol. 33, p. 317 (1978).

[87] M. H. Sadd, *Elasticity — Theory, Applications, and Numerics*, 2 ed., Amsterdam: Elsevier (2009).

[88] J. J. Shi and J. Z. Wu, *Philos. Mag.* **92**, 4205 (2012).

[89] J. J. Shi and J. Z. Wu, *Philos. Mag.* **92**, 2911 (2012).

[90] Judy Z. Wu, Javier Baba, Jack J. Shi, Rose Emergo, Timothy J. Haugan,
      Boris Maiorov and Terry Holesinger, *Supercond. Sci. Technol.*, asap (2014).

[91] K. Y. Arutyunov, D. S. Golubev and A. D. Zaikin, *Physics Reports-Review
      Section of Physics Letters* **464**, 1 (2008).

[92] M. D. Croitoru, A. A. Shanenko, C. C. Kaun and F. M. Peeters, *Phys. Rev.
      B* **80** 024513 (2009).

[93] A. Del Maestro, B. Rosenow, N. Shah and S. Sachdev, *Phys. Rev. B* **77**
      180501 (2008).

[94] R. Tidecks, *Springer Tracts Mod. Phys.* **121**, 1 (1990).

[95] J. Meyer and Minniger. Gv, *Phys. Lett. A* **38**, 529 (1972).

[96] M. Tinkham, J. U. Free, C. N. Lau and N. Markovic, *Phys. Rev. B* **68**, 7
      (2003).

[97] W. W. Webb and Warburto. Rj, *Phys. Rev. Lett.* **20**, 461 (1968).

[98] S. Michotte, S. Matefi-Tempfli and L. Piraux, *Appl. Phys. Lett.* **82**, 4119
      (2003).

[99] D. Y. Vodolazov, F. M. Peeters, L. Piraux, S. Matefi-Tempfli and
      S. Michotte, *Phys. Rev. Lett.* **91**, 4 (2003).

[100] A. Bezryadin, C. N. Lau and M. Tinkham, *Nature* **404**, 971 (2000).

[101] A. Rogachev and A. Bezryadin, *Appl. Phys. Lett.* **83**, 512 (2003).

[102] Y. F. Zhang, Y. H. Tang, Y. Zhang, C. S. Lee, I. Bello and S. T. Lee, *Chem.
      Phys. Lett.* **330**, 48 (2000).

[103] Y. Y. Wu, B. Messer and P. D. Yang, *Adv. Mater.* **13**, 1487 (2001).

[104] J. Hua, Z. L. Xiao, A. Imre, S. H. Yu, U. Patel, L. E. Ocola, R. Divan, A.
      Koshelev, J. Pearson, U. Welp and W. K. Kwok, *Phys. Rev. Lett.* **101**, 4
      (2008).

[105] A. Falk, M. M. Deshmukh, A. L. Prieto, J. J. Urban, A. Jonas and H. Park,
      Phys. Rev. B **75**, 4 (2007).

[106] A. Ohtomo and H. Y. Hwang, *Nature* **427**, 423 (2004).

[107] H. B. Heersche, P. Jarillo-Herrero, J. B. Oostinga, L. M. K. Vandersypen
      and A. F. Morpurgo, *Nature* **446**, 56 (2007).

[108] M. Tarasov, N. Lindvall, L. Kuzmin and A. Yurgens, *JEPT Lett.* **94**, 329
      (2011).

[109] B. M. Kessler, C. O. Girit, A. Zettl and V. Bouchiat, *Phys. Rev. Lett.* **104**,
      4 (2010).

[110] Mazin, II and A. V. Balatsky, *Philos. Mag. Lett.* **90**, 731 (2010).

[111] N. B. Hannay, T. H. Geballe, B. T. Matthias, K. Andres, P. Schmidt and
      D. Macnair, *Phys. Rev. Lett.* **14**, 225 (1965).

[112] N. Emery, C. Herold, M. d'Astuto, V. Garcia, C. Bellin, J. F. Mareche, P.
      Lagrange and G. Loupias, *Phys. Rev. Lett.* **95**, 4 (2005).

[113] T. E. Weller, M. Ellerby, S. S. Saxena, R. P. Smith and N. T. Skipper, *Nat.
      Phys.* **1**, 39 (2005).

[114] H. Mousavi, "Doped graphene as a superconductor", *Phys. Lett. A* **374**,
      2953 (2010).

[115] I. T. Belash, A. D. Bronnikov, O. V. Zharikov and A. V. Palnichenko, *Solid State Commun.* **69**, 921 (1989).

[116] I. T. Belash, A. D. Bronnikov, O. V. Zharikov and A. V. Palnichenko, *Synth. Met.* **36**, 283 (1990).

[117] N. Rey, P. Toulemonde, D. Machon, L. Duclaux, S. Le Floch, V. Pischedda, J. P. Itie, A. M. Flank, P. Lagarde, W. A. Crichton, M. Mezouar, T. Strassle, D. Sheptyakov, G. Montagnac and A. San-Miguel, *Phys. Rev. B* **77**, 125433 (2008).

[118] P. G. De Gennes, *Superconductivity of Metals and Alloys*, New York: Benjamin (1966).

[119] J. Z. Wu, C. S. Ting, W. K. Chu and X. X. Yao, *Phys. Rev. B* **44**, 411 (1991).

[120] J. Z. Wu, X. X. Yao, C. S. Ting and W. K. Chu, *Phys. Rev. B* **46**, 14059 (1992).

[121] S. De Franceschi, L. Kouwenhoven, C. Schonenberger and W. Wernsdorfer, *Nat. Nanotechnol.* **5**, 703 (2010).

[122] A. Y. Kasumov, R. Deblock, M. Kociak, B. Reulet, H. Bouchiat, Khodos, II, Y. B. Gorbatov, V. T. Volkov, C. Journet and M. Burghard, *Science* **284**, 1508 (1999).

[123] P. Jarillo-Herrero, J. A. van Dam and L. P. Kouwenhoven, *Nature* **439**, 953 (2006).

[124] J. P. Cleuziou, W. Wernsdorfer, V. Bouchiat, T. Ondarcuhu and M. Monthioux, *Nat. Nanotechnol.* **1**, 53 (2006).

[125] E. Pallecchi, M. Gaass, D. A. Ryndyk and C. Strunk, *Appl. Phys. Lett.* **93**, 3 (2008).

[126] G. Liu, Y. Zhang and C. N. Lau, *Phys. Rev. Lett.* **102**, 4 (2009).

[127] J. D. Pillet, C. H. L. Quay, P. Morfin, C. Bena, A. L. Yeyati and P. Joyez, *Nat. Phys.* **6**, 965 (2010).

[128] X. Du, I. Skachko and E. Y. Andrei, *Phys. Rev. B* **77**, 184507 (2008).

[129] D. Jeong, J. H. Choi, G. H. Lee, S. Jo, Y. J. Doh and H. J. Lee, *Phys. Rev. B* **83**, 094503 (2011).

[130] G. H. Lee, D. Jeong, J. H. Choi, Y. J. Doh and H. J. Lee, *Phys. Rev. Lett.* **107**, 146605 (2011).

[131] B. M. Kessler, C. O. Girit, A. Zettl and V. Bouchiat, *Phys. Rev. Lett.* **104**, 047001 (2010).

5

Improvement of Critical Current Density and Study on Flux Pinning Mechanism in MgB$_2$

W. X. Li^{1,2,3,4} and S. X. Dou^4

5.1 Introduction

According to the BCS theory, high critical transition temperatures T_c can be found in the low-mass elements due to their high frequency phonon modes [1]. Hydrogen under high pressure is predicted to show the highest superconducting temperature because of its lightest mass [2]. The discovery of superconductivity in MgB$_2$ confirms the prediction of high T_c existing in compounds containing light elements [3]. The unit cell of MgB$_2$ has a hexagonal crystal structure with space group $P6/mmm$ [4], which is common among diborides and known as AlB$_2$ type. The boron atoms form a graphite-like honeycomb network and the Mg atoms are located at the pores of these hexagons up and down the boron layer. The lattice parameters are $a = 0.3084$ nm and $c = 0.3524$ nm. MgB$_2$ exhibits a strong anisotropy in the B-B lengths, because the distance between the boron planes is almost twice as long as in the in-plane B-B distance. The metallic B layers are believed to play a crucial role in the superconductivity of MgB$_2$ [5]. Furthermore, two-gap superconducting behavior has also been discovered in MgB$_2$. The

^{1}Institute of Materials, Shanghai University, Shanghai 200072, China.
^{2}Institute for Sustainable Energy, Shanghai University, Shanghai 200444, China.
^{3}Shanghai Key Laboratory of High Temperature Superconductors, Shanghai 200444, China.
^{4}Institute for Superconducting and Electronic Materials, University of Wollongong, Australia.

mechanism behind the 40 K superconducting transition is under investigation. The MgB_2 critical temperature is close to or above the theoretical value predicted by the BCS theory [6]. This may be a strong argument to consider MgB_2 as a non-conventional superconductor. On the other hand, MgB_2 is easy to process into different types according to the application purposes, in various forms such as bulk, wire, tape, and thin film. The most competitive advantage is the low price of MgB_2 compared to those of the high temperature superconductors. However, research into both the application potential and the mechanism of high superconducting performance is underway to make MgB_2 the most affordable superconductor.

5.1.1   Application of MgB_2

MgB_2 is a promising superconductor for high-magnetic-field applications [7] because of its high critical current density, J_c. The grain boundaries in MgB_2 do not significantly degrade J_c and even serve as pinning centers [8, 9], which is different from the weak-link effects in high-T_c superconductors. Doping or alloying with carbon or SiC has been shown to significantly enhance H_{c2} [10] and the pinning force [11, 12]. Wires and tapes of MgB_2 have been made using the powder-in-tube (PIT) technique with encouraging results [13, 14]. The T_c at 40 K allows practical operation above 20 K using cryocoolers, allowing for cryogen-free operation of magnets, such as in magnetic resonance imaging (MRI) systems. Commercial companies around the world have made impressive progress in producing long length wires and magnets, and 5 years after the discovery of superconductivity in MgB_2, an MRI system using MgB_2 magnets cooled by cryocoolers was manufactured, and able to produce brain images [15]. The interest in MgB_2 for electronic applications, such as in superconducting digital circuits, is mainly due to its high T_c as compared with the currently used Nb-based superconductors, which have to be cooled to 4.2 K [16].

   MgB_2 Josephson junctions and superconducting quantum interference devices (SQUIDs) have been demonstrated to work well at temperatures over 20 K [17]. Sandwich-type all-MgB_2 junctions have been reported [18]. Other MgB_2 devices, such as bolometers and neutron detectors, have also been explored [19]. The low resistivity (0.1 $\mu\Omega\cdot$cm), in combination with an energy gap of 2 meV or 7 meV, promises a BCS microwave surface resistance much lower than that of Nb, which is currently the best material for the superconducting radio-frequency (RF) cavities used in accelerators [20]. The small value of the Ginzburg–Landau parameter, κ, also leads to high values of the thermodynamic critical field H_c ($\sim$800 mT as compared with 200 mT for Nb) or the superheating field, H_{sh}. Low BCS surface resistance is important for high quality factor, Q, of a cavity, and high H_c or H_{sh} is important for high ultimate RF critical field. The great potential of MgB$_2$ for superconducting RF cavities has been recognized by researchers in the field [21].

5.1.2 Critical fields of MgB$_2$

As a type II superconductor, MgB$_2$ shows both an upper critical field H_{c2} and a lower critical field H_{c1} in magnetic field. H_{c1} will not be discussed, as it has a weak influence on the superconductivity and transportation. Instead, an irreversibility field, H_{irr}, is defined and explored for practical purposes.

For single-gap dirty limit superconductors, the upper critical field $H_{c2}(0) = 0.69T_c(dH_{c2}/dT)_{T_c}$ and $(dH_{c2}/dT)_{T_c} \propto \rho_n$; therefore H_{c2} increases with normal-state resistivity ρ_n, which can be achieved by adding impurities and defects into the superconductor. Gurevich pointed out that the two-band superconductor MgB$_2$ can be understood as a weak-coupling bi-layer in which two thin films corresponding to the σ and π bands are in contact through Josephson coupling [22]. Using the dirty-limit weak-coupling multi-band BCS model and taking into account both interband and intraband scattering by non-magnetic impurities, Gurevich showed that the temperature dependence of $H_{c2}(T)$ depends on whether the σ bands or π bands are dirtier and can be very different from that in the one-band theory [23]. The global $H_{c2}(T)$ of the bi-layer is dominated by the layer with the higher H_{c2}. If the π layer is dirtier, it will have higher H_{c2} at low temperature, even though its T_c is much lower. As a result, an upturn in the global $H_{c2}(T)$ occurs at low temperature. $H_{c2}(0)$ of MgB$_2$ can exceed $0.69T_c(dH_{c2}/dT)_{T_c}$ considerably because of the existence of the two bands. Considering the electron–phonon coupling effect, Gurevich argued that the strong coupling paramagnetic limit in MgB$_2$ can be as high as 130 T; thus, there is still room for further enhancement of H_{c2} by engineering the σ- and π-band scattering [22]. The high H_{c2} in MgB$_2$ is very attractive for high-magnetic-field applications. The H_{c2} behaviors described by Gurevich have been observed in experimental results. For example, Braccini et al. observed different types of temperature dependence of H_{c2}, including an anomalous upturn at low temperature, reflecting different multi-band scattering in thin films from various groups, with disorder introduced in different ways.

The value of H_{c2} in carbon-doped thin films has reached over 60 T at low temperature, approaching the BCS paramagnetic limit of 65 T [10].

Knowledge of the irreversibility field, H_{irr}, is important in potential applications, as non-zero critical currents are confined to magnetic fields below this line. The irreversibility fields extrapolated at 0 K range between 6 T and 12 T for MgB_2 bulks, films, wires, tapes, and powders. A substantial enhancement of the H_{irr}, accompanied by a significantly large critical current density between 10^6 A/cm^2 and 10^7 A/cm^2 at 4.2 K and 1 T, have been reported in MgB_2 thin films with lower T_c [24, 25]. These results give further encouragement to the development of MgB_2 for high current applications.

5.1.3 *Critical current density*

The critical current density (J_c) of MgB_2 superconductors has been optimized through many different kinds of dopants or additives [26]. Carbon sources in particular are the most effective, with great promise for application [11, 27–30]. J_c values as high as 1×10^4 A/cm^2 at $\sim$12.5 T, 4.2 K have been reported for 10 *wt%* SiC doped *in-situ* wires. However, the self-field values, $J_c(0)$, of these samples are much lower than the depairing current density, J_d. The J_d value can be estimated from the Ginzburg–Landau (GL) formula:

$$J_d = \Phi_0/[3/(\sqrt{3})\pi\mu_0\lambda^2(T)\xi(T)], \qquad (5.1)$$

where Φ_0 is the flux quantum, μ_0 the permeability of vacuum, λ the penetration depth, and ξ the coherence length [31]. However, it is not the theoretical maximum [32]. With the help of optimized pinning, about 15% of J_d can be obtained at low magnetic fields in superconductors [33]. J_d at 0 K is about 1.3×10^8 A/cm^2 in MgB_2, assuming the high field values of $\lambda = 80$ nm and $\xi = 12$ nm. This is the relevant value for fields above about 1 T. The π-band charge carriers contribute only about 10% to the depairing current density [32], and most of the difference from the high field value is caused by the interaction energy. The depairing current decreases with disorder, since λ increases.

It was pointed out soon after the discovery of MgB_2 that clean grain boundaries are, in principle, no obstacle for supercurrents [34, 35], although they are known as weak links in the high temperature superconductors. Nevertheless, the connections between the grains remain delicate, since dirty grain boundaries potentially reduce the critical current [36]. Insulating phases have been found at the grain boundaries, consisting of MgO, boron oxides [37], or boron carbide [38]. Cracks [39], porosity, or normal conducting phases [40] can further reduce the cross-section over which supercurrents effectively flow. The density of *in-situ* prepared MgB_2 is typically only about half (or less) of its theoretical value [41], which leads to high porosity.

The concept of connectivity, A_{con}, has been introduced to quantify this reduction of the effective cross-section, σ_{eff}, for supercurrents [36, 42]: $A_{con} = \sigma_{eff}/\sigma_0$, where σ_0 is the geometrical cross-section. The connectivity can be estimated from the phonon contribution to the normal state resistivity by $A_{con} \approx \Delta\rho_{theo}/\Delta\rho_{exp}$. A practical quantity to evaluate the connectivity is the active area fraction, A_F [36]:

$$A_F = \frac{\Delta\rho_{ideal}}{\Delta\rho(300 \text{ K})}, \qquad (5.2)$$

where

$$\Delta\rho_{ideal} = \rho_{deal}(300 \text{ K}) - \rho_{ideal}(40 \text{ K}) \approx 9 \ \mu\Omega \cdot cm \qquad (5.3)$$

is the resistivity of fully-connected MgB_2 without any disorder [43], and

$$\Delta\rho(300 \text{ K}) = \rho(300 \text{ K}) - \rho(T_c). \qquad (5.4)$$

This estimate is based on the assumption that the effective cross-section is reduced equivalently in the normal and superconducting state, which is a severe simplification. The supercurrents are limited by the smallest effective cross-section along the conductor and the resistivity is given more or less by the average effective cross-section. A single large transverse crack strongly reduces J_c and only slightly increases the resistivity of a long sample. Unreacted magnesium decreases $\Delta\rho_{exp}$ [44] and the cross-section for supercurrents. Thin insulating layers on the grain boundaries strongly increase $\Delta\rho_{exp}$, but might be transparent to supercurrents. Finally, $\Delta\rho_{theo}$ within the grains can change due to disorder. Even a negative $\Delta\rho_{exp}$ has been reported in highly resistive samples [45]. Despite these objections, A_{con} is very useful, at least if the resistivity is not too high. A clear correlation between the resistivity and the critical current was found in thin films [42]. Nevertheless, one should be aware of the fact that this procedure is not really reliable, but just offers a possible way to get an idea about the connectivity.

The rather large upper critical field anisotropy leads to a new phenomenon in polycrystalline superconductors. The randomly oriented grains of the conductor attain different properties according to their orientation, when a magnetic field is applied, and the conductor can be considered as highly inhomogeneous [46]. This 'field induced inhomogeneity' is unique to MgB_2 conductors and is responsible for a rather rapid decrease of the critical current in magnetic fields. MgB_2 grains with their c-axis oriented parallel to the magnetic field become normal conducting at rather low fields, leading to a reduction of the superconducting cross-section. The number of superconducting grains decreases with increasing magnetic field, and finally, a continuous superconducting current path is no longer present, when the fraction p of superconducting grains becomes too small. This minimum fraction is called the percolation threshold, p_c, in percolation theory [47]. It mainly depends on the number of connections between the grains [48]. The resistivity increases with decreasing p until it meets the normal state resistivity, when all particles are normal conducting ($p = 0$).

Macroscopic superconductivity exists above the percolation threshold ($p > p_c$), but the maximum (critical) current remains reduced. This reduction, A_p, can be modeled by introducing a percolation cross-section $\sigma_p = A_p\sigma_0$, over which the current effectively flows. A_p can be easily defined in binary systems, which consist of only two different types of particles, i.e. superconducting particles with fraction p and normal conducting particles ($1 - p$). A_p is the ratio between J_c of the mixed and pure ($p = 1$) systems. Polycrystalline MgB_2 is, of course, not a binary system, but the material properties are continuously distributed according to the orientation of the grains with respect to the applied field. The resulting critical current density can be calculated from the percolation cross-section, if the distribution function of the critical current density within the grains is known [49]:

$$J_c = \int_0^\infty A_p(p(J))dJ, \qquad (5.5)$$

where $p(J)$ denotes the fraction of grains with critical current densities above J. Good agreement with experimental data was obtained with a simple power law for $p > p_c$ and $A_p = 0$, otherwise:

$$A_p(p, p_c) = \left(\frac{p - p_c}{1 - p_c}\right)^\alpha. \qquad (5.6)$$

Such power laws are typical within percolation theory, but in principle restricted to the vicinity of p_c. The exponent α depends on the dimensionality of the system and on the exponent of the current voltage law ($U \propto I^n$) in nonlinear resistor networks [50]. α is predicted to be 1.79 in three-dimensional systems with $n \rightarrow \infty$. The percolation threshold might be around 0.2 in dense samples [49], but it is higher in samples with secondary phases or voids. p_c increases with a decreasing number of connections between the grains, and thus, a low connectivity is expected to result in a larger p_c. Voids or non-superconducting particles can be treated as substitutions for MgB_2 grains, if these defects have a similar size to the superconducting grains.

5.1.4   Experimental results on H_{c2} and J_c in MgB_2

The *in-situ* route seems to be the most promising method to improve the H_{c2} and J_c performance of MgB_2. Magnesium or MgH_2 [51–60] reacts with boron after mixing and compacting of the precursor powders. MgB_2 samples with small grains [39, 61–73] of poor crystallinity can be obtained with low processing temperatures, resulting in strong pinning and high H_{c2}. The stoichiometry can be modified to yield samples with magnesium deficiency, which induces lattice strain, decreases T_c, and increases H_{c2} [65, 74–77]. An excess of magnesium in the starting powders may compensate the loss of magnesium due to evaporation or due to a reaction with other elements (e.g. with oxygen or with the sheath material). The precursor powders are very important for the properties of the final samples [78]. They should be clean to ensure good grain connectivity. The grain size is strongly influenced by the grain size of the precursor powders, especially of the boron powders. Ball milling or mechanical alloying of the precursor powders reduces the grain size and improves the critical current [76, 79–81].

Chemical or compound doping changes the reaction kinetics [82] and therefore influences the grain growth [83, 84], the formation of secondary phases [85], the density [86], or the stoichiometry [87]. Carbon-doping can be easily performed by the addition of B_4C [88, 89], carbon [73, 82, 83, 90], carbon nanotubes [27, 91–93], nanodiamonds [93, 94], NbC [95], SiC [11, 62, 72, 78, 82, 88, 96–103], or organic compounds [60, 68, 104]. SiC is by far the most popular dopant, because carbon can be doped into MgB_2 at low temperatures ($600°C$) according to the dual reaction model [82]. Higher processing temperatures are necessary for most of the other carbon sources, leading to grain growth and worse pinning. However, comparable results to those with SiC have also obtained with nanoscale carbon powder [105], stearic acid [106], and carbon nanotubes [91].

The main disadvantage of the *in-situ* techniques is the low mass density, which is caused by the process, because the precursor powder has a lower density than MgB_2. High pressure synthesis increases the density [107, 108]. The density of *ex-situ* materials is usually significantly higher [108] and can be further improved by hot isostatic pressing (HIP) [105, 109, 110]. The grain size of *ex-situ* produced materials is comparatively large and inhomogeneous [111, 112], but small grains ($\sim$100–200 nm) have also been reported [113]. The grain size is also correlated to the grain size of the precursor powders [114–118], which can be decreased by ball milling [115, 118–121].

Starting with fine alloyed powder seems to be the most promising approach for the *ex-situ* route [121]. A high-temperature heat treatment is necessary after sintering in the *ex-situ* process to improve the connectivity [14, 113, 122, 123]. This heat treatment leads to recrystallization [39], which increases the crystallinity and reduces H_{c2}. Thus, disorder cannot be introduced simply by low-temperature processing in the *ex-situ* process, and thermally stable defects (e.g. carbon atoms) are needed to enhance H_{c2} for high-field performance.

5.1.5   Contents in this chapter

$J_c(0)$ values have been reported as high as 3.5×10^7 A/cm^2 at 4.2 K and 1.6×10^8 A/cm^2 at 2 K in highly-connected thin films made by hybrid physical-chemical vapor deposition (HPCVD) [124, 125]. The connectivity is poor in *in-situ* wires, tapes, and even bulks because the *in-situ* technique in wire fabrication involves a phase reaction process with considerable shrinkage due to the high theoretical density of MgB_2 compared to the initial mixture of Mg and B [126, 127]. It is critical to discover how to make a highly-connected sample in order to obtain high $J_c(0)$ and then introduce proper disorder to keep J_c dropping as slowly as possible with increasing magnetic field. Several attempts have been undertaken to achieve enhancement of the connectivity of MgB_2 bulk samples and *in-situ* filamentary tapes. High pressure sintering, especially hot isostatic pressing (HIP) treatment, is used in making high density bulks and wires [110]. Cold high pressure densification (CHPD) is also effective in improving the density to as high as 73% in wires [128, 129]. However, the equipment for the high pressure processes is quite complicated, and the sample densities are still lower than needed for practical application. To improve the connectivity and flux pinning force, we tried several techniques, such as using excess Mg, combined *in-situ/ex-situ*, and diffusion, as discussed in the following sections.

5.2 Effect of excess Mg in the nano-SiC doped Mg-B system

In this work, the effects of excess Mg combined with nano-SiC co-doping on the current transport properties and flux pinning mechanisms of MgB_2/Fe wires were studied systematically. To confirm the cooperation between the connectivity and lattice distortion, Raman scattering measurements were employed in this study. Raman scattering can disclose the crystal strain caused by lattice distortion and chemical pressure, which is related to the flux pinning force (F_p) [130].

5.2.1 *Experimental details*

The PIT process was employed to make practical MgB_2 wires from starting powders of Mg (99%), B (99%, amorphous), and SiC (< 30 nm). The starting powders were weighed out according to the molecular formulae of MgB_2 and $Mg_{1.15}B_2$, and of $Mg_xB_2 + 10wt\%SiC$ with $x = 1.0$, 1.15, 1.20, 1.25, and 1.30. The thoroughly ground powder mixture of micron-size Mg powder (99%), nanosize amorphous B powder (99.99%), and SiC powder ($\leqslant$30 nm) was packed into pure iron tubes with an 8 mm outer diameter and a 6 mm inner diameter, and cold-drawn into wires 1.4 mm in diameter. The mixing of the powders and packing of the mixed powder into iron tubes were carried out under a high-purity argon gas atmosphere in a glove box in order to avoid oxidation of the powders. Sintering was carried out at 650°C for half an hour under a flowing argon gas atmosphere.

The MgB_2 cores extracted from the sintered wires were ground and subjected to X-ray diffraction (XRD) measurements. The microstructures were observed by field emission gun scanning electron microscopy (FEG-SEM: JSM-6700F). The magnetization hysteresis loops $M(H)$ of the cores were measured at 5 K and 20 K using a Physical Properties Measurement System (PPMS: Quantum Design) in a time-varying magnetic field with sweep rate of 50 Oe/s and amplitude of 8.5 T. The magnetic field was parallel to the wire axes. The magnetic J_c was derived from the height of the magnetization loop ΔM using the Bean model: $J_c = 15\Delta M/[\pi a^3 h)]$, where a and h are the radius and height of a cylindrical sample, respectively. The magnetic T_c could be deduced from the temperature dependence curve of the magnetic susceptibility. The magnetic field and temperature-dependent resistivity, $\rho(H, T)$, were measured with H applied perpendicular to the current direction, using the four-probe method in the temperature range from 15 K to 300 K and the field range from 0 T to 8.7 T. H_{c2} and H_{irr} were defined as the magnetic field values at 90% and 10% of the superconducting transition on the $\rho(H, T)$ curve, respectively. The sample dimensions are $\Phi \sim 0.7 \times 4$ mm^3 for magnetization measurements and $\Phi \sim 0.7 \times 8$ mm^3 with 2 mm distances of two inner probes for resistivity measurements. The Raman scattering was measured by a confocal laser Raman spectrometer (Renishaw inVia plus) with a 100× microscope. The 514.5 nm line of an Ar$^+$ laser was used for excitation, with the laser power maintained at about 20 mW, measured on the laser spot on the samples, in order to avoid laser heating effects on the materials studied. Several spots were selected on the same sample to collect the Raman signals in order to make sure that the results were credible.

5.2.2   Phase structure and electromagnetic properties

The Mg excess and nanosize SiC-doping effects on the phases and superconducting properties were first studied to explore their different functions in the behavior of the critical temperature, T_c, and the critical current density, J_c. Figure 5.1 shows the XRD patterns for the wire cores of MgB$_2$, Mg$_{1.15}$B$_2$, and Mg$_x$B$_2$ + 10wt%SiC. These XRD patterns reveal that nearly single-phase MgB$_2$ was obtained in the pure, stoichiometric MgB$_2$ samples along with a small amount of MgO. A small amount of residual Mg could be found in Mg$_{1.15}$B$_2$, while the Mg$_2$Si content was obvious in MgB$_2$ + 10wt%SiC.

Figure 5.2 shows the field dependence of J_c at 5 K and 20 K in fields of 0–8.5 T. The pure MgB$_2$ shows a gradual decrease in J_c with increasing measurement field at 5 K and 20 K. The J_c values of Mg$_{1.15}$B$_2$ in low magnetic field, $< \sim 2.5$ T at 5 K and $< \sim 1.1$ T at 20 K, are the highest. Although this behavior cannot be observed directly at 5 K, because the J_c values at 5 K are difficult to calculate due to the flux jumping, it is

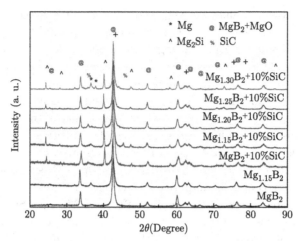

Fig. 5.1   X-ray diffraction patterns of the MgB_2 cores in the iron-clad wires of MgB_2, $Mg_{1.15}B_2$, and $Mg_xB_2 + 10wt\%SiC$ ($x = 1.00$, 1.15, 1.20, 1.25, and 1.30) sintered at $650°C$ for 30 min.

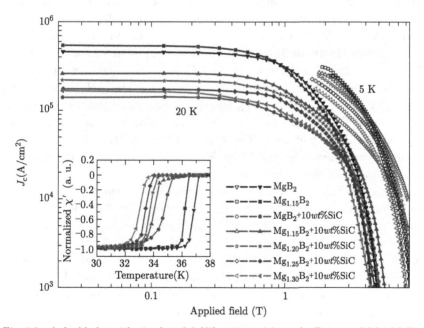

Fig. 5.2   A double-logarithmic plot of $J_c(H)$ estimated from the Bean model for MgB_2, $Mg_{1.15}B_2$, and $Mg_xB_2 + 10wt\%SiC$ ($x = 1.00$, 1.15, 1.20, 1.25, and 1.30) wires at 5 K and 20 K. The magnetic transition behavior around the critical temperature is shown in the inset.

obvious that the J_c values for $Mg_{1.15}B_2$ wires at 20 K exceed those of all the other samples. The nano-SiC addition has improved the J_c over that of pure stoichiometric MgB_2 in high measurement fields, $>\sim 7$ T at 5 K and $>\sim 4$ T at 20 K. However, the J_c values are very low at low field, as in previous results [11]. The superconducting transitions are shown in the inset of Fig. 5.2. The additional Mg degrades the superconductivity of MgB_2, and the T_c value of $Mg_{1.15}B_2$ is 0.7 K lower than that of the stoichiometric MgB_2, because the residual Mg is a source of impurity scattering effects. The T_c value, 35.3 K, for $MgB_2 + 10wt\%SiC$ is even lower than that of $Mg_{1.15}B_2$ because of both the C substitution (estimated as $\sim 2.2\%$, as shown in Table 5.1 [131]) and the impurity scattering.

To take advantage of the doping effects of Mg and SiC, $Mg_xB_2 + 10wt\%SiC$ samples were studied with $x = 1.15$, 1.20, 1.25, and 1.30. The residual Mg contents are similar in all the samples, as indicated by the XRD patterns in Fig. 5.1. As estimated from the XRD patterns, the C substitution content is $\sim 2.3\%$ in $Mg_{1.15}B_2 + 10wt\%SiC$, which is similar to that in $MgB_2 + 10wt\%SiC$, and then it drops to 2.1% in $Mg_{1.20}B_2 + 10wt\%SiC$ and 1.8% in both $Mg_{1.25}B_2 + 10wt\%SiC$ and $Mg_{1.30}B_2 + 10wt\%SiC$. It seems that the MgB_2 lattice excludes C atoms due to the excess Mg, as in the trend shown in Table 5.1. The most impressive advantage of the Mg excess samples is the improved $J_c(H)$ in low magnetic field, especially at 5 K. However, the J_c values drop quickly for $Mg_{1.20}B_2 + 10wt\%SiC$, $Mg_{1.25}B_2 + 10wt\%SiC$, and $Mg_{1.30}B_2 + 10wt\%SiC$ because of the high impurity content. The decreased T_c values with increasing Mg content are responsible for the J_c drop at 20 K. The J_c performance of $Mg_{1.15}B_2 + 10wt\%SiC$ improves over the whole range of applied magnetic field due to the optimum ratio of the component elements.

Although the T_c values decrease with the dopant content, the different superconducting transition processes suggest quality differences between the MgB_2 crystals according to the normalized AC susceptibility, as shown in the inset of Fig. 5.2. The transition width, ΔT_c, is defined as the temperature difference between the onset value of the transition (the point where the AC susceptibility deviates from zero), $T_{c,onset}$, and the termination point of the transition (the point where the AC susceptibility reaches -1), $T_{c,ter}$. All the values of $T_{c,onset}$ and ΔT_c are listed in Table 5.1. The superconducting transition of MgB_2 is quite narrow, with a ΔT_c of 0.98 K. However, it is even sharper for $Mg_{1.15}B_2$, with a ΔT_c of 0.85 K, which is attributed to the improved quality and connectivity of the MgB_2 because

Table 5.1   Comparison of the onset transition temperature ($T_{c,\text{onset}}$), the transition width (ΔT_c), the lengths of the a-axis and c-axis parameters, the C substitution content, the density of states $N(0)$, the resistivity values at 40 K and 300 K, $\Delta\rho(300)$, and the active area fraction (A_F) for iron-clad wires of MgB_2, $Mg_{1.15}B_2$, and $Mg_xB_2 + 10wt\%SiC$ ($x = 1.15$, 1.20, 1.25, and 1.30) sintered at 650°C for 30 min.

| Samples | MgB_2 | $Mg_{1.15}B_2$ | With 10$wt\%$SiC addition | | | | | |
			MgB_2	$Mg_{1.15}B_2$	$Mg_{1.20}B_2$	$Mg_{1.25}B_2$	$Mg_{1.30}B_2$
$T_{c,\text{onset}}$ (K)	37.18	36.50	35.32	34.55	34.28	34.04	33.60
ΔT_c (K)	0.98	0.85	3.36	1.69	1.57	1.53	1.54
a-axis (Å)	3.0869	3.0881	3.0771	3.0766	3.0773	3.0786	3.0788
c-axis (Å)	3.5244	3.5238	3.5203	3.5201	3.5218	3.5224	3.5227
C content (%)	N/A	N/A	2.2	2.3	2.1	1.8	1.8
$N(0)$	0.354	0.354	0.332	0.331	0.333	0.336	0.336
$\rho_{40K}(\mu\Omega \cdot \text{cm})$	86.8	70.0	277.2	152.2	168.6	191.3	213.3
$\rho_{300K}(\mu\Omega \cdot \text{cm})$	171.6	125.5	422.5	246.3	269.8	307.7	347.3
$\Delta\rho(300)(\mu\Omega \cdot \text{cm})$	84.8	55.5	145.3	94.1	101.2	116.4	134.0
A_F	0.106	0.162	0.062	0.096	0.089	0.077	0.067

the flux is easy to expel, as shown by the decreased temperature transition width. The transition of $MgB_2 + 10wt\%SiC$ is very broad, with a ΔT_c of 3.36 K, because of the strong disorder. Excess Mg in $MgB_2 + 10wt\%SiC$ is of benefit to the transition: the ΔT_c of $Mg_{1.15}B_2 + 10wt\%SiC$ drops significantly to 1.69 K. More Mg addition decreases the value of ΔT_c to ~ 1.55 K in $Mg_xB_2 + 10wt\%SiC$ ($x > 1.20$). The stable ΔT_c values indicate that the quality cannot be improved with more Mg addition in $MgB_2 + 10wt\%SiC$.

5.2.3   *Connectivity*

Based on the collective pinning model [132], the disorder-induced spatial fluctuations in the vortex lattice can be clearly divided into different regimes according to the strength of the applied field: single-vortex, small-bundle, large-bundle, and charge-density-wave type relaxation of the vortex lattice. For high current purposes, the practicable J_c for MgB_2 is in the regions of single-vortex pinning and small-bundle pinning in the phase diagram. J_c value is independent of applied magnetic field in single-vortex region and it depends exponentially on applied magnetic field in small-bundle pinning region. The vortex flux deforms greatly and merges in higher magnetic field and moves at last when the magnetic field is as high as H_{c2}. It is believed that the quality of the connectivity is responsible for the J_c performance in the single-vortex pinning regime (low magnetic field region) due to its weak field dependence, while disorder is responsible for the J_c performance in the small-bundle regime (high magnetic field) due to the strong F_p.

To explore the effect on connectivity of Mg excess, microstructures of all samples were observed with SEM, as shown in Fig. 5.3. The grains in MgB_2 samples show an independent growth process, which is responsible for their isolated distribution. The grains in $Mg_{1.15}B_2$ have clearly melted into big clusters because the additional Mg can extend the liquid reaction time. The grain shapes in $MgB_2 + 10wt\%SiC$ are different from that in pure, stoichiometric MgB_2 because the crystals are grown under strain due to the C substitution effect. The strain is also strong in $Mg_{1.15}B_2 + 10wt\%SiC$, as long bar grains can be observed under SEM. The strain is released in the high Mg content samples ($x > 1.20$), judging from the homogeneous grain sizes and shapes. Compared with $MgB_2 + 10wt\%SiC$, the grain connectivity improved greatly with increasing Mg addition. The grains are merged into big particles. This means that grain boundaries have replaced the gaps

Fig. 5.3   SEM images of MgB$_2$ (a), Mg$_{1.15}$B$_2$ (b), MgB$_2$ + 10wt%SiC (c), Mg$_{1.15}$B$_2$ + 10wt%SiC (d), Mg$_{1.20}$B$_2$ + 10wt%SiC (e), Mg$_{1.25}$B$_2$ + 10wt%SiC (f), and Mg$_{1.30}$B$_2$ + 10wt%SiC (g).

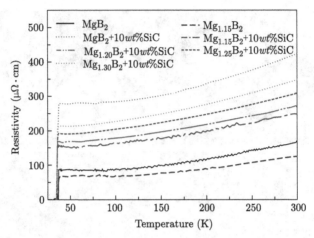

Fig. 5.4 Temperature dependence of resistivity curves of MgB_2, $Mg_{1.15}B_2$, and $Mg_xB_2 + 10wt\%SiC$ ($x = 1.00$, 1.15, 1.20, 1.25, and 1.30).

between grains. However, more impurities are induced in forms such as residual Mg and MgO.

Compared with the microstructure observations, the resistivity is a more reliable and quantitative method to estimate the connectivity. The resistivity dependences on temperature are shown in Fig. 5.4. The $\rho_{300\ K}$ and $\rho_{40\ K}$ of MgB_2 wires are always high compared with thin films and bulks due to the low density of the wire samples. Although residual Mg exists in $Mg_{1.15}B_2$, its resistivity is decreased due to the highly-connected grain boundaries. The resistivity increases greatly in $MgB_2 + 10wt\%SiC$ because of the strong band scattering effects due to the C substitution on B sites [133]. The strong localization of the boron p_{xy} orbital on the boron plane, which is sensitive to the C substitution for boron, is responsible for the great resistivity enhancement. Additionally, the reduction in charge carriers has to be considered, because the conducting holes are annihilated by the electrons introduced into the system by the substitutional carbon atoms. The most amazing effect of Mg excess is that the resistivity of $MgB_2 + 10wt\%SiC$ is reduced by $\sim$45% with 15% Mg addition. This is because the grain boundary scattering effects are greatly weakened. Although a greater Mg excess ($x > 1.15$) ought to further reduce the resistivity of $MgB_2 + 10wt\%SiC$, the increased scattering effects of the impurity phases make the resistivity a little higher than that of $Mg_{1.15}B_2 + 10wt\%SiC$.

The A_F values for all the samples are listed in Table 5.1. It should be noted that the connectivity is far removed from that found in ideal crystals, as reflected by the low A_F values. Although the A_F values of pure and 10% SiC-doped MgB_2 are just 0.106 and 0.062, the additional Mg can improve them to 0.162 and 0.096 for $15wt\%$ Mg excess samples, respectively. High A_F values are the reflection of a broad channel of supercurrents, while impurities reduce the connectivity in large x samples, as shown in Table 5.1. Considering the weak field dependence of J_c in the applied field, H_A, ranges of $H_A <\sim 0.7$ T, 20 K and $H_A <\sim 2$ T, 5 K, as shown in Fig. 5.2, it is believed that the high connectivity improves the supercurrent channels because the currents can easily meander through the well-connected grains.

5.2.4 H_{irr}, H_{c2}, and F_p

In contrast with the high connectivity mechanism in the single-vortex region, a strong flux pinning force is the dominant feature that keeps the J_c high in the small-bundle region. Most defects, lattice distortions, and grain boundaries are believed to act as flux pinning centers [126, 130, 134]. Carbon substitution for boron in the MgB_2 lattice can cause great lattice distortion because of the shorter C-B bonds [133]. Furthermore, H_{c2} and H_{irr} are significantly enhanced by the increased band scattering due to carbon substitution. The competitive J_c in high field for $MgB_2 + 10wt\%SiC$ and $Mg_{1.15}B_2 + 10wt\%SiC$ should be ascribed to the enhanced H_{c2} and H_{irr}. Figure 5.5 compares the H_{c2} and H_{irr} for all the samples. The H_{c2} and H_{irr} values of pure, stoichiometric MgB_2 increase gently as the temperature is decreased, while the enhancement is a little slower in $Mg_{1.15}B_2$. $MgB_2 + 10wt\%SiC$ has high H_{c2} and H_{irr} values in the range of $T <\sim 24$ K, and the additional Mg in $Mg_{1.15}B_2 + 10wt\%SiC$ does not degrade the good performance of H_{c2} and H_{irr}. The high H_{c2} and H_{irr} of $Mg_{1.15}B_2 + 10wt\%SiC$ are attributed to both the carbon substitution and the high amount of impurity phase.

The H_{irr} is responsible for the J_c performance in high magnetic field because the field dependence of F_p is proportional to $h^n(1 - h)^m$, with $h = H/H_{irr}$, where the values of n and m depend on the different flux pinning mechanisms [135]. The h dependence of $F_p/F_{p,max}$ is shown in Fig. 5.6. The x-axis is not normalized to 1.0 because the values of H_{irr} are higher than those in Refs. [15] and [16] due to the different standards

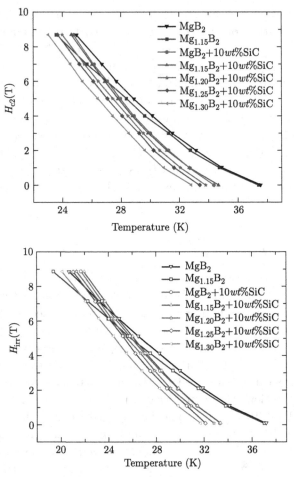

Fig. 5.5   Comparisons of H_{c2} (a) and H_{irr} (b) of MgB$_2$, Mg$_{1.15}$B$_2$, and Mg$_x$B$_2$ + 10$wt\%$SiC (x = 1.00, 1.15, 1.20, 1.25, and 1.30).

for H_{irr}. The most striking phenomena in Fig. 5.6 are: (i) the $F_p/F_{p,max}$ for Mg$_{1.15}$B$_2$ + 10$wt\%$SiC is effective both in the single-vortex region and in the small-bundle region; (ii) the $F_p/F_{p,max}$ for MgB$_2$ + 10$wt\%$SiC is the most effective in the small-bundle region; (iii) the $F_p/F_{p,max}$ values for MgB$_2$ and Mg$_{1.15}$B$_2$ are uncompetitive in the small-bundle region; and (iv) the $F_p/F_{p,max}$ values for Mg$_x$B$_2$ + 10$wt\%$SiC (x = 1.20, 1.25, and 1.30) are also effective in the small-bundle region. These phenomena are in agreement with the J_c behavior shown in Fig. 5.2: (i) the J_c performance for

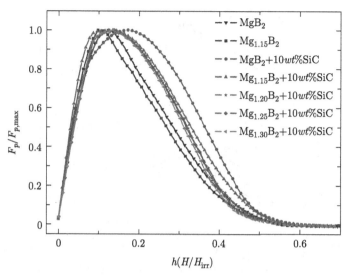

Fig. 5.6   Comparisons of the h($h = H/H_{\text{irr}}$) dependence of $F_p/F_{p,\text{max}}$ for MgB$_2$, Mg$_{1.15}$B$_2$, and Mg$_x$B$_2$ + 10wt%SiC ($x = 1.00$, 1.15, 1.20, 1.25, and 1.30).

Mg$_{1.15}$B$_2$+10wt%SiC is promising over the whole range of the H_A; (ii) the J_c values for MgB$_2$ + 10wt%SiC drop slowly with the H_A, and it can be imagined that the J_c values are higher than those of Mg$_{1.15}$B$_2$ + 10wt%SiC in the stronger H_A; (iii) the $J_c(0)$ values of MgB$_2$ and Mg$_{1.15}$B$_2$ are out-standing, however, they drop quickly with H_A; and (iv) the high impurity content degrades the J_c performance in low H_A for Mg$_x$B$_2$ + 10wt%SiC ($x = 1.20$, 1.25, and 1.30), whereas these impurities are also efficient flux pinning centers in high H_A. The high J_c of Mg$_{1.15}$B$_2$+10wt%SiC is ascribed to the cooperation of the high A_F, high H_{c2}, and high H_{irr}. The H_{c2} and H_{irr} values of Mg$_x$B$_2$ + 10wt%SiC ($x \geqslant 1.20$) are lower than those of Mg$_{1.15}$B$_2$ + 10wt%SiC, which is attributed to the high impurity content of the samples.

5.2.5   Raman spectra

Raman scattering is employed to study the combined influence of connectivity and lattice distortion. Chemical substitution [126] and lattice distortion [136] are expected to modify the phonon spectrum, by changing the phonon frequency and the electron-phonon interaction. The effects of C substitution include an increase in impurity scattering and band filling, which reduces

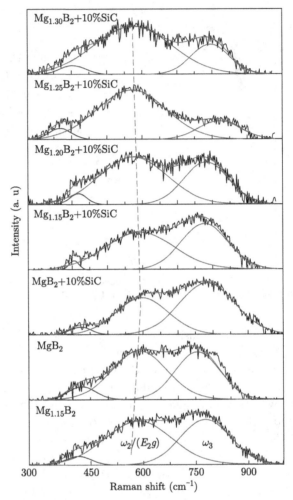

Fig. 5.7   Ambient Raman spectra of MgB_2, $Mg_{1.15}B_2$, and $Mg_xB_2 + 10wt\%SiC$ ($x = 1.00$, 1.15, 1.20, 1.25, and 1.30) fitted with three peaks: ω_1, ω_2, and ω_3. The dashed line indicates the vibration of the E_{2g} mode (ω_2) with different components.

the density of states and alters the shape of the Fermi surface. The E_{2g} phonon peak shifts to the higher energy side, and the peak is narrowed with increasing x in $Mg(B_{1-x}C_x)_2$ [126]. As a carbon source, nano-SiC shows a similar influence, due to its C atoms, on the J_c, H_{irr}, H_{c2}, and even Raman spectra in MgB_2. Figure 5.7 shows the Raman spectra fitted with three peaks: ω_1, ω_2, and ω_3. The ω_1 and ω_3 peaks are understood to arise from sampling of the phonon density of states (PDOS) due to disorder, while ω_2 is associated with the E_{2g} mode, which is the only Raman active mode for MgB$_2$ [137]. A reasonable explanation for the appearance of ω_1 and ω_3 is the violation of Raman selection rules induced by disorder. All three peaks are broad, as in previous results, due to the strong electron-phonon coupling. The influence of ω_1 on the superconducting performance is negligible compared with those of ω_2 and ω_3, because of its weak contribution to the Raman spectrum. The frequency and full width at half maximum (FWHM) of ω_2 and ω_3 are shown in Fig. 5.8. Both ω_2 and ω_3 are hardened with SiC addition. The ω_2 frequency is reduced with further Mg addition, whereas the ω_3 frequency remains almost stable. The frequencies

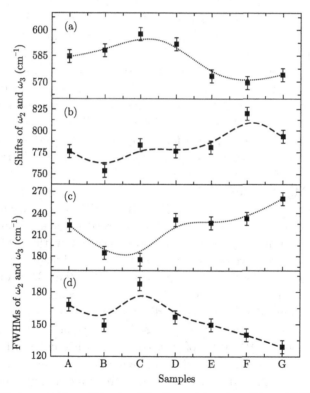

Fig. 5.8   Fitted parameters of Raman shifts for ω_2 (a) and ω_3 (b), and FWHMs for ω_2 (c) and ω_3 (d). The sample labels are defined as: A for Mg$_{1.15}$B$_2$, B for MgB$_2$, C for MgB$_2$ + 10wt%SiC, D for Mg$_{1.15}$B$_2$ + 10wt%SiC, E for Mg$_{1.20}$B$_2$ + 10wt%SiC, F for Mg$_{1.25}$B$_2$ + 10wt%SiC, and G for Mg$_{1.30}$B$_2$ + 10wt%SiC.

of ω_2 for the $x \geqslant 1.20$ samples are even lower than that of the pure, stoichiometric MgB_2. The FWHM of ω_2 decreases with SiC-doping, while the Mg excess remedies this trend. On the contrary, the ω_3 FWHM increases with SiC addtion and becomes narrow with further addition of Mg.

The Raman scattering properties are the direct reflection of the phonon behaviors of MgB_2. The parameters of Raman spectra vary with the composition of MgB_2 crystals and the influences of their surroundings, which depend on both the connectivity and the disorder of the samples. Furthermore, the disorder should be considered to be divided into intrinsic and extrinsic parts based on their different sources. The crystallinity and chemical substitution are believed to be responsible for the intrinsic disorder effects, while the grain boundaries and impurities are treated as responsible for the extrinsic disorder effects. The influences of intrinsic disorder on the basic characteristics of Raman spectra are significant because the physical properties of MgB_2 depend on the intrinsic disorder. The Raman parameters can also be tuned by the extrinsic disorder. Especially in the samples with good connectivity, the influences of grain boundaries and impurities on the Raman spectra are taken into account because of their strain effects on the MgB_2 crystals [138]. The differences between shifts and FWHMs in Raman spectra for MgB_2, $Mg_{1.15}B_2$, $MgB_2 + 10wt\%SiC$, and $Mg_{1.15}B_2 + 10wt\%SiC$ are mostly attributable to the intrinsic characteristics because of their different chemical compositions. The Raman spectra of $Mg_xB_2 + 10wt\%SiC$ ($x > 1.20$) can be considered as the gradual modification of that of $Mg_{1.15}B_2 + 10wt\%SiC$. The weakened C substitution effects are reponsible for the decreased frequencies and slightly increased FWHMs of ω_2 with Mg addition. Accordingly, the FWHMs of ω_3 decrease due to the weakened lattice distortion. Although the A_F values are quite low for $Mg_xB_2 + 10wt\%SiC$ ($x > 1.20$), the effects of extrinsic disorder on Raman parameters are considerable through the MgB_2-MgB_2 and MgB_2-impurities interfaces, and the connectivity degrades with the increased x values due to the decreased numbers of MgB_2-MgB_2 interfaces. The high FWHM value of ω_2 is correlated with high self-field J_c due to the high carrier density, while the high FWHM value of ω_3 is correlated with the high-field J_c because of the strong flux-pinning force due to the large disorder. The FWHM behaviors show that high connectivity and strong disorder are best combined in $Mg_{1.15}B_2 + 10wt\%SiC$ among all the samples.

5.2.6   Electron-phonon coupling

Not only are the Raman parameters related to the superconducting properties, but they are also related to the electron-phonon coupling (EPC) strength. As for the superconductivity, a broad FWHM of ω_2 and a narrow FWHM of ω_3 in Raman spectra imply a strong electron-E_{2g} coupling in MgB$_2$. According to the frequency and the linewidth of the E_{2g} mode reflected in the Raman spectra, direct evaluation of the contribution of the E_{2g} mode to the EPC is possible due to the negligible anharmonic effects in the system. The constant of the electron-E_{2g} coupling is given by the Allen equation [139]:

$$\Gamma_2 = 2\pi \lambda_{E_{2g}} N(0)\omega_2^2, \qquad (5.7)$$

where $\lambda_{E_{2g}}$ is the strength of the electron-E_{2g} coupling and $N(0)$ is the density of states (per spin per unit energy per unit cell) on the Fermi surface, and is the only electronic property explicitly occurring in this equation. The measured phonon frequency and phonon linewidth, in the absence of anharmonic contributions, are simply and directly related to the EPC constant, $\lambda_{E_{2g}}$. The total DOS at E_F in pure MgB$_2$ is taken as $N(0)$ = 0.354 states/eV/cell/spin, with the contribution from the σ-band being 0.15 states/eV/cell/spin and that from the π-band being 0.204 states/eV/cell/spin, respectively [140]. The $N(0)$ is assumed to be constant for the small changes of electron and holes in Mg$_{1.15}$B$_2$ and MgB$_2$. The factor of $N^\pi(0)$ dependence on the carbon concentration is about -1 in the range of 0–0.04, while $N^\sigma(0)$ remains constant [141]. Taking the C substitution contents shown in Table 5.1 into account, the $\lambda_{E_{2g}}$ values for the different samples can be obtained from the Allen equation, as shown in Fig. 5.9. These values are in agreement with 2.5 ± 1.1 obtained for the $q = 0.2\Gamma - AE_{2g}$ mode from inelastic X-ray scattering measurements [142]. It is unexpected that the $\lambda_{E_{2g}}$ of MgB$_2$ + 10$wt\%$SiC is just a little smaller than that of the pure, stoichiometric MgB$_2$. A possible explanation is that both the samples are imperfect because of the low-temperature sintering and impurity phases. The crystallinity is improved for all the Mg excess samples according to the higher $\lambda_{E_{2g}}$ values. The $\lambda_{E_{2g}}$ values in Mg$_x$B$_2$ + 10$wt\%$SiC ($x > 1.20$) are even higher than that for Mg$_{1.15}$B$_2$, which is attributed to the low $N(0)$ values and the large amount of Mg excess. This phenomenon is in agreement with the strong FWHM values of ω_2 and the low FWHM values of ω_3.

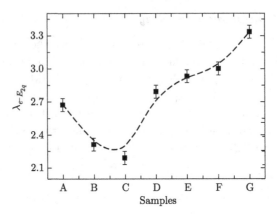

Fig. 5.9   Calculated coupling constants of electron-E_{2g}, $\lambda_{e\text{-}E_{2g}}$, based on the Raman shifts and FWHMs of ω_2 according to the Allen equation. The sample labels are defined as: A for $Mg_{1.15}B_2$, B for MgB_2, C for $MgB_2 + 10wt\%SiC$, D for $Mg_{1.15}B_2 + 10wt\%SiC$, E for $Mg_{1.20}B_2 + 10wt\%SiC$, F for $Mg_{1.25}B_2 + 10wt\%SiC$, and G for $Mg_{1.30}B_2 + 10wt\%SiC$.

Although their E_{2g} mode contribution to the EPC is higher, the total EPC constants are degraded by the high impurity contents, which can be estimated with the McMillan formula [6], as modified by Allen and Dynes [143]:

$$T_c = \frac{\langle\omega_{\log}\rangle}{1.2}\exp\left[\frac{-1.04(1+\lambda)}{\lambda - \mu^*(1 + 0.62\lambda)}\right],\tag{5.8}$$

where $\langle\omega_{\log}\rangle = (390\times\omega_{E_{2g}}^2\times690)$ is the averaged phonon frequency [5], with 690 and 390 cm^{-1} being the phonon frequencies of the other modes in the MgB_2 system (taken from Ref. [144]), μ^* is the Coulomb pseudopotential, taken as equal to 0.13 [145], and λ is the EPC constant. Taking these values, the λ for each sample is calculated and is shown in the inset of Fig. 5.10. The λ of MgB_2 is the highest one because of its high T_c and low impurity content. Both Mg and SiC addition depress λ, which is in agreement with the high impurity contents shown in the XRD patterns.

5.2.7   Section summary

In conclusion, connectivity and disorder are both important factors in improvement of the J_c performance of MgB_2. The connectivity is respon-sible for the high J_c performance in the single-vortex regime, while strong disorder is responsible for the promising J_c in the small-bundle pinning

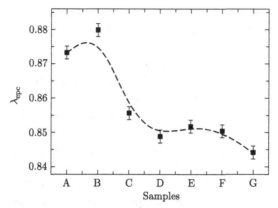

Fig. 5.10   Calculated total electron-phonon coupling constants, λ_{epc}, according to the McMillan formula. The sample labels are defined as: A for $Mg_{1.15}B_2$, B for MgB_2, C for $MgB_2 + 10wt\%SiC$, D for $Mg_{1.15}B_2 + 10wt\%SiC$, E for $Mg_{1.20}B_2 + 10wt\%SiC$, F for $Mg_{1.25}B_2 + 10wt\%SiC$, and G for $Mg_{1.30}B_2 + 10wt\%SiC$.

regime. The promising J_c values in $Mg_{1.15}B_2 + 10wt\%SiC$ are the result of optimized connectivity and disorder, which are reflected in the Raman spectrum, with both a strong E_{2g} mode and a strong PDOS. The Raman scattering measurements imply that excess Mg is effective in improving the connectivity of MgB_2 grains, while nano-SiC is responsible for the great lattice distortion in the SiC-doped samples. The superconductivity transition is advanced in Mg excess samples according to the decreased ΔT_c. The EPC analysis shows that the excess Mg can also improve the electron-E_{2g} coupling. However, the impurity phases depress the total EPC strengths.

5.3   The combined in-situ/ex-situ technique

The *ex-situ* technique is promising for making high J_c MgB_2 superconductors. The problem with the *ex-situ* technique is the low connectivity due to the low quality connections among the MgB_2 grains. In this work, a mixed *in-situ/ex-situ* technique was employed to develop MgB_2 wires with high connectivity and strong disorder to increase both the low- and high-field J_c properties. The evolution of microstructures for the samples sintered at different temperatures was observed and analyzed systematically to demonstrate the J_c dependence on microstructure, connectivity, and disorder in MgB_2 wires.

The PIT process was employed to make practical MgB_2 wires from a ball-milled mixture of Mg (99%) and B (99%, amorphous). A part of the mixture was made into MgB_2/Fe wires with a diameter of 1.4 mm and sintered in high purity Ar at temperatures of 750°C, 850°C, 950°C, and 1050°C for 30 min as standard *in-situ* samples, which were marked as 750 in, 850 in, 950 in, and 1050 in, respectively. The other part of the mixture was sintered at 650°C for 30 min in pure argon flow and then ball-milled to yield precursor MgB_2 powder. Then *in-situ/ex-situ* combined MgB_2/Fe wires were made by the PIT method using precursor powder and a mixture of Mg and B powders in a 1:3 ratio. All the green wires were annealed in high purity Ar at temperatures of 750°C, 850°C, 950°C, and 1050°C for 30 min, yielding samples which were marked as 750 in/ex, 850 in/ex, 950 in/ex, and 1050 in/ex, respectively. The characterization of all the samples were similar with those in Sec. 5.2.

5.3.1 *Superconducting transition properties*

According to the indexed XRD patterns, most of the samples show quite high purity of MgB_2 with a little amount of MgO. The MgO content is quite high in 950 in, 1050 in, and 1050 in/ex, which is in agreement with the broad transition from the normal state to the superconducting state, as shown in the insets of Fig. 5.11. In particular, the transition width of 1050 in is as broad as ∼7 K compared with the width of ∼2 K in the other samples because the high MgO content has destroyed the connectivity of the magnetic flux. The T_c values of *in-situ* sintered samples are 0.5–1 K higher than those of the *in-situ/ex-situ* ones. The T_c of 1050 in drops as low as that of 1050 in/ex, as shown in Fig. 5.11.

The low T_c of *in-situ/ex-situ* samples is attributed to the strong disorder induced by both ball-milling of raw materials and the defects among the grains [146, 147]. Furthermore, the transition widths for *in-situ/ex-situ* samples are a little broader than those of 750 in, 850 in, and 950 in, which means that the connectivity of these samples is lower than that of most *in-situ* samples. Although there are MgO impurities in 1050 in/ex, its transition width is still quite narrow compared with 1050 in due to the low amount of Mg evaporation. The precursor MgB_2 is stable at 1050°C, and the low quantity of Mg in the raw materials of 1050 in/ex has only a weak influence on the phases.

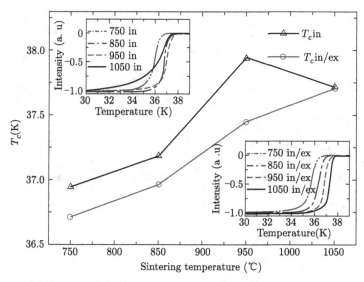

Fig. 5.11 Comparison of the dependence of the critical transition temperature T_c on sintering temperature for *in situ* and combined *in situ/ex situ* samples. The insets show the normalized magnetic moment dependence on temperature for all *in situ* samples (upper) and for all *in situ/ex situ* samples (lower). Wide transition widths imply higher impurity samples.

5.3.2 Microstructure

Figure 5.12 shows typical SEM images of all *in situ* sintered samples. The grains in 750 in show an independent growth process with a grain size of about 300 nm, which is responsible for their isolated distribution. The grains have clearly melted into big clusters for 850 in, 950 in, and 1050 in because the increased sintering temperature has extended the crystal growth time. Actually, besides the high sintering temperature, long sintering time is required for this as well. The crystal quality is obviously improved with the increased sintering temperature. Some clusters of 1 μm in size can be found in 1050 in. The high sintering temperature accelerates the Mg evaporation in the raw materials. Insulating MgO impurities can be observed in 1050 in under SEM, shown as small white particles without contrasts.

The *in situ/ex situ* sintered samples show irregular crystal shapes and a wide distribution of grain size, as shown in Fig. 5.13. The grain size of 750 in/ex is quite small compared with those of 850 in/ex, 950 in/ex, and 1050 in/ex because the low sintering temperature is not high enough to

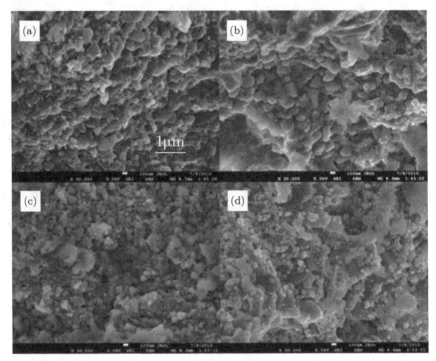

Fig. 5.12   SEM images for the *in situ* samples sintered at (a) 750°C, (b) 850°C, (c) 950°C, and (d) 1050°C. The crystal growth was improved gradually with increasing sintering temperature.

trigger recrystallization. The small crystals grow quickly into big ones in 850 in/ex, 950 in/ex, and 1050 in/ex, and grain sizes as large as 1 μm can be found in these samples. It should be noted that the microstructures are very similar for 1050 in and 1050 in/ex. The high sintering temperature provides enough energy, and there is a long sintering time for the crystal growth of 1050 in/ex.

5.3.3   *Critical current density*

The J_c dependence on the applied field is shown in Fig. 5.14 for all the samples, which were measured at 5 K and 20 K, respectively. 750 in, 850 in, 750 in/ex, and 1050 in/ex display quite similar performance over the whole field range. The J_c properties deteriorated in 950 in and 1050 in in high magnetic fields, and the J_c value of 1050 in is the worst among all the

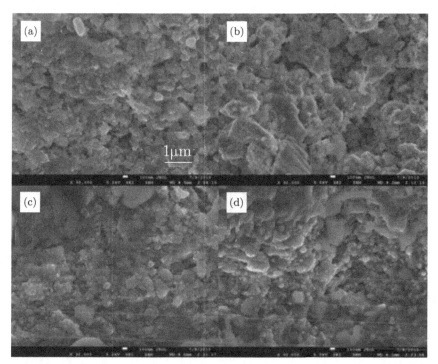

Fig. 5.13   SEM images for the combined *in situ/ex situ* samples sintered at (a) 750°C, (b) 850°C, (c) 950°C, and (d) 1050°C. The grain size is inhomogeneous in all the samples due to the presintered MgB$_2$ in the raw materials and recrystallization effects.

samples in this work due to the high MgO content. 850 in/ex and 950 in/ex show amazingly high J_c performances among all the samples, which are about five times higher than the best *in situ* sintered samples at 5 K under 8 T magnetic field. To avoid the influence of the flux jumping effect, the low-field performances were compared at 20 K in detail, as shown in the inset of Fig. 5.14. All *in situ* sintered samples, except 1050 in, show very competitive J_c in lower magnetic field. 950 in shows quite high self-field J_c performance because of the high crystal quality. The self-field J_c values of 750 in/ex and 850 in/ex are lower than those of *in situ* sintered samples. The increased sintering temperature for 950 in/ex and 1050 in/ex improved their self-field J_c to a little higher than those of 750 in and 850 in. However, the higher MgO contents prevent their self-field J_c enhancement compared with 950 in.

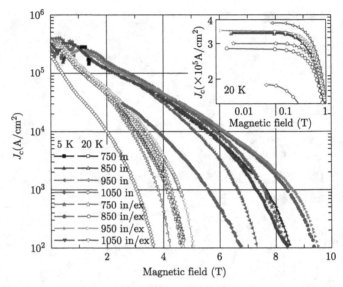

Fig. 5.14 The dependence of critical current density, J_c, on applied magnetic field at 5 K and 20 K for all the samples. The inset shows the J_c performance in low magnetic field at 20 K to demonstrate the self-field J_c behavior without the influence of flux jumping effect.

5.3.4 *Connectivity and critical fields*

The A_F values for all the samples are displayed in Fig. 5.15. It should be noted that the connectivity is much lower than those of ideal crystals, as reflected by the low A_F values. Although the A_F value of 750 in is just 0.169, increased sintering temperature improves it to $\sim$0.26. The A_F values are stable when the sintering temperature is higher than 850°C, which means that the connectivity easily reaches its peak value for *in situ* sintered samples. Despite the high MgO content, 1050 in shows a high A_F value due to sufficient crystallization. However, its self-field J_c was destroyed by MgO. The *in situ/ex situ* samples show lower A_F values compared with those *in situ* ones, which is in agreement with the resistivity values shown in the insets of Fig. 5.15. The resistivity decreases with increased sintering temperature for all the samples. In particular, the resistivity values of 1050 in and 1050 in/ex reach the same level after high temperature sintering. The gradually increased A_F values for *in situ/ex situ* samples are attributed to the improved solid state reaction at high sintering temperature, which is responsible for the increased self-field J_c performance in

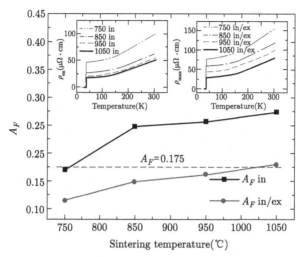

Fig. 5.15   The active area fraction, A_F, as a function of sintering temperature for all the samples. The short dashed line indicates $A_F = 0.175$. The upper left inset shows the resistivity dependence on temperature of all the *in situ* samples in the range 0–305 K. The upper right inset shows the resistivity dependence on temperature of all the *in situ/ ex situ* samples in the range of 0–305 K.

950 in/ex and 1050 in/ex. In contrast to the high connectivity mechanism in the single-vortex region, a strong flux-pinning force is the dominant feature that keeps the J_c high in the small-bundle region. Most defects, lattice distortion, and grain boundaries are believed to act as flux-pinning centers [130, 134]. The strength of the pinning force can be reflected by the dependence of H_{c2} and H_{irr} on temperature, as shown in Fig. 5.16. All the H_{c2} and H_{irr} behavior is in agreement with the high-field J_c performance. It should be noted that 1050 in shows both the lowest H_{c2} and the lowest H_{irr} among all the samples. 750 in and 750 in/ex show quite similar H_{c2} and H_{irr} dependences on temperature. 850 in/ex and 950 in/ex show the best H_{c2} and H_{irr} performances, which are responsible for their high J_c values under high magnetic field. The high H_{c2} and H_{irr} values are attributed to the strong disorder introduced by the *ex situ* powder and proper crystallization.

5.3.5   *Section summary*

In conclusion, connectivity and disorder are both important factors in improvement of the J_c performance of MgB_2. The connectivity is

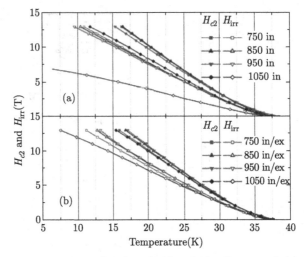

Fig. 5.16   Comparisons of H_{c2} (solid symbols) and H_{irr} (open symbols) of (a) *in situ* MgB$_2$ and (b) *in situ/ex situ* MgB$_2$ sintered at 750°C, 850°C, 950°C, and 1050°C, respectively.

responsible for the high J_c performance in the single-vortex regime, where J_c is magnetic field independent. Strong flux-pinning force is responsible for the promising J_c in the small-bundle pinning regime, where the defects are believed to act as flux-pinning centers. The promising J_c values in 850 in/ex and 950 in/ex are the result of optimized connectivity and disorder. This work shows the importance of strong disorder in high density MgB$_2$ for high J_c values over the whole range of applied field. The J_c values in bulk MgB$_2$ samples are far below the J_d value, $\sim8.7 \times 10^8$ A/cm^2 in pure MgB$_2$. There is still huge room for J_c improvement in bulk MgB$_2$. Based on the current work, chemically doped MgB$_2$/Fe wires fabricated by a combined *in situ/ex situ* process are worth ongoing research for practical purposes.

5.4   Diffusion technique for high density MgB$_2$ superconductors

Based on the above discussion, the connectivity can be improved by the density of the samples. In this work, the effects of the stress field on the flux pinning and electron-phonon coupling are discussed to explain the behavior of high density SiC-MgB$_2$ composite made by the diffusion process.

Crystalline B (99.999%) powders, with and without $10wt\%$ SiC particles, were mixed and pressed into pellets. The pellets were then put into iron tubes filled with Mg powder (99.8%). The atomic ratio of Mg to B was 1.15:2.0. Considering the time dependence of the diffusion process, sintering was conducted at 850°C for 10 hours under a flow of high purity argon gas to achieve fully reacted MgB_2 bulks.

5.4.1 Phase composition

The density of the pure MgB_2 sample is about 1.86 g/cm^3, which is about 80% of the theoretical density. This value is much higher than those of the samples made from *in situ* process, less than 50%. The SiC-MgB_2 composite shows an even higher density of 1.91 g/cm^3 due to the SiC addition. Figure 5.17 shows the Rietveld refinement XRD patterns of the pure MgB_2 and the $10wt\%$ SiC-doped MgB_2 samples. It is interesting to note that the SiC particles remained unreacted and formed a composite with the MgB_2 in the SiC-doped sample. The Rietveld refinement analysis results show that the unreacted SiC was about $9.3wt\%$, which is about the same as in the precursor. This is consistent with the absence of Mg_2Si in the XRD pattern as shown in Fig. 5.17. The result is different with the SiC-doped

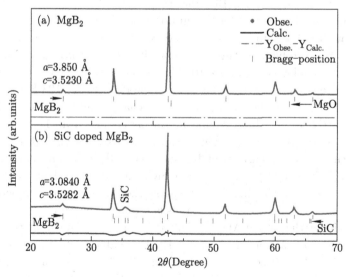

Fig. 5.17   Rietveld refined XRD patterns of pure MgB_2 and $10wt\%$ SiC-doped MgB_2 samples made from diffusion process at 850°C for 10 hours.

MgB$_2$ prepared by the *in situ* technique [62, 82], in which a very small amount of SiC remained, while Mg$_2$Si is always present due to the reaction of Mg with SiC. The a- and c-axis lattice parameters were 3.0850 Å and 3.5230 Å for pure MgB$_2$, and 3.0840 Å and 3.5282 Å for SiC-doped MgB$_2$, respectively. The a-axis parameters are virtually equivalent for the two samples, whereas the c-axis parameter is slightly enlarged in the SiC-MgB$_2$ composite. In contrast, the a-axis parameter for *in situ* processed SiC-doped MgB$_2$ is reduced while the c-axis parameter should remain unchanged, as reported by a number of groups [82, 103].

5.4.2   Strain effect

To explain the abnormal c-axis enlargement of SiC-MgB$_2$ composites, the thermal expansion coefficients, α, of MgB$_2$ and SiC are considered. It is reasonable to assume that both the MgB$_2$ and the SiC are in a stress-free state at the sintering temperature of 850°C due to the relatively high sintering temperature over a long period of time. However, the lattice parameters are determined by the thermal strain during the cooling process. Especially, the temperature dependences of the α values for MgB$_2$ and SiC are different. Figure 5.18(a) plots the $\alpha(T)$ for MgB$_2$ and SiC along the a- and c-axes, based on the data of Refs. [134, 148–150]. It clearly shows the weak temperature dependence $\alpha(T)$ for SiC in both directions, whereas the changes are great for MgB$_2$ and are characterized by high anisotropy. The averaged $\alpha(T)$ is also huge in MgB$_2$ as shown in the inset of Fig. 5.18(a). The α_{SiC} decreases slightly from 5×10^{-6}/K at 850°C to 2.5×10^{-6}/K at 0 K, whereas the α_{MgB_2} drops quickly from 1.7×10^{-5}/K at 1123 K to zero at 0 K. Based on $\alpha(T)$, the normalized lattice change and lattice strain in the MgB$_2$ matrix of SiC-MgB$_2$ composite during cooling from 1123 K to 0 K can be derived, as shown in Fig. 5.18(b). An assumption of Fig. 5.18(b) is that the two phases are strongly connected and the volume shrinkage of MgB$_2$ is confined by the relative stable SiC. The normalized lattice strain is estimated to be -0.55% in SiC-MgB$_2$ along the c-axis at room temperature. The negative value corresponds to tensile strain in the MgB$_2$. The large c-axis strain in the doped MgB$_2$ resulted in an enlargement in the c-axis by 0.15% in comparison with pure MgB$_2$. Estimated from Williamson-Hall model [151], the lattice strain is 0.208 and 0.306 in a-axis, and 0.292 and 1.13 in c-axis for pure and SiC-MgB$_2$, respectively. The lattice strain in c-axis in the SiC-MgB$_2$ increased from pure MgB$_2$ by a factor of 4, which

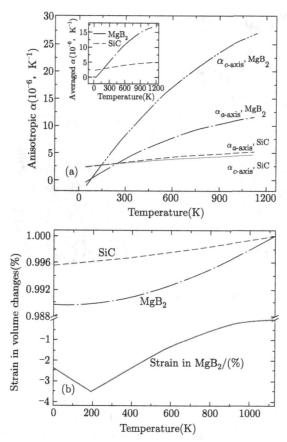

Fig. 5.18   (a) The thermal expansion coefficients [134, 148–150] (α) along the a-axis and c-axis for MgB$_2$ and SiC as a function of temperature. The averaged α(T) for MgB$_2$ and SiC are ploted in the inset. (b) Plots of the normalized lattice changes for MgB$_2$ and SiC, and the thermal strain in the matrix during cooling from 1123 K to 0 K.

is attributed to the high anisotropy in the thermal expansion coefficient of MgB$_2$.

Figure 5.19 shows a transmission electron microscope (TEM) image of an interface between SiC and MgB$_2$. Based on the fast Fourier transform (FFT) analysis, the interface is marked by a dashed line on the image. The left side is a SiC grain parallel to the [101] plane and the right side is an MgB$_2$ grain parallel to the [001] plane. This kind of interface will impose tensile stress along the c-axis in MgB$_2$, which is responsible for the enlarged c-axis parameter of MgB$_2$.

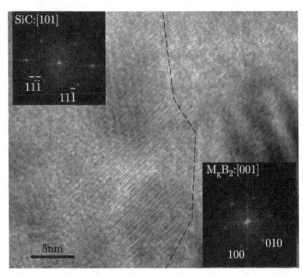

Fig. 5.19 High resolution TEM image of interface in SiC-MgB$_2$ and (inset) FFT
patterns of SiC and MgB$_2$ from each side of the interface. The dashed line indicates
the interface of SiC and MgB$_2$.

5.4.3 Superconducting performance

Based on the collective pinning model [132], J_c is independent of the
applied field in the single-vortex pinning regime (low magnetic field region:
$H < H_{sb}$), where H_{sb} is the crossover field from single-vortex to small-
bundle pinning. The J_c decreases exponentially in the small-bundle regime
(high magnetic field: $H_{sb} < H < H_{irr}$). According to the dual model [82],
the significant effect of SiC-doping on J_c comes from the high level of C
substitution on the B plane, which is responsible for the reduction of the
self-field J_c [127, 152]. However, the SiC-MgB$_2$ composite sample shows
not only an improved in-field J_c, but also no degradation in self-field J_c, as
indicated in Fig. 5.20. The approximate H_{sb} values are also indicated on
the J_c curves for 20 K and 30 K, although H_{sb} has not been detected at 5 K
due to the relatively high supercurrents. The *in situ* processed SiC-doped
MgB$_2$ normally shows a decrease in T_c of 1.5–2 K [82, 127, 152], but this
present SiC-MgB$_2$ composite sample shows a small drop of 0.6 K, as shown
in the inset of Fig. 5.20. This phenomenon is attributed to the absence of
any reaction between Mg and SiC, as well as the stretched MgB$_2$ lattice,
as indicated by the XRD pattern [153].

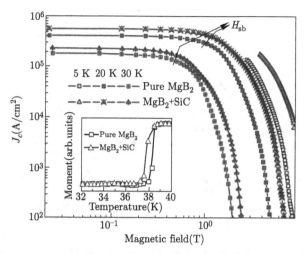

Fig. 5.20   The magnetic J_c versus field at 5 K, 20 K, and 30 K for pure and nano-SiC doped samples. The inset shows the superconducting transition of the two samples.

5.4.4   *Raman spectra and electron-phonon coupling*

To investigate whether the lattice strain is significant in SiC-MgB$_2$ during low temperature measurements to obtain $M(H)$ and $M(T)$ curves, Raman spectra were collected before and after the measurements with a confocal laser Raman spectrometer (Renishaw inVia plus) under a 100× microscope. The 514.5 nm line of an Ar$^+$ laser was used for excitation. The Raman spectra for pure MgB$_2$ are shown in Figs. 5.21(a) and (b) to compare the cooling effects on the matrix. Both the spectra have been fitted with three peaks: ω_1, ω_2, and ω_3 [126, 136, 154]. Based on the previous results, ω_2 is the reflection of the E_{2g} mode at the Γ point of the Brillouin zone in the simple hexagonal MgB$_2$ structure (space group: $P6/mmm$), while ω_1 and ω_3 come from the lattice distortion. The effects of ω_1 are not discussed in the following analysis because of its small influence on the spectra. As indicated by the fitting parameters that are shown in Fig. 5.21, both the peak centers and the FWHM values show negligible differences before and after the low-temperature measurements because of synchronic volume fluctuation. The weak temperature dependence of the Raman spectra for pure MgB$_2$ is in agreement with the results of Shi *et al.* [155]. The ω_2 peak of the Raman spectrum of SiC-MgB$_2$, before low temperature measurement, has shifted to the low frequency of 585 cm^{-1}, as shown in Fig. 5.22(a). The FWHM of the ω_2 peak increases from $\sim$200 cm^{-1} to 210 cm^{-1}. Furthermore, the

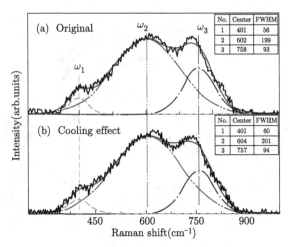

Fig. 5.21   Fitting and experimental results for the normalized ambient Raman spectrum of MgB$_2$ sintered at 850°C for 10 hrs (a), and the cooling effect on the Raman spectrum (b).

FWHM of the ω_3 peak increases from ~93 cm^{-1} to 125 cm^{-1}. The variations of both the Raman shift and the FWHM indicate the strong lattice strain in the SiC-MgB$_2$ composite. Figure 5.22(b) shows the cooling effect on the Raman spectrum of SiC-MgB$_2$. The FWHM of the ω_2 peak further increases to 228 cm^{-1}, and the frequency of ω_3 peak shifts to 770 cm^{-1}. These results suggest that the stress field is very strong during the low temperature measurements in the SiC-MgB$_2$ composite. Considering the stable defect structures in the sample at room temperature and the measurement temperatures, the high J_c performance is attributed to the thermal strain. Although the interface or grain boundaries themselves are effective flux-pinning centers, the thermal strain provides more efficient flux-pinning force, based on the comparison of the J_c values in pure MgB$_2$ and SiC-MgB$_2$ composite.

It should be noted that the broadened ω_2 peak in SiC-MgB$_2$ is a signal of strong electron-E_{2g} coupling, which is responsible for the high T_c in MgB$_2$. The electron-E_{2g} coupling constant is estimated by Eq. (5.7) and the $\lambda_{E_{2g}}$ values for the pure MgB$_2$ and SiC-MgB$_2$ are 2.327 and 2.706, respectively. The $\lambda_{E_{2g}}$ of SiC-MgB$_2$ is just slightly higher than that of the pure MgB$_2$. However, the T_c of SiC-MgB$_2$ is decreased slightly compared to that of the pure MgB$_2$. The total EPC constants are degraded by the

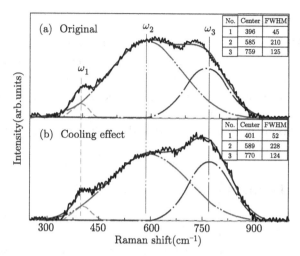

Fig. 5.22   Fitting and experimental results for the normalized ambient Raman spectrum of SiC-MgB$_2$ sintered at 850°C for 10 hrs (a), and the cooling effect on the Raman spectrum (b).

scattering effects of SiC impurities in the MgB$_2$ matrix, which can be estimated with Eq. (5.8). λ is calculated as 0.888 in pure MgB$_2$ and 0.886 in SiC-MgB$_2$, respectively. Although the values are very similar, the λ of MgB$_2$ is a little higher because of its low impurity scattering effects. The residual resistivity of SiC-MgB$_2$ is 16 $\mu\Omega \cdot$ cm, but it is just 12 $\mu\Omega \cdot$ cm for pure MgB$_2$, due to the weak impurity scattering effects.

5.4.5   *Section summary*

In summary, the thermal strain originating from the interface of SiC and MgB$_2$ is one of the most effective sources of flux-pinning centers to improve the supercurrent critical density. The weak temperature dependence of the thermal expansion coefficient of SiC stretches the MgB$_2$ lattice as the temperature decreases. The thermal strain supplies much more effective flux-pinning force than the interfaces and grain boundaries themselves. The low-temperature effects on Raman spectra show very strong lattice stretch at the application temperature of MgB$_2$, which benefits both the J_c and the T_c behaviors.

5.5 Conclusion

Based on the systematic study of the MgB_2 with high quality grain connectivity and strong flux-pinning centers, it is concluded that the techniques to optimize the solid state reaction of the *in situ* or *ex situ* processing of MgB_2 superconductors. The increased connectivity provides broad superconducting channels for the transfer of cooper pairs. On the other hand, the strong flux-pinning force restrains the decrease of J_c with improved magnetic field.

Acknowledgements

This work is financially supported by National Natural Science Foundation fo China (Grant No. 51572166) and the Shanghai Key Laboratory of High Temperature Superconductors (No. 14DZ226D700). The authors thank the Analysis and Research Center of Shanghai University for their technical support. W.X. Li acknowledges research support from the Program for Professors with Special Appointments (xxxxxxxxx) at Shanghai Institutions of Higher Learning.

References

[1] J. Bardeen, L. N. Cooper and J. R. Schrieffer, *Phys. Rev.* **108**, 1175 (1957).

[2] C. F. Richardson and N. W. Ashcroft, *Phys. Rev. Lett.* **78**, 118 (1997).

[3] J. Nagamatsu, N. Nakagawa, T. Muranaka, Y. Zenitani and J. Akimitsu, *Nature* **410**, 63 (2001).

[4] M. E. Jones and R. E. Marsh, *J. Am. Chem. Soc.* **76**, 1434 (1954).

[5] J. Kortus, I. I. Mazin, K. D. Belashchenko, V. P. Antropov and L. L. Boyer, *Phys. Rev. Lett.* **86**, 4656 (2001).

[6] W. L. McMillan, *Phys. Rev.* **167**, 331 (1968).

[7] D. Larbalestier, A. Gurevich, D. M. Feldmann, and A. Polyanskii, *Nature* **414**, 368 (2001).

[8] D. C. Larbalestier, L. D. Cooley, M. O. Rikel, A. A. Polyanskii, J. Jiang, S. Patnaik, X. Y. Cai, D. M. Feldmann, A. Gurevich, A. A. Squitieri, M. T. Naus, C. B. Eom, E. E. Hellstrom, R. J. Cava, K. A. Regan, N. Rogado, M. A. Hayward, T. He, J. S. Slusky, P. Khalifah, K. Inumaru and M. Haas, *Nature* **410**, 186 (2001).

[9] Y. Bugoslavsky, G. K. Perkins, X. Qi, L. F. Cohen and A. D. Caplin, *Nature* **410**, 563 (2001).

[10] V. Braccini, A. Gurevich, J. E. Giencke, M. C. Jewell, C. B. Eom, D. C. Larbalestier, A. Pogrebnyakov, Y. Cui, B. T. Liu, Y. F. Hu, J. M. Redwing, Q. Li, X. X. Xi, R. K. Singh, R. Gandikota, J. Kim, B. Wilkens,

N. Newman, J. Rowell, B. Moeckly, V. Ferrando, C. Tarantini, D. Marre, M. Putti, C. Ferdeghini, R. Vaglio and E. Haanappel, *Phys. Rev. B* **71**, 012504 (2005).

[11] S. X. Dou, S. Soltanian, J. Horvat, X. L. Wang, S. H. Zhou, M. Ionescu, H. K. Liu, P. Munroe and M. Tomsic, *Appl. Phys. Lett.* **81**, 3419 (2002).

[12] J. Chen, V. Ferrando, P. Orgiani, A. V. Pogrebnyakov, R. H. T. Wilke, J. B. Betts, C. H. Mielke, J. M. Redwing, X. X. Xi and Q. Li, *Phys. Rev. B* **74**, 174511 (2006).

[13] S. Jin, H. Mavoori, C. Bower and R. B. van Dover, *Nature* **411**, 563 (2001).

[14] G. Grasso, A. Malagoli, C. Ferdeghini, S. Roncallo, V. Braccini, A. S. Siri and M. R. Cimberle, *Appl. Phys. Lett.* **79**, 230 (2001).

[15] Webpage: http://www.columbussuperconductors.com/press.htm.

[16] J. Rowell, *Nat. Mater.* **1**, 5 (2002).

[17] K. Chen, Y. Cui, Q. Li, X. X. Xi, S. A. Cybart, R. C. Dynes, X. Weng, E. C. Dickey and J. M. Redwing, *Appl. Phys. Lett.* **88**, 222511 (2006).

[18] H. Shim, K. S. Yoon, J. S. Moodera and J. P. Hong, *Appl. Phys. Lett.* **90**, 212509 (2007).

[19] T. Ishida, M. Nishikawa, Y. Fujita, S. Okayasu, M. Katagiri, K. Satoh, T. Yotsuya, H. Shimakage, S. Miki, Z. Wang, M. Machida, T. Kano and M. Kato, *J. Low Temp. Phys.* **151**, 1074 (2008).

[20] H. Padamsee, *2004 Applied Superconductivity Conference* **15**, 2432 (2004).

[21] E. W. Collings, M. D. Sumption and T. Tajima, *Supercond. Sci. Technol.* **17**, S595 (2004).

[22] A. Gurevich, *Physica C* **456**, 160 (2007).

[23] A. Gurevich, *Phys. Rev. B* **67**, 184515 (2003).

[24] S. Patnaik, L. D. Cooley, A. Gurevich, A. A. Polyanskii, J. Jing, X. Y. Cai, A. A. Squitieri, M. T. Naus, M. K. Lee, J. H. Choi, L. Belenky, S. D. Bu, J. Letteri, X. Song, D. G. Schlom, S. E. Babcock, C. B. Eom, E. E. Hellstrom and D. C. Larbalestier, *Supercond. Sci. Techn.* **14**, 315 (2001).

[25] C. B. Eom, M. K. Lee, J. H. Choi, L. J. Belenky, X. Song, L. D. Cooley, M. T. Naus, S. Patnaik, J. Jiang, M. Rikel, A. Polyanskii, A. Gurevich, X. Y. Cai, S. D. Bu, S. E. Babcock, E. E. Hellstrom, D. C. Larbalestier, N. Rogado, K. A. Regan, M. A. Hayward, T. He, J. S. Slusky, K. Inumaru, M. K. Haas and R. J. Cava, *Nature* **411**, 558 (2001).

[26] E. W. Collings, M. D. Sumption, M. Bhatia, M. A. Susner and S. D. Bohnenstiehl, *Supercond. Sci. Techn.* **21**, 103001 (2008).

[27] S. X. Dou, W. K. Yeoh, J. Horvat and M. Ionescu, *Appl. Phys. Lett.* **83**, 4996 (2003).

[28] W. X. Li, Y. Li, R. H. Chen, W. K. Yeoh and S. X. Dou, *Physica C* **460**, 570 (2007).

[29] W. K. Yeoh, J. Horvat, S. X. Dou and P. Munroe, *IEEE Trans. Appl. Supercond.* **15**, 3284 (2005).

[30] W. X. Li, Y. Li, M. Y. Zhu, R. H. Chen, X. Xu, W. K. Yeoh, J. H. Kim and S. X. Dou, *IEEE Trans. Appl. Supercond.* **17**, 2778 (2007).

[31] M. Tinkham, *Introduction to Superconductivity*, 2nd ed, New York: McGraw-Hill (1996).

[32] E. J. Nicol and J. P. Carbotte, *Phys. Rev. B* **72**, 014520 (2005).

[33] D. H. Arcos and M. N. Kunchur, *Phys. Rev. B* **71**, 184516 (2005).

[34] D. K. Finnemore, J. E. Ostenson, S. L. Bud'ko, G. Lapertot and P. C. Canfield, *Phys. Rev. Lett.* **86**, 2420 (2001).

[35] K. Kawano, J. S. Abell, M. Kambara, N. H. Babu and D. A. Cardwell, *Appl. Phys. Lett.* **79**, 2216 (2001).

[36] J. M. Rowell, *Supercond. Sci. Technol.* **16**, R17 (2003).

[37] R. F. Klie, J. C. Idrobo, N. D. Browning, K. A. Regan, N. S. Rogado and R. J. Cava, *Appl. Phys. Lett.* **79**, 1837 (2001).

[38] A. V. Pogrebnyakov, X. X. Xi, J. M. Redwing, V. Vaithyanathan, D. G. Schlom, A. Soukiassian, S. B. Mi, C. L. Jia, J. E. Giencke, C. B. Eom, J. Chen, Y. F. Hu, Y. Cui and Q. Li, *Appl. Phys. Lett.* **85**, 2017 (2004).

[39] C. Fischer, W. Hassler, C. Rodig, O. Perner, G. Behr, M. Schubert, K. Nenkov, J. Eckert, B. Holzapfel and L. Schultz, *Physica C* **406**, 121 (2004).

[40] W. Hassler, B. Birajdar, W. Gruner, M. Herrmann, O. Perner, C. Rodig, M. Schubert, B. Holzapfel, O. Eibl and L. Schultz, *Supercond. Sci. Technol.* **19**, 512 (2006).

[41] W. K. Yeoh, J. H. Kim, J. Horvat, S. X. Dou and P. Munroe, *Supercond. Sci. Technol.* **19**, L5 (2006).

[42] J. M. Rowell, S. Y. Xu, H. Zeng, A. V. Pogrebnyakov, Q. Li, X. X. Xi, J. M. Redwing, W. Tian and X. Q. Pan, *Appl. Phys. Lett.* **83**, 102 (2003).

[43] M. Eisterer, *Supercond. Sci. Technol.* **20**, R47 (2007).

[44] K. H. Kim, J. B. Betts, M. Jaime, A. H. Lacerda, G. S. Boebinger, C. U. Jung, H. J. Kim, M. S. Park and S. I. Lee, *Phys. Rev. B* **66**, 020506 (2002).

[45] P. A. Sharma, N. Hur, Y. Horibe, C. H. Chen, B. G. Kim, S. Guha, M. Z. Cieplak and S. W. Cheong, *Phys. Rev. Lett.* **89**, 167003 (2002).

[46] M. Eisterer, C. Krutzler and H. W. Weber, *J. Appl. Phys.* **98**, 033906 (2005).

[47] D. Stauffer and A. Aharony, *Introduction to Percolation Theory*, London: Taylor and Francis (1992).

[48] S. C. van der Marck, *Phys. Rev. E* **55**, 1514 (1997).

[49] M. Eisterer, M. Zehetmayer and H. W. Weber, *Phys. Rev. Lett.* **90**, 247002 (2003).

[50] S. W. Kenkel and J. P. Straley, *Phys. Rev. Lett.* **49**, 767 (1982).

[51] H. Kumakura, H. Kitaguchi, A. Matsumoto and H. Yamada, *Supercond. Sci. Technol.* **18**, 1042 (2005).

[52] T. Nakane, C. H. Jiang, T. Mochiku, H. Fujii, T. Kuroda and H. Kumakura, *Supercond. Sci. Technol.* **18**, 1337 (2005).

[53] S. Hata, T. Yoshidome, H. Sosiati, Y. Tomokiyo, N. Kuwano, A. Matsumoto, H. Kitaguchi and H. Kumakura, *Supercond. Sci. Technol.* **19**, 161 (2006).

[54] C. H. Jiang, H. Hatakeyama and H. Kumakura, *Supercond. Sci. Technol.* **18**, L17 (2005).

[55] H. Fujii, K. Togano and H. Kumakura, *Supercond. Sci. Technol.* **15**, 1571 (2002).

[56] C. H. Jiang, T. Nakane, H. Hatakeyama and H. Kumakura, *Physica C* **422**, 127 (2005).

[57] C. H. Jiang, H. Hatakeyama and H. Kumakura, *Physica C* **423**, 45 (2005).

[58] A. Matsumoto, H. Kumakura, H. Kitaguchi and H. Hatakeyama, *Supercond. Sci. Technol.* **16**, 926 (2003).

[59] W. Pachla, A. Morawski, P. Kovac, I. Husek, A. Mazur, T. Lada, R. Diduszko, T. Melisek, V. Strbik and M. Kulczyk, *Supercond. Sci. Technol.* **19**, 1 (2006).

[60] H. Yamada, M. Hirakawa, H. Kumakura and H. Kitaguchi, *Supercond. Sci. Technol.* **19**, 175 (2006).

[61] W. Goldacker, S. I. Schlachter, B. Obst, B. Liu, J. Reiner and S. Zimmer, *Supercond. Sci. Technol.* **17**, S363 (2004).

[62] A. Matsumoto, H. Kumakura, H. Kitaguchi, B. J. Senkowicz, M. C. Jewell, E. E. Hellstrom, Y. Zhu, P. M. Voyles and D. C. Larbalestier, *Appl. Phys. Lett.* **89**, 132508 (2006).

[63] A. Yamamoto, J. Shimoyama, S. Ueda, I. Iwayama, S. Horii and K. Kishio, *Supercond. Sci. Technol.* **18**, 1323 (2005).

[64] O. Perner, W. Habler, R. Eckert, C. Fischer, C. Mickel, G. Fuchs, B. Holzapfel and L. Schultz, *Physica C* **432**, 15 (2005).

[65] C. H. Jiang, T. Nakane and H. Kumakura, *Appl. Phys. Lett.* **87**, 252505 (2005).

[66] Y. F. Wu, Y. F. Lu, G. Yan, J. S. Li, Y. Feng, H. P. Tang, S. K. Chen, H. L. Xu, C. S. Li and P. X. Zhang, *Supercond. Sci. Technol.* **19**, 1215 (2006).

[67] J. H. Kim, W. K. Yeoh, M. J. Qin, X. Xu and S. X. Dou, *J. Appl. Phys.* **100**, 013908 (2006).

[68] J. H. Kim, S. Zhou, M. S. A. Hossain, A. V. Pan and S. X. Dou, *Appl. Phys. Lett.* **89**, 142505 (2006).

[69] S. K. Chen, Z. Lockman, M. Wei, B. A. Glowacki and J. L. MacManus-Driscoll, *Appl. Phys. Lett.* **86**, 242501 (2005).

[70] S. Ueda, J. Shimoyama, I. Iwayama, A. Yamamoto, Y. Katsura, S. Horii and K. Kishio, *Appl. Phys. Lett.* **86**, 222502 (2005).

[71] X. P. Zhang, Z. S. Gao, D. L. Wang, Z. G. Yu, Y. W. Ma, S. Awaji and K. Watanabe, *Appl. Phys. Lett.* **89**, 132510 (2006).

[72] O. Shcherbakova, S. X. Dou, S. Soltanian, D. Wexler, M. Bhatia, M. Sumption and E. W. Collings, *J. Appl. Phys.* **99**, 08M510 (2006).

[73] Y. W. Ma, X. P. Zhang, G. Nishijima, K. Watanabe, S. Awaji and X. D. Bai, *Appl. Phys. Lett.* **88**, 072502 (2006).

[74] R. A. Ribeiro, S. L. Bud'ko, C. Petrovic and P. C. Canfield, *Physica C* **385**, 16 (2003).

[75] X. Z. Liao, A. Serquis, Y. T. Zhu, D. E. Peterson, F. M. Mueller and H. F. Xu, *Supercond. Sci. Technol.* **17**, 1026 (2004).

[76] O. Perner, J. Eckert, W. Hassler, C. Fischer, J. Acker, T. Gemming, G. Fuchs, B. Holzapfel and L. Schultz, *J. Appl. Phys.* **97**, 056105 (2005).

[77] A. Serquis, Y. T. Zhu, E. J. Peterson, J. Y. Coulter, D. E. Peterson and F. M. Mueller, *Appl. Phys. Lett.* **79**, 4399 (2001).

[78] H. Yamada, M. Hirakawa, H. Kumakura, A. Matsumoto and H. Kitaguchi, *Appl. Phys. Lett.* **84**, 1728 (2004).

[79] H. Fang, S. Padmanabhan, Y. X. Zhou and K. Salama, *Appl. Phys. Lett.* **82**, 4113 (2003).

[80] C. Fischer, C. Rodig, W. Hassler, O. Perner, J. Eckert, K. Nenkov, G. Fuchs, H. Wendrock, B. Holzapfel and L. Schultz, *Appl. Phys. Lett.* **83**, 1803 (2003).

[81] N. M. Strickland, R. G. Buckley and A. Otto, *Appl. Phys. Lett.* **83**, 326 (2003).

[82] S. X. Dou, O. Shcherbakova, W. K. Yeoh, J. H. Kim, S. Soltanian, X. L. Wang, C. Senatore, R. Flukiger, M. Dhalle, O. Husnjak and E. Babic, *Phys. Rev. Lett.* **98**, 097002. (2007).

[83] Y. Zhao, Y. Feng, T. M. Shen, G. Li, Y. Yang and C. H. Cheng, *J. Appl. Phys.* **100**, 123902 (2006).

[84] X. Xu, J. H. Kim, W. K. Yeoh, Y. Zhang and S. X. Dou, *Supercond. Sci. Technol.* **19**, L47 (2006).

[85] S. Haigh, P. Kovac, T. A. Prikhna, Y. M. Savchuk, M. R. Kilburn, C. Salter, J. Hutchison and C. Grovenor, *Supercond. Sci. Technol.* **18**, 1190 (2005).

[86] Y. W. Ma, H. Kumakura, A. Matsumoto, H. Hatakeyama and K. Togano, *Supercond. Sci. Technol.* **16**, 852 (2003).

[87] D. Kumar, S. J. Pennycook, J. Narayan, H. Wang and A. Tiwari, *Supercond. Sci. Technol.* **16**, 455 (2003).

[88] A. Yamamoto, J. Shimoyama, S. Ueda, Y. Katsura, I. Iwayama, S. Horii and K. Kishio, *Appl. Phys. Lett.* **86**, 212502 (2005).

[89] P. Lezza, C. Senatore and R. Flukiger, *Supercond. Sci. Technol.* **19**, 1030 (2006).

[90] W. K. Yeoh, J. H. Kim, J. Horvat, X. Xu, M. J. Qin, S. X. Dou, C. H. Jiang and T. Nakane, *Supercond. Sci. Technol.* **19**, 596 (2006).

[91] J. H. Kim, W. K. Yeoh, M. J. Qin, X. Xu, S. X. Dou, P. Munroe and H. Kumakura, *Appl. Phys. Lett.* **89**, 122510 (2006).

[92] P. Kovac, I. Husek, V. Skakalova, J. Meyer, E. Dobrocka, M. Hirscher and S. Roth, *Supercond. Sci. Technol.* **20**, 105 (2007).

[93] C. H. Cheng, Y. Yang, P. Munroe and Y. Zhao, *Supercond. Sci. Technol.* **20**, 296 (2007).

[94] C. H. Cheng, H. Zhang, Y. Zhao, Y. Feng, X. F. Rui, P. Munroe, H. M. Zeng, N. Koshizuka and M. Murakami, *Supercond. Sci. Technol.* **16**, 1182 (2003).

[95] Z. S. Gao, X. P. Zhang, D. L. Wang, X. Liu, X. H. Li, Y. W. Ma and E. Mossang, *Supercond. Sci. Technol.* **20**, 57 (2007).

[96] S. Soltanian, X. L. Wang, J. Horvat, S. X. Dou, M. D. Sumption, M. Bhatia, E. W. Collings, P. Munroe and M. Tomsic, *Supercond. Sci. Technol.* **18**, 658 (2005).

[97] S. K. Chen, K. S. Tan, B. A. Glowacki, W. K. Yeoh, S. Soltanian, J. Horvat and S. X. Dou, *Appl. Phys. Lett.* **87**, 182504 (2005).

[98] S. X. Dou, V. Braccini, S. Soltanian, R. Klie, Y. Zhu, S. Li, X. L. Wang and D. Larbalestier, *J. Appl. Phys.* **96**, 7549 (2004).

[99] Y. W. Ma, X. P. Zhang, A. X. Xu, X. H. Li, L. Y. Xiao, G. Nishijima, S. Awaji, K. Watanabe, Y. L. Jiao, L. Xiao, X. D. Bai, K. H. Wu and H. H. Wen, *Supercond. Sci. Technol.* **19**, 133 (2006).

[100] M. D. Sumption, M. Bhatia, M. Rindfleisch, M. Tomsic and E. W. Collings, *Appl. Phys. Lett.* **86**, 102501 (2005).

[101] H. Kumakura, H. Kitaguchi, A. Matsumoto and H. Hatakeyama, *Appl. Phys. Lett.* **84**, 3669 (2004).

[102] S. Li, T. White, K. Laursen, T. T. Tan, C. Q. Sun, Z. L. Dong, Y. Li, S. H. Zhou, J. Horvat and S. X. Dou, *Appl. Phys. Lett.* **83**, 314 (2003).

[103] M. D. Sumption, M. Bhatia, M. Rindfleisch, M. Tomsic, S. Soltanian, S. X. Dou and E. W. Collings, *Appl. Phys. Lett.* **86**, 092507 (2005).

[104] M. S. A. Hossain, J. H. Kim, X. L. Wang, X. Xu, G. Peleckis and S. X. Dou, *Supercond. Sci. Technol.* **20**, 112 (2007).

[105] X. Z. Liao, A. Serquis, Y. T. Zhu, L. Civale, D. L. Hammon, D. E. Peterson, F. M. Mueller, V. F. Nesterenko and Y. Gu, *Supercond. Sci. Technol.* **16**, 799 (2003).

[106] Z. S. Gao, Y. W. Ma, X. P. Zhang, D. L. Wang, Z. G. Yu, K. Watanabe, H. A. Yang and H. H. Wen, *Supercond. Sci. Technol.* **20**, 485 (2007).

[107] H. H. Wen, S. L. Li, Z. W. Zhao, H. Jin, Y. M. Ni, Z. A. Ren, G. C. Che and Z. X. Zhao, *Supercond. Sci. Technol.* **15**, 315 (2002).

[108] T. A. Prikhna, W. Gawalek, Y. M. Savchuk, V. E. Moshchil, N. V. Sergienko, T. Habisreuther, M. Wendt, R. Hergt, C. Schmidt, J. Dellith, V. S. Melnikov, A. Assmann, D. Litzkendorf and P. A. Nagorny, *Physica C* **402**, 223 (2004).

[109] A. Serquis, X. Z. Liao, Y. T. Zhu, J. Y. Coulter, J. Y. Huang, J. O. Willis, D. E. Peterson, F. M. Mueller, N. O. Moreno, J. D. Thompson, V. F. Nesterenko and S. S. Indrakanti, *J. Appl. Phys.* **92**, 351 (2002).

[110] A. Serquis, L. Civale, D. L. Hammon, X. Z. Liao, J. Y. Coulter, Y. T. Zhu, M. Jaime, D. E. Peterson, F. M. Mueller, V. F. Nesterenko and Y. Gu, *Appl. Phys. Lett.* **82**, 2847 (2003).

[111] D. Eyidi, O. Eibl, T. Wenzel, K. G. Nickel, S. I. Schlachter and W. Goldacker, *Supercond. Sci. Technol.* **16**, 778 (2003).

[112] A. V. Pan, S. H. Zhou, H. K. Liu and S. X. Don, *Supercond. Sci. Technol.* **16**, 639 (2003).

[113] A. Serquis, L. Civale, D. L. Hammon, J. Y. Coulter, X. Z. Liao, Y. T. Zhu, D. E. Peterson and F. M. Mueller, *Appl. Phys. Lett.* **82**, 1754 (2003).

[114] R. Flukiger, H. L. Suo, N. Musolino, C. Beneduce, P. Toulemonde and P. Lezza, *Physica C* **387**, 419 (2003).

[115] W. Goldacker, S. I. Schlachter, B. Liu, B. Obst and E. Klimenko, *Physica C* **401**, 80 (2004).

[116] T. Nakane, H. Kitaguchi and H. Kumakura, *Appl. Phys. Lett.* **88**, 022513 (2006).

[117] W. Pachla, A. Presz, R. Diduszko, P. Kovac and I. Husek, *Supercond. Sci. Technol* **15**, 1281 (2002).

[118] R. Flukiger, P. Lezza, C. Beneduce, N. Musolino and H. L. Suo, *Supercond. Sci. Technol.* **16**, 264 (2003).

[119] R. Flukiger, H. L. Suo, N. Musolino, C. Beneduce, P. Toulemonde and P. Lezza, *Physica C* **385**, 286 (2003).

[120] C. R. M. Grovenor, L. Goodsir, C. J. Salter, P. Kovac and I. Husek, *Supercond. Sci. Technol.* **17**, 479 (2004).

[121] B. J. Senkowicz, J. E. Giencke, S. Patnaik, C. B. Eom, E. E. Hellstrom and D. C. Larbalestier, *Appl. Phys. Lett.* **86**, 202502 (2005).

[122] A. Serquis, L. Civale, D. L. Hammon, X. Z. Liao, J. Y. Coulter, Y. T. Zhu, D. E. Peterson and F. M. Mueller, *J. Appl. Phys.* **94**, 4024 (2003).

[123] H. L. Suo, C. Beneduce, M. Dhalle, N. Musolino, J. Y. Genoud and R. Flukiger, *Appl. Phys. Lett.* **79**, 3116 (2001).

[124] C. G. Zhuang, S. Meng, C. Y. Zhang, Q. R. Feng, Z. Z. Gan, H. Yang, Y. Jia, H. H. Wen and X. X. Xi, *J. Appl. Phys.* **104**, 013924 (2008).

[125] X. H. Zeng, A. V. Pogrebnyakov, M. H. Zhu, J. E. Jones, X. X. Xi, S. Y. Xu, E. Wertz, Q. Li, J. M. Redwing, J. Lettieri, V. Vaithyanathan, D. G. Schlom, Z. K. Liu, O. Trithaveesak and J. Schubert, *Appl. Phys. Lett.* **82**, 2097 (2003).

[126] W. X. Li, Y. Li, R. H. Chen, R. Zeng, S. X. Dou, M. Y. Zhu and H. M. Jin, *Phys. Rev. B* **77**, 094517 (2008).

[127] W. X. Li, R. Zeng, L. Lu, Y. Li and S. X. Dou, The combined influence of connectivity and disorder on J_c and T_c performances in $Mg_xB_2 + 10$ $wt\%$ SiC, *J. Appl. Phys.* **106**, 093906 (2009).

[128] R. Flukiger, M. S. A. Hossain and C. Senatore, Strong enhancement of J_c and B_{irr} in binary in situ MgB_2 wires after cold high pressure densification, *Supercond. Sci. Technol.* **22**, 085002 (2009).

[129] M. S. A. Hossain, C. Senatore, R. Flukiger, M. A. Rindfleisch, M. J. Tomsic, J. H. Kim and S. X. Dou, *Supercond. Sci. Technol.* **22**, 095004 (2009).

[130] W. X. Li, R. H. Chen, Y. Li, M. Y. Zhu, H. M. Jin, R. Zeng, S. X. Dou and B. Lu, *J. Appl. Phys.* **103**, 013511 (2008).

[131] S. Lee, T. Masui, A. Yamamoto, H. Uchiyama and S. Tajima, *Physica C* **397**, 7 (2003).

[132] G. Blatter, M. V. Feigelman, V. B. Geshkenbein, A. I. Larkin and V. M. Vinokur, *Rev. Mod. Phys.* **66**, 1125 (1994).

[133] C. Buzea and T. Yamashita, *Supercond. Sci. Technol.* **14**, R115 (2001).

[134] D. Dew-Hughes, *Philos. Mag.* **30**, 293 (1974).

[135] W. X. Li, Y. Li, R. H. Chen, R. Zeng, M. Y. Zhu, H. M. Jin and S. X. Dou, *J. Phys. Condens. Matter* **20**, 255235 (2008).

[136] K. Kunc, I. Loa, K. Syassen, R. K. Kremer and K. Ahn, *J. Phys. Condens. Matter* **13**, 9945 (2001).

[137] R. Zeng, S. X. Dou, L. Lu, W. X. Li, J. H. Kim, P. Munroe, R. K. Zheng and S. P. Ringer, *Appl. Phys. Lett.* **94**, 042510 (2009).

[138] P. B. Allen, *Phys. Rev. B* **6**, 2577 (1972).

[139] J. Kortus, O. V. Dolgov, R. K. Kremer and A. A. Golubov, *Phys. Rev. Lett.* **94**, 027002 (2005).

[140] G. A. Ummarino, D. Daghero, R. S. Gonnelli and A. H. Moudden, *Phys. Rev. B* **71**, 134511 (2005).

[141] A. Shukla, M. Calandra, M. d'Astuto, M. Lazzeri, F. Mauri, C. Bellin, M. Krisch, J. Karpinski, S. M. Kazakov, J. Jun, D. Daghero and K. Parlinski, *Phys. Rev. Lett.* **90**, 095506 (2003).

[142] P. B. Allen and R. C. Dynes, *Phys. Rev. B* **12**, 905 (1975).

[143] R. Osborn, E. A. Goremychkin, A. I. Kolesnikov and D. G. Hinks, Phonon density of states in MgB_2, *Phys. Rev. Lett.* **87**, 017005 (2001).

[144] A. Brinkman, A. A. Golubov, H. Rogalla, O. V. Dolgov, J. Kortus, Y. Kong, O. Jepsen and O. K. Andersen, *Phys. Rev. B* **65**, 180517 (2002).

[145] X. Xu, J. H. Kim, S. X. Dou, S. Choi, J. H. Lee, H. W. Park, M. Rindfleish and M. Tomsic, *J. Appl. Phys.* **105**, 103913 (2009).

[146] G. Romano, M. Vignolo, V. Braccini, A. Malagoli, C. Bernini, M. Tropeano, C. Fanciulli, M. Putti and C. Ferdeghini, *IEEE Trans. Appl. Supercond.* **19**, 2706 (2009).

[147] J. J. Neumeier, T. Tomita, M. Debessai, J. S. Schilling, P. W. Barnes, D. G. Hinks and J. D. Jorgensen, *Phys. Rev. B* **72**, 220505 (2005).

[148] J. D. Jorgensen, D. G. Hinks and S. Short, *Phys. Rev. B* **6322**, 224522 (2001).

[149] Z. Li and R. C. Bradt, *J. Appl. Phys.* **60**, 612 (1986).

[150] G. K. Williamson and W. H. Hall, *Acta Metall.* **1**, 22 (1953).

[151] W. X. Li, R. Zeng, L. Lu, Y. Zhang, S. X. Dou, Y. Li, R. H. Chen and M. Y. Zhu, *Physica C*, **469**, 1519 (2009).

[152] A. V. Pogrebnyakov, J. M. Redwing, S. Raghavan, V. Vaithyanathan, D. G. Schlom, S. Y. Xu, Q. Li, D. A. Tenne, A. Soukiassian, X. X. Xi, M. D. Johannes, D. Kasinathan, W. E. Pickett, J. S. Wu and J. C. H. Spence, *Phys. Rev. Lett.* **93**, 147006. (2004).

[153] W. X. Li, R. Zeng, C. K. Poh, Y. Li and S. X. Dou, *J. Phys. Condens. Matter* **22**, 135701 (2010).

[154] L. Shi, H. R. Zhang, L. Chen and Y. Feng, *J. Phys. Condens. Matter* **16**, 6541 (2004).

Part III

Applications

6

High-Temperature Superconductive Maglev Vehicle

J. S. Wang[1] and S. Y. Wang[1]

6.1 Introduction

The levitation of a NdFeB permanent magnet of volume 0.7 cm^3 above a disk of YBCO bulk superconductor of 2.5 cm diameter, 0.6 cm thickness bathed in liquid nitrogen was observed by Hellman *et al.* [1]. Peter *et al.* [2] observed the very stable suspension of YBCO samples in the magnetic field below the permanent magnet. Soon after, Brandt [3] presented that the high-temperature superconducting (HTS) bulk magnetic levitation (maglev) could realize stable levitation without any active control. Since the discovery of the HTS maglev phenomenon, a lot of progress has been made in the theory and application of HTS maglev, especially in the fields of energy storage [4] and transportation [5]. There are some comprehensive review papers on HTS maglev which can be found elsewhere [6–9]. Special comments on the HTS maglev vehicle can be found in literature [10–13] written by the present authors.

After the development of high speed information technology, the human race dream off "door to door" high speed transport system just like modern information technology. The conventional automobile-centric transport system has many disadvantages such as traffic jam, traffic accidents, energy consumption, pollution etc. Therefore, transportation using tracks is considered one of the mainstream transport modes in the 21st century. Traditional railway transport, normal conductor maglev trains in Germany and the low-

---

[1]Applied Superconductivity Laboratory, Southwest Jiaotong University, China.

temperature superconducting (LTS) maglev trains in Japan did not solve the future problems of ground transportation at high speed or super-high speed. In the constant pursuit of the "perfect" mode of transportation, the HTS maglev train is one of the best candidates.

The HTS maglev train is a mode of transportation whose suspension and guidance do not require on-board power, and it is a self-stabilization system that does not require active control. The HTS maglev system can integrate levitation, guidance and superconducting linear motor propulsion. The vehicle in the evacuate tube transportation (ETT) can achieve super-high speeds (thousands of kilometers per hour) and it is economical in energy consumption (5% of an airplane's energy consumption), environmentally friendly (no chemical nor noise pollution), safe and comfortable.

In 1997, a cooperation project between China and Germany had led to the development of a small HTS maglev model of 20 kg with a levitation gap of 7 mm [14]. International research zest on the man-loading HTS maglev vehicle was aroused after the first man-loading HTS maglev test vehicle in the world, "Century", was built in the Applied Superconductivity Laboratory (ASCLab) of the Southwest Jiaotong University (SWJTU) at 2:26 pm on December 31st, 2000 [5]. The second and third man-loading HTS maglev vehicles were developed in Germany [15] and Russia [16] respectively in 2004. The full scale HTS maglev train, as a substitute for urban light track train, is under development in Brazil [17, 18]. Moreover, research groups in Japan [19] and Italy [20] etc. also developed HTS maglev prototypes.

The technology of electromagnetic launch "will have a significant and far-reaching impact for a country's national defense, space planning, and energy". The self-stability of the HTS maglev made it the best candidate for repeatable (cold) launch systems compared to conventional guns and normal conducting maglev. Especially, the traits of low energy loss, high energy density, quick discharge, small size, and light weight superconducting maglev FESS (flywheel energy storage system), will solve the problems to do with high power pulse supply in the electromagnetic launch system. The speed of the launch system using this high efficiency FESS pulse power can reach upto several kilometers per second.

The project of the HTS maglev vehicle was ratified by the National High-Technology R&D program of China (863 Program) in 1997. The first man-loading HTS maglev test vehicle "Century" in the world was successfully developed in 2000. Since then several other prototypes have been built all around the world, making this research field extremely exciting.

Thereafter, both experimental and theoretical studies of the HTS maglev have been made in detail. A series of important achievements have been obtained. On the foundation of these research results, a new full-scale HTS maglev prototype is under development.

This chapter will introduce the key properties of HTS maglev vehicle at first, the HTS maglev vehicle project as part of the 863 Program, as well as support and attention from various experts and organizations. Then, the research and development process on "Century" vehicle before and after its creation are reviewed. Lastly, research on the moderated/low-speed HTS maglev train, super-high-speed HTS maglev train, and HTS maglev launch system are presented.

More books and review articles about HTS maglev vehicles can be found in Refs. [10, 11, 21–23].

6.2   Rail transit–the mainstream of 21st century transportation

Automobiles produce pollution in amounts 30 times higher than that of trains. The total area of the highway is four times that of the railway. The numbers of car and airplane accidents are, respectively, 8.6 times and 7.1 times the number of railway accidents in Japan in 1985. The mortality rate of the Japanese Shinkansen is zero in its 34 years of operation. In 1991, in 37 American cities each with a population over one million, 91% of cars experienced traffic jams, leading to an economic loss of $100 billion. The mortality rates (number of people per ten billion person kilometers) of cars, planes, and trains were 501, 5.31, and 0.78, respectively [24]. Stretching back to the beginning of records on automobile road traffic fatalities, the total number of people worldwide killed in road traffic accidents is more than 22 million, which is close to the 26 million soldiers killed in World War II [25]. Transportation tax is based on environmental (pollution, noise) cost, thus it is competitive to increase railway usage relative to highway and aviation usage.

More than 200 countries worldwide will construct an urban railway transit, and the vast majorities are funded by the governments. Rail transportation is the mainstay of 21st century transportation.

When the velocity of ground railway transportation is higher than 350 km/h, the train consumes a large amount of energy and produces great noise. Therefore, in March 1998, the International Railway Association of Innovation proposed that the maximum speed of ground transportation should be 350 km/h. High-speed ground transport by rail should use ETT.

The energy consumption of maglev vehicles is 28% of that of airplanes at the speed of 480 km/h, but only 5% at the speed of 3200 km/h. Hence, super-high speed transportation is necessary in order to increase the energy efficiency. As wheel-rail high-speed railway trains did not solve the problem of the super-high speed ground traffic, the maglev can be the best choice. In 1998, the United States announced the Transportation Equity Act TEA-21 for the 21st Century to provide the legal form for the development of maglev transportation technology in the United States, and a maglev line of 969 km was planned and reviewed.

6.3   Maglev train

Magnetic levitation includes electromagnetic suspension (EMS) and electrodynamic suspension (EDS). Maglev trains can be superconducting or non-superconducting. There are two types of superconducting maglev trains: the low-temperature superconducting maglev train (with liquid helium, 4.2 K) and the high-temperature superconducting (with liquid nitrogen, 77 K) maglev train.

The German "Transrapid" is a normal conducting (non-super-conducting) maglev train. Both the levitation forces and the guidance forces are provided by the attraction between the ground rail track and the on-board winding magnets, which requires on-board power supply and complex active control. The active control (100 ms delay control) is responsible for its stability of levitation and guidance. The shortcomings of this maglev train include a heavier vehicle body, higher energy consumption, lower magnetic field (0.4 T), lower levitation height (about 8 mm), and the higher precision requirement of roadbed and track. As a result, it is not suitable for super-high speed.

The LTS maglev train was developed in Japan. It uses the winding magnet of LTS Nb-Ti wire. The LTS magnets produce a stronger magnetic field (5.0 T) and higher levitation height (100 mm). However, its shortcomings include higher manufacturing and operational costs due to its super-low operational temperatures (4.2 K, liquid helium), and the train is not suspended in an equilibrium state (similar to the take off of an aircraft).

Permanent magnets can also provide levitation, but this levitation is not stable.

A 1 G (bismuth (Bi) series) HTS wire generates relatively low magnetic fields at liquid nitrogen temperature (77 K), so it is not suitable for use in maglev train systems. A 2 G (yttrium (Y) series) HTS wire maglev magnet [26] generates high magnetic field at liquid nitrogen temperature (77 K), but both the levitation and guidance must be the same as the normal conducting maglev train: the complex active control system is indispensable. Similarly, a hybrid HTS maglev vehicle assembled with HTS bulk, wire, HTS permanent magnets, etc. needs a more complex active control system.

Levitation methods that need active control are not suitable for the high speed maglev train system, especially for super-high speed ground transportation, and they are not suitable for the launch system.

The principle of maglev vehicles using HTS bulk is completely different from all of the methods mentioned above, the self-stable levitation system spares automatic control systems. The HTS maglev system is compatible with the HTS linear motor, therefore, it can integrate levitation, guidance and propulsion. The power of the HTS linear motor is provided by the HTS maglev flywheel energy storage system (FESS), which will be fully applied in HTS maglev launch systems with ultra-high speed, low power and high efficiency.

Table 6.1 lists the research and development history of the HTS maglev vehicle in the world.

6.4 Advantages of inherent self-stable maglev systems

The interaction between a HTS bulk and magnetic field provided by a magnet can generate a levitation force. The magnetic levitation of the HTS bulks is a self-stable levitation system, which does not require active control. It is a unique, inherently self-stable system. The HTS maglev vehicle is truly an energy-efficient, environmentally friendly, safe, and comfortable mode of ground transportation. Specific advantages include:

(1) Energy efficiency

Both levitation and guidance do not require active control, nor an onboard power supply system. The total operational cost is almost negligible due to the affordability of liquid nitrogen (77 K, 78% abundance in air and is easy to liquefy). The price of liquid nitrogen is about 1/100 of that of

Table 6.1　R&D history of HTS maglev vehicle in the world.

Years	Progress
1934	Patent of normal conducting maglev train in Germany
1966	Patent of LTS maglev train in U.S.A.
1988	Discovery of the HTS suspension phenomena in U.S.A.
1996	Suspension of a 220 kg sumo by a superconducting disk in Japan
1997	A small HTS maglev mode with 20 kg weight and 7 mm levitation height, as a China-Germany joint project
1997	HTS maglev test vehicle research project at SWJTU approved by the 863 Program in China
2000	The world's first man-loading HTS maglev test vehicle "Century" was successfully developed by ASCLab, SWJTU, China
2004	More man-loading HTS maglev demonstration vehicles were developed by Germany and Russia
2009	Man-loading HTS maglev test vehicle "Cobra" in Brazil
2011	Man-loading HTS maglev test vehicle "SupraTrans II" in Germany (80 m loop track, 2.2 million Euros)
2013	A man-loading HTS maglev experimental vehicle system, 'Super-Maglev', in ASCLab, SWJTU, China (45 m loop track, 2.5 million RMB)
2014	A full-scale HTS maglev project 'Maglev Cobra' in Brazil (200 m track, no budget details yet)

liquid helium (4.2 K). Compared with the LHS maglev train in Japan and the normal conducting maglev vehicle in Germany, the HTS maglev train system is lightweight, simple in structure, and has low manufacturing and operational cost. As mentioned, the energy consumption of the HTS maglev train with super-high speed in the ETT is only 1/20 of the airplane's.

(2) Environmentally friendly

The HTS maglev train in the ETT has no noise, nor does it produce electromagnetic and chemical pollution.

(3) Safe

The HTS maglev train is an inherently self-stable levitation system. The levitation force will increase exponentially when levitation height decreases. Consequently, it is impossible to contact the permanent magnetic guideway in all operation circumstances. The safe operation of the guidance system in the horizontal direction can be achieved without control.

(4) Comfortable

The HTS maglev train is different from all the other modes of transportation. It provides a very high level of comfort to passengers because of its inherent self-stability.

(5) Super-high speed

Operation speed can be greater than 1000 km/h in ETT.

(6) Low cost

Construction and operating costs are lower than for the urban light track.

6.5   Typical application of the HTS maglev technology

(1) HTS maglev train

As mentioned earlier, the principle of the HTS maglev train is completely different from the normal conducting (non-superconducting) train in Germany and the LTS maglev train in Japan. The HTS maglev train is a self-stable maglev, economic in energy (1/20 of the aircraft), environmentally friendly (no chemical nor noise pollution), and can achieve super-high speed (thousands of kilometers per hour). Thus, it is an ideally safe and comfortable mode of ground transportation.

(2) HTS maglev launch system (inexpensive reusable cold launch system)

The reusable cold launch system has no ignition nor explosion, and can be used repeatedly. It has many advantages such as no noise production, no pollution, allowing for all-electric control, mobility, reusability and so on. The HTS maglev rocket is only 20% of the weight of a conventional propellant rocket, and the cost of the space launch will be greatly reduced. With no fuel on the vehicle, super-high speed can be achieved with only 200 kW power. The launch cost can be reduced from $10000/lb to $1000/lb. Therefore, the HTS maglev is the right choice to best meet the requirements of the reusable cold launch technology when compared to other maglev alternatives.

(3) HTS maglev bearings

HTS maglev bearings have a friction coefficient of 10^{-7}, which is 10,000 times lower than conventional bearings.

(4) HTS maglev FESS

FESS can store both electrical and mechanical energy. It can directly store mechanical energy without any energy conversion loss, and the mechanical energy storage efficiency is 10 times higher than the traditional storage system. Energy density has already reached 230 Wh/kg. In five years it could reach 2700 Wh/kg, which is 3 times higher than the hydrogen fuel cell.

6.6   HTS maglev in the 863 Program

In 1996, the authors discussed the problems about the man-loading HTS maglev vehicle with Prof. Hong-Tao Ren of the GRINM (General Research Institute for Nonferrous Metals) in the Dalian conference of information net, of Cryogenics and Superconductivity in China. Prof. Ren finally acceded to the big, challenging project.

During the period of the ninth national five-year plan, ASCLab directed one of the national 863 programs in China. The research project to develop the man-loading HTS maglev vehicle (serial number: 863-CD080000) was approved by "National 863 Program" in 1997. The magnetic levitation performances of the HTS YBCO bulk over permanent magnet guideway (PMG) are experimentally investigated by using the special measurement equipment developed by ASCLab, and the first to reach a number of theoretical and basic technological achievements. The first man-loading HTS maglev test vehicle in the world, "Century", was created in ASCLab on December 31, 2000. All the on-board melt-textured YBaCuO bulk superconductors were purchased from Prof. Hong-Tao Ren's group in GRINM in Beijing, China, which are fixed in rectangle-shaped vessels with very thin soleplates. The maglev vehicle [5, 27, 32] is 3.5 m long, 1.1 m wide, 0.9 m high, and the PMG has a length of 15.5 m. The net levitation height is more than 20 mm when five people sit on the vehicle (about 630 kg in total, including the vehicle's weight). Thereafter, the R&D of the HTS maglev vehicle continually received investment from the National 863 Program. The completed project was reviewed and approved by the National Committee of Experts in Superconductivity of "863 Program" on February 11, 2001. The comments from the National Committee of Experts in Superconductivity of 863 Program were, "This is a creative achievement via a long period of time of hard work. It has expanded the possibility of leap frogging mode development about HTS maglev technology."

"By the hard working of the group, China made an important breakthrough in basic research of HTS maglev application and other related technology. Now we can play a leading role in this field in the world."

"This original work also laid the foundation in applying HTS maglev technology in transportation and other fields and provided guidance for long-term development in the subject".

"Century" was awarded "Top 10 Advancements of Science and Technology of Chinese Colleges" in 2001, and the comment was that "it possesses tremendous development opportunity and commercial potential". AsCLab is also the recipient of the National Certificate of Merit for the May Day by All-China Federation of Trade Unions in 2001. The International Workshop on HTS maglev (ISMAGLEV 2002) was held on June 25–27, 2002 in AsCLab of SWJTU, Chengdu, China. The topic centered around HTS maglev technology and its application.

On November 5th, 2003, fourteen academicians of the Chinese Academy of Sciences co-signed a letter to the national council to petition for higher priority on further researching and developing HTS maglev train technology and relating it to the domestic high-technology industry.

During the period of the national 9th five-year plan, the SCML-01 HTS maglev measurement system was developed. Thereafter, the SCML-02 HTS maglev measurement system, with high precision and multi-functions, was developed in 2004. The SCML-03 HTS maglev dynamic test system with the linear speed of 300 km/h was developed in 2005. A laboratory prototype of the HTS maglev launch system was completed in 2006.

The project proposal review meeting of the super-high-speed ETT-HTS maglev train was held in the Southwest Jiaotong University on January 12, 2004. According to research results on the long-term stability (less than 5% of the levitation force decrease), this author produced the report of a ETT-HTS maglev train project which presented a high speed of more than 600 kilometers per hour. A meeting of more than 50 persons discussing the project lasted all day long. The meeting resulted in approval of the author's project and suggested starting R&D on the low-speed HTS maglev vehicle, but temporarily, to not carry out investigations in high-speed experiments.

In December 2004, the First Academicians Forum Meeting about the ETT-HTS high-speed transportations was held in the Sichuan Academy Advisory Centre. The academicians and experts unanimously recommended strong support for the development and industrialization of China's first HTS maglev vehicle after Jia-Su Wang's report on high-speed HTS maglev vehicle technology.

"National 863 Program Strategy Development Report (Superconducting Materials and Technology)" pointed out: China has developed the man-loading HTS maglev system technology, and it has expanded the possibility for crossover of development in HTS maglev technology. HTS maglev system technology has bright Chinese characteristics. This maglev technology results in a new transport vehicle which has high energy efficiency, environmental friendliness, safety and comfort, among other attractive features and will be expected to occupy more than 10 billion RMB/y in the market.

6.7 The special advantage of developing HTS maglev train in China

(1) The first man-loading HTS maglev test vehicle in the world, "Century", was developed successfully in China in 2000, and all intelligent property rights are owned by China.

(2) In 2007, one of over 30 patents owned by ASCLab (ZL:01128867.1) was selected by the State Intellectual Property Office (SIPO) of China as one of three most important patents in China to be added to the permanent exhibition in the SIPO's IP Exhibit Hall.

(3) The comment from National Committee of Experts in Superconductivity of "863 Program" was that "It has expanded the possibility to leap frogging mode development of HTS maglev technology".

(4) "Century" was awarded one of the Top 10 Advancements of Science and Technology of Chinese Colleges in 2001, and the comment was that "it possesses tremendous development opportunity and commercial potential".

(5) Hitherto, China still keeps more obvious technique advantages.

(6) A new PMG had been developed in China, and the cost of this PMG is lower.

(7) Massive research achievements were obtained by using several self-developed HTS maglev measurement equipment; it provided solid technique foundations on the development of the super-high speed HTS maglev train.

(8) Both rare-earth materials yttrium (Y) and neodymium (Nd) are used in the HTS maglev train, and these resources are abundant in China.

(9) Both maglev and superconductivity have been included in long-term science and technology development programs in China.

6.8 Development of "Century"

High-temperature superconductors are highly attractive because they can operate at liquid nitrogen temperature. The stable levitation of a PM above a YBCO bulk superconductor bathed in liquid nitrogen was observed by Hellman et al. [1]. The forces of YBCO bulks over a PMG have been

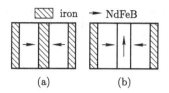

Fig. 6.1   Cross-sectional view of the PMG.

reported [28]. The onboard HTS maglev equipment is the key component of the HTS maglev vehicle that was developed [29]. The first man-loading HTS maglev vehicle in the world was tested successfully on December 31, 2000 at the Applied Superconductivity Laboratory of Southwest Jiaotong University in China [5, 30–32].

6.8.1   *PMG*

Figure 6.1 shows the two construction cross-sectional drawings of the PMG. The PMG is composed of normal permanent magnets and iron plate. The arrows represent magnetic poles where the arrowheads represent North. The length of the PMG is 920 mm, and the concentrating magnetic induction of the PMG is up to 1.2 T at its surface.

In Fig. 6.2 [33], the magnetic field in the center of the PMG is stronger than that of any other position, and it decreases rapidly with the increase of the gap from the surface of the PMG. The surface magnetic field of single normal PM is about 0.45 T, while the surface concentrating magnetic flux density of the PMG is up to 1.2 T (Fig. 6.2(a)). The magnetic flux density is 0.4 T at 20 mm above the surface of the PMG and is equivalent to about surface magnetic field of a single PM. The levitation and guidance forces of the HTS bulk materials are measured by PMG in Fig. 6.1(a).

6.8.2   *Liquid nitrogen vessel with thin bottom*

One of the important features in the onboard HTS maglev equipment is the liquid nitrogen vessel with a thin bottom. Generally, the wall thickness of liquid nitrogen vessels is not limited, and its main specification is the lower vapor rate. However, YBaCuO bulks are levitated over a magnetic guideway in the HTS maglev vehicle system, therefore a liquid nitrogen vessel with a thin bottom is very necessary in order to get a maximum net levitation gap, i.e. the gap between the outside wall of the vessel and the surface of

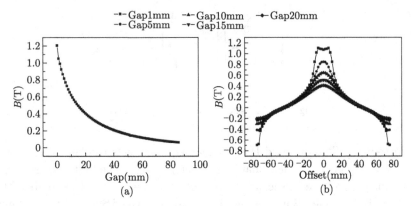

Fig. 6.2   Measurement results of the PMG's magnetic field along the transverse (a) and vertical (b) directions.

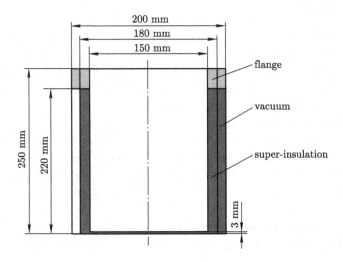

Fig. 6.3   Schematic of the columnar liquid nitrogen vessel with a thin bottom.

the guideway. A small columnar liquid nitrogen vessel with a thin bottom has been developed [28, 34] in order to verify the possibility of producing a large-size liquid nitrogen vessel with thin wall, with a bottom thickness of only 3.5 mm. The schematic and vapor rate of the small columnar liquid nitrogen vessel are shown in Figs. 6.3 and 6.4, respectively. The outside and inside diameters of the vessel are 200 and 150 mm, and the height is 250 mm. The liquid nitrogen vessel can operate continuously for over

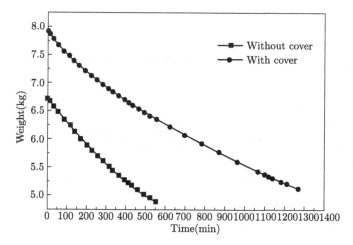

Fig. 6.4   Vapor rate of the columnar liquid nitrogen vessel with thin bottom.

16 hours. After 14 years, the performance of the liquid nitrogen vessel is still excellent with constant temperature performance. The vessel has been used successfully to measure the levitation forces of YBaCuO bulk over a magnetic guideway. During the experiment, the YBaCuO was fixed on the bottom of the columnar liquid nitrogen vessel.

6.8.3   Onboard HTS maglev equipments [29]

Based on the experimental outcome and experience with the small columnar liquid nitrogen vessel, an onboard rectangle-shaped liquid nitrogen vessel was developed. The bottom thickness of the rectangle-shaped vessel is thinner than that of the small columnar vessel, which is only 3 mm. The schematic of the rectangle-shaped liquid nitrogen vessel is shown in Fig. 6.3. Its outer and inner outline size is 150 mm×516 mm×168 mm and 102 mm × 470 mm × 168 mm, respectively. The rectangle-shaped liquid nitrogen vessel can operate continuously over 6 h. The vessel was successfully used in the HTS maglev test vehicle.

The onboard HTS maglev equipment is a core component of the HTS maglev vehicle system. The photograph of the two sets of onboard HTS maglev equipment are shown in Fig. 6.6.

The onboard maglev equipment (Fig. 6.5) [29] is composed of YBaCuO bulks and rectangle-shaped liquid nitrogen vessel. There are 43 pieces of YBaCuO bulks in each vessel (Fig. 6.6). The YBaCuO bulks are 30 mm

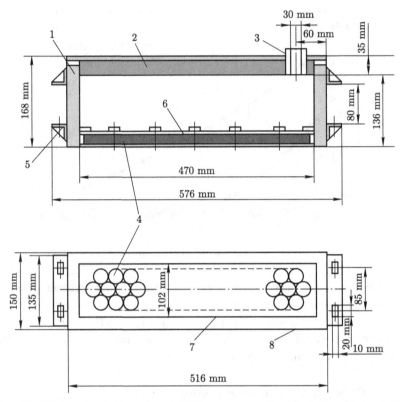

Fig. 6.5   Schematic of the onboard HTS maglev equipment. 1–thermal insulation, 2–vessel cover, 3–liquid nitrogen entrance, 4–YBaCuO, 5–pedestal of connecting with vehicle, 6–press board of YBaCuO, 7–inner vessel container, 8–outer vessel container.

in diameter and 17–18 mm in thickness. The YBaCuO bulks are arranged on the bottom of the liquid nitrogen vessel, which are sealed by a special technology free from erosion. All YBaCuO bulks are fixed firmly in the liquid nitrogen vessel. The onboard maglev equipment are connected rigidly to the two sides of the vehicle body.

6.8.4   *SCLM-01 HTS maglev measurement system* [12, 35, 36]

A HTS maglev measurement system was developed [35, 36] in order to investigate the magnetic levitation properties of the HTS YBCO bulk above a PMG in 1999. A series of properties, for example, the levitation

Fig. 6.6   The photograph of the two sets of onboard HTS maglev equipment.

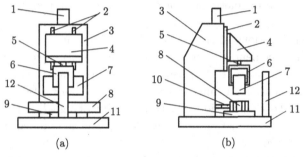

(a)   (b)

Fig. 6.7   Schematic of the HTS Maglev measurement system. 1–Servo motor, 2–Vertical guided way, 3–Vertical column, 4–Cantlever, 5–Vertical sensor, 6–Fix frame of vessel, 7–Liquid nitrogen vessel, 8–Permanent magnet guideway (PMG), 9–Horizontal drive platform, 10–Horizontal sensor, 11–Base, 12–Drive device of three dimension.

force, guidance force, levitation stiffness, etc., of YBCO bulk HTS over a PMG were investigated with the SCLM-01 HTS measurement system. The measurement system includes liquid nitrogen vessels (circular and rectangular-shaped), a permanent magnet guideway (PMG), data collection and processing, a mechanical drive and control system, and scanning of the magnetic flux.

Figure 6.7 shows the schematic diagram of the HTS maglev measurement system. During the experiment, the YBCO is placed in the columnar liquid nitrogen vessel which is positioned above the PMG. The YBCO is zero-field cooled (ZFC) and the vessel is allowed to move up and down

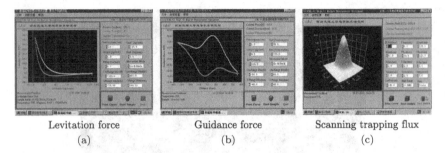

Levitation force   Guidance force   Scanning trapping flux
(a)   (b)   (c)

Fig. 6.8   Main interfaces of the measurement results.

at different speeds. The horizontal drive platform is used to measure the guidance force (stable equilibrium force along the longitudinal orientation of guideway). The drive device of three dimensions can make scanning measurements of the magnetic field of the PMG and trapped flux of an HTS bulk.

The specifications of the SCML-01 measurement system are: vertical maximal displacement of 200 mm, ±0.1 mm precision, ±2000 N vertical maximal support force, 0.2% precision; 100 mm guideway horizontal maximal displacement, ±0.1 mm precision, 1000 N of horizontal maximal support force, and 0.1% precision. The trapping flux of high T_c superconductors and the magnetic induction of the guideway can be scanned in the range 100 mm × 100 mm.

In the measurement, the YBCO HTS bulk sample is fixed at the bottom of the thin-walled liquid nitrogen vessel and cooled to the superconducting state in zero magnetic field. Secondly, the vessel is fixed at a connecting fixture with a servo electromotor. In order to avoid collision between the bottom of the vessel and the surface of the PMG, there is a gap of 1.5 mm left between the bottom of the vessel and the surface of the PMG. When the vessel is lowered to the lowest point, the minimum gap is 5 mm between the bottom of the sample and the surface of the PMG. The vessel first moves downward, reaches the lowest point, then moves upward at a speed of 2 mm/s, and the computer samples a data point every 0.5 s. The system can make real-time measurements of one or many superconductors. A computer controls the measurement process. The main interfaces of the measurement provide results of the magnetic levitation force, guidance force; these and the scanning magnetic field of a HTS trapping flux are shown in Fig. 6.8.

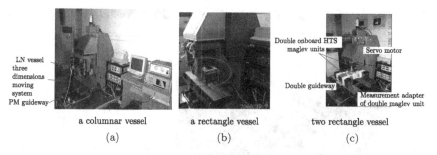

a columnar vessel   a rectangle vessel   two rectangle vessel

(a)   (b)   (c)

Fig. 6.9   Photos of HTS maglev measurement system SCML-01.

The SCML-01 HTS maglev measurement system is capable of making real-time measurements of maglev properties with one or many YBCO pieces with a PM or PMGs. This setup is especially employed in onboard HTS maglev equipment over one or two PMGs (Fig. 6.9). The onboard maglev equipment includes a rectangle-shaped liquid nitrogen vessel and an array of YBCO bulk superconductors (Fig. 6.6). Figure 6.9 is a picture of the HTS maglev measurement system. Three types of liquid nitrogen vessels are shown in Fig. 6.9: a columnar vessel, a rectangle vessel, and two rectangle vessels.

Many of these research results [10, 11, 33, 36–40] were obtained by the SCML-01 HTS maglev measurement system. Based on the original research results from SCML-01, the first man-loading HTS maglev test vehicle in the world was successfully developed in 2000 [5].

6.8.5   *Experimental results of the single onboard maglev equipment*

Figure 6.9(b) shows the measuring system of the levitation forces of single onboard HTS maglev equipment [29] over a single NdFeB guideway. Figure 6.9(c) shows the measuring system of the levitation forces of two sets of onboard HTS maglev equipment over two corresponding NdFeB guideways.

Figure 6.10 shows the measurement results of levitation forces of the best single onboard HTS maglev equipment over the NdFeB guideway (surface magnetic flux density is 1.2 T). The levitation force of the single onboard HTS maglev equipment over the PMG was 1202 N at the levitation gap of 15 mm, and the levitation force was 1724 N at the levitation gap of 8 mm. This was first measured in Dec. 2000. The main data of the levitation

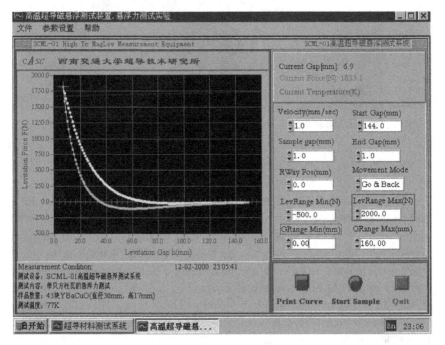

Fig. 6.10   The measurement result of the best single onboard HTS maglev equipment over a single NdFeB guideway.

forces of single onboard HTS maglev equipment are listed in Table 6.2. As shown in Fig. 6.10, the levitation force of a single piece of onboard HTS maglev equipment over the PMG is 1202 N at the levitation gap (between guideway surface and superconductors) of 15 mm, and the levitation force is 1724 N at the levitation gap of 8 mm. Moreover, the levitation force is up to 1823 N at the levitation gap of 7 mm. The total levitation force has reached 13792 N of eight sets of onboard maglev equipment at the levitation gap of 8 mm. The eight sets of onboard HTS maglev equipment have been successfully used in a manned HTS maglev vehicle.

6.8.6   The first man-loading HTS maglev test vehicle in the world [5]

The PMG consists of two parallel PM tracks, whose concentrated magnetic field at the height of 20 mm is about 0.4 T. The total length of the PMG is 15.5 m. The HTS maglev provides inherent stability along both the vertical

Table 6.2   Levitation forces of the best single onboard
HTS maglev equipment.

Gaps (mm)	Forces (N)	Gaps (mm)	Forces (N)
60	44.9	8	1591.5
50	122.5	9	1422.1
40	265.1	10	1277.6
30	506.3	11	1152.2
25	685.2	12	1031.6
20	913.0	13	928.3
15	1202.0	14	830.7
14	1267.1	15	746.4
13	1341.7	20	421.1
12	1409.7	25	216.2
11	1484.3	30	87.0
10	1561.8	35	3.8
9	1638.4	40	−46.8
8	1723.6	50	−96.6
7	1823.1	60	−105.2

Fig. 6.11   To summarize measurement results of 8 sets of HTS maglev equipment.

and lateral directions, thus there is no need to control the vehicle along
these two directions. The only control systems used are linear motors as
driving and breaking devices.

The eight sets of onboard HTS maglev equipment [29] are connected
rigidly on the two sides of the vehicle body, with four sets of maglev
equipment assembled on each side. Figure 6.11 was the photograph taken

Fig. 6.12   The HTS maglev test vehicle body.

while the measurement results of eight sets of HTS maglev equipment were summarized. The vehicle body (Fig. 6.12) is 2268 mm long, 1038 mm wide and 120 mm high. Both the linear motor of the vehicle and the PMG are under the vehicle body.

The vehicle body was lifted by hydraulic equipment until the gap between the bottom of the liquid nitrogen vessels and the surface of the PMG was larger than 75 mm. Then, the HTS bulks were cooled with liquid nitrogen. Figure 6.13(a) shows the historic moment of the first pouring of liquid nitrogen into the onboard vessels. In Figure 6.12, we can see the two tracks and the vehicle body.

The net levitation gap of the HTS maglev test vehicle body is more than 20 mm when five people stood on the vehicle, and the levitation height of the vehicle body was 33 mm when the same five people got off the vehicle (Fig. 6.13(b)). Figure 6.13(c) shows the photograph of the HTS maglev test vehicle with an outer shell. The external size of the vehicle with a shell is 3.5 m long, 1.2 m wide, and 0.8 m high. There are four seats in the HTS maglev vehicle. The total levitation force of the entire maglev vehicle was measured to be 6351 N and 7850 N at the net levitation gap of 20 mm and at 15 mm, respectively, in July 2001. The net levitation gap is the distance between the PMG's upper surface and the liquid nitrogen vessel's bottom. Figure 6.13(d) is a photograph of the people who took the first

The first pouring of liquid nitrogen	The net levitation gap of the HTS maglev test vehicle body was more than 20 mm when five people stood on the vehicle
(a)	(b)

The researchers' gratification of experiencing the first-time success
(c)

The first man-loading HTS, maglev test vehicle in the world "Century"
(d)

Fig. 6.13   The historic photographs of the first man-loading HTS maglev test vehicle "Century" in the world.

man-loading HTS maglev test vehicle "Century" in the 15th (1986–2000) anniversary exhibition of 863 Program of China in Beijing, 2001.

Essential parameters of the first man-loading HTS maglev test vehicle in the world, "Century", are listed in Table 6.3.

6.8.7   Levitation forces of the HTS maglev test vehicle [27, 30–32]

Figure 6.14(a) shows that the levitation forces of eight sets of onboard maglev equipment above the PMG have a small difference. Those measured differences of result among the equipment assemblies perhaps originated from the integration of YBCO bulks and the accuracy of the measured

Table 6.3   Essential parameters of the first man-loading
HTS maglev test vehicle in the world — "Century".

Specifications	Units	Value
Passengers	seats	4
Length of guideway	m	15.5
Vehicle body size	mm	2268 × 1038 × 120
Exterior size of vehicle	m	3.5 × 1.2 × 0.8
Total levitation force	N	6350 at gap of 20 mm
Levitation gap	mm	20
Total guidance force	N	1980 at gap of 20 mm
Acceleration	m/s^2	1

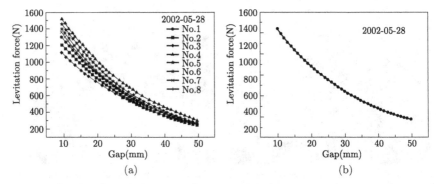

Fig. 6.14   (a) Measured results of eight onboard HTS magnetic levitation equipment
assemblies (rectangle-shaped vessels include 43 pieces of YBCO bulks), and (b) The total
levitation force of eight onboard HTS maglev equipment assemblies above the PMG.

device assembly. For example, the levitation force of maglev equipment
No. 7 is 1493 N at the levitation gap of 10 mm, and 1227 N at that of
15 mm. The levitation force of maglev equipment No. 3 is 1091 N at the
levitation gap of 10 mm, and 902 N at that of 15 mm. Figure 6.14(b)
shows the total levitation forces of the eight onboard maglev equipment
assemblies. The total levitation force of eight onboard maglev equipments
assemblies was 10431 N at the levitation gap of 10 mm, and 8486 N at
15 mm.

6.8.8   Guidance forces of the HTS maglev vehicle [41, 42]

The guidance force defines the lateral stability of the maglev vehicle when it
is either at a standstill or moving. The lateral guidance force is dependent on

the trapped flux in the bulk superconductors, i.e. the greater the amount of trapped flux, the stronger the guidance force. This is a unique characteristic of the bulk HTS maglev. This sort of maglev vehicle with bulk HTS does not need any lateral stability control system, which makes it better than all the other conventional maglev vehicle systems. The guidance forces are large and suffice to guide the vehicle while large levitation forces are also ensured.

The measuring equipment of the guidance force of the entire HTS maglev vehicle is shown in Fig. 6.15(a). The set up includes two horizontal

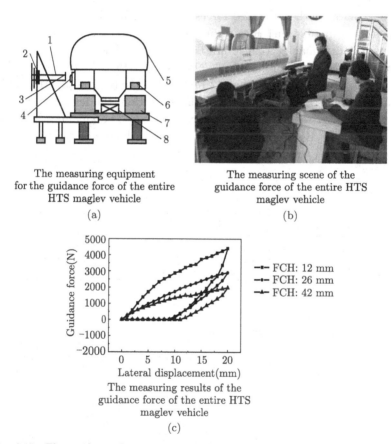

The measuring equipment
for the guidance force of the entire
HTS maglev vehicle

(a)

The measuring scene of the
guidance force of the entire HTS
maglev vehicle

(b)

The measuring results of the
guidance force of the entire HTS
maglev vehicle

(c)

Fig. 6.15 The guidance force measurement of the entire HTS maglev vehicle. In (a), 1–Horizontal propulsion system, 2–Vertical propulsion system, 3–Screw of adjusting zero, 4–Force sensor, 5–Vehicle body, 6–HTS, 7–Permanent magnetic railway, 8–Linear motor.

propulsion systems and two sets of force sensors which are fixed on the vehicle. Each set of the propulsion system can move in both the horizontal and vertical directions so that they can measure the guidance force of the entire vehicle at different levitation gaps. A chain with a synchronization precision of 0.5 mm connects two sets of propulsion systems. The moving range of the propulsion system along the horizontal direction is 0 to 20 cm and the moving precision is 1 mm; and along the vertical direction the range is from 0 to 10 cm, with a vertical moving precision of 1 mm. A photograph of the measuring equipment for the guidance force is shown in Fig. 6.15(b).

The data of the guidance forces are listed in Table 6.4. It can be seen from Table 6.4 and Fig. 6.15(c) that the lateral guidance forces of the entire HTS maglev vehicle are large enough. The guidance forces at displacement 20 mm are 4407 N, 2908 N, and 1980 N for field cooling heights of 12 mm, 26 mm, and 42 mm, respectively. The practical guidance force should be 1980 N, for in this condition, the levitation force is not significantly reduced, guiding force is larger. The guidance force can be restored to the original central position without additional external force.

Table 6.4   Guidance force (unit: N) under different field cooling height (Disp: displacement, FCH: field cooling height).

Disp (mm)	FCH (mm) 12	26	42	Disp (mm)	FCH (mm) 12	26	42
1	215	207	250	16	3834	2517	1740
2	592	414	450	17	3941	2659	1785
3	1026	603	600	18	4117	2744	1845
4	1391	802	725	19	4246	2851	1930
5	1713	987	835	**20**	**4407**	**2908**	**1980**
6	2033	1122	945	19	3364	2422	1515
7	2330	1289	1085	18	2682	2120	1295
8	2521	1465	1165	17	2361	1834	1070
9	2685	1610	1275	16	1932	1533	885
10	2883	1734	1355	15	1580	1301	665
11	3003	1913	1420	14	1244	1073	470
12	3220	2038	1510	13	886	767	295
13	3350	2151	1475	12	590	606	120
14	3418	2263	1565	11	354	375	0
15	3726	2399	1660	10	110	190	0

6.8.9 Long-term stability of the HTS maglev vehicle in 2001–2003 [21, 32]

The results were measured by the SCML-01 HTS maglev measurement system in July 2001, December 2001, May 2002, and March 2003, respectively. Figure 6.16 shows the total levitation force of eight liquid nitrogen vessels over the PGM at different levitation gaps. In July 2001, the levitation force was measured to be 8940 N at the levitation gap of 15 mm and 7271 N at the levitation gap of 20 mm. The total levitation force of eight liquid nitrogen vessels over the PMG in March 2003 was 8630 N at the levitation gap (subtract 3 mm of bottom thickness of liquid nitrogen vessel) of 15 mm.

Figure 6.16 and Table 6.5 show that the levitation force of the entire HTS maglev vehicle was 8486 N at the levitation gap of 15 mm and 6908 N at the levitation gap of 20 mm in May 2002. At a gap of 30 mm, there was a 46% decrease in the levitation force compared to the gap of 15 mm.

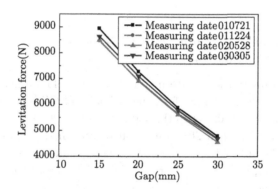

Fig. 6.16   Change of total levitation force of the entire HTS maglev vehicle.

Table 6.5   Total levitation force of the entire HTS maglev vehicle.

Gap (mm)	Levitation force (N)			
	07/21 2001	12/24 2001	05/28 2002	03/05 2003
15	8940	8457	8486	8633
20	7271	6927	6908	7094
25	5890	5656	5618	5782
30	4791	4685	4561	4693

The data set of levitation forces for March 2003 was slightly higher than the data set of May 2002, it was due to the fact that one of the PMGs was not laid to overlap accurately.

The comparison of total levitation force of the entire HTS maglev vehicle during the 10-month period from July 2001 with May 2002 showed levitation force decreases of 5.1%, 5.0%, 4.6%, and 4.8% at the levitation gaps of 15 mm, 20 mm, 25 mm, and 30 mm, respectively. All data are nearly the same, i.e., the levitation force decreases were about 5.0% at different levitation gaps.

The levitation forces have became small after two years. Up to September 2009, over 27000 passengers had taken a ride on the maglev vehicle, and the total recorded shuttling mileage was about 400 km. Experimental results indicate that the long-term stability of the HTS maglev vehicle is better when in the static mode than when the HTS maglev vehicle was operating at low speeds. The behavior of a HTS maglev vehicle is very different at high speeds. Therefore, the investigation of the dynamic properties of the HTS maglev vehicle at high-speed operation is extremely important.

6.8.10   Long-term stability of YBCO bulks in 2001–2009 [43]

In order to evaluate the long-term stability of the HTS bulks for practical application, the levitation performance of bulk YBCO samples was investigated by the SCML-02 setup throughout 2001 to 2009. The same YBCO bulk pieces (A5) used in 2000 were still placed on the man-loading HTS maglev vehicle and had been running for nearly nine years. The levitation force (Fig. 6.17), hysteresis, and relaxation (Fig. 6.18) of the YBCO bulks, which were cooled at the liquid nitrogen temperature (77 K), were measured by the SCML-02 HTS maglev measurement system in June 2009. A batch of YBCO samples was synthesized in 2007 (07-1) and in 2008 (08-1), and some of these samples were loaded in 2007 while others were, at the latter time. This was to analyze the variation of levitation performances of different bulk YBCO samples during different times after fabrication of the YBCO.

Figure 6.17 is the comparison of the levitation force of different YBCO bulks during 2001–2008, and Fig. 6.18 is the comparison of the levitation force relaxation of different YBCO bulks in 2001–2008. A5, old-1, 07-1, and 08-1 in the graphs represented onboard YBCO bulk in 2000, YBCO bulk which was not loaded in 2000, new YBCO bulk in 2007, and new YBCO

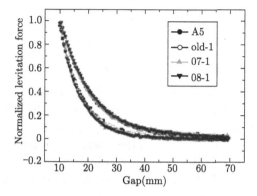

Fig. 6.17   Measurement results of levitation force.

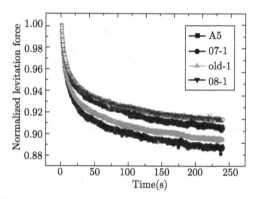

Fig. 6.18   Measurement results of levitation force relaxation. A5 is onboard bulk YBCO in 2000; Old-1 is bulk YBCO that was not loaded in 2000; 07-1 is new in 2007; 08-1 is new in 2008.

bulk in 2008, respectively. All measurement results are normalized with respect to the results of A5 and all are aligned along the centerline of the PMG. In Figs. 6.17 and 6.18, the measurement results of the levitation force and relaxation time of the A5, 07-1, 08-1, and old-1 samples remained almost constant.

These experimental results validate the fact that the performances of the bulk YBCO samples above the PMG did not change after nearly nine years. Our research results can be applied to not only the man-loading HTS maglev vehicle, but also to other superconducting applications such as HTS maglev bearings and superconducting flywheel energy storage devices.

6.8.11 *Brief summary of "Century"*

Based on a series of parameters, such as levitation forces, guidance forces, levitation stiffness, etc. of single- and multi-bulk YBCO, have been measured by the SCML-01 HTS maglev measurement system, the first man-loading HTS maglev test vehicle in the world, "Century", was built in the ASCLab of the SWJTU at 2:26 pm on December 31st, 2000. The "Century" HTS maglev vehicle is 3.5 m × 1.1 m × 0.9 m with 4 passenger seats. The PMG track has a length of 15.5 meters, and they have the concentrating field of 1.2 T at its center surface and 0.4 T at 20 mm above the surface, respectively. The onboard HTS maglev equipment, with an external size of 150 mm × 516 mm × 168 mm, is composed of a rectangle-shaped liquid nitrogen vessel and 43 pieces of YBaCuO bulks, and the vessel bottom has a very small thickness of only 3 mm. The linear motor acceleration is 1 m/s². Here are the key observations from our HTS maglev system:

(1) The total levitation force and guidance force of the whole vehicle are 6350 N and 1980 N at the net levitation gap of 20 mm, respectively.

(2) The levitation force of a single set of onboard HTS maglev equipment over the PMG are 1202 N and 1724 N at the levitation gap of 15 mm and 8 mm, respectively. The total levitation force of eight onboard maglev equipment assemblies was 10431 N at the levitation gap of 10 mm.

(3) The levitation forces of two onboard HTS maglev equipment sets above two corresponding PMGs, i.e. one-fourth of the whole vehicle, are measured by the simulation of a vehicle running system.

(4) During the ten-month period from July 2001 to May 2002, the levitation force only decreased by about 5.0% at the levitation gap of 20 mm.

(5) By September 2009, more than 27000 passengers had taken a ride on the maglev vehicle, and the total recorded mileage was about 400 km.

(6) Experimental results verified that the man-loading running performances of the single bulk YBCO samples above the PMG barely changed after about nine years.

6.9   Research advances after "Century"

6.9.1   *SCML-02 HTS maglev measurement systems* [12, 44]

Although the first man-loading HTS maglev vehicle in the world [5] was developed successfully with the research results from the SCML-01 measurement system [35, 36], the improvement of measurement functions and precision of the SCML-01 platform is necessary. Therefore, to make more thorough and careful investigations on the magnetic levitation properties of the HTS maglev vehicle over a PMG, a SCML-02 HTS maglev measurement system [44] with several special functions and high precision was successfully developed.

The experiment system is shown in Fig. 6.19. Four vertical support posts labeled 1 are fixed on an optical bedplate. The upper liquid nitrogen vessel 6 with HTS bulks is placed above a PM or PMG. The liquid nitrogen vessel with HTS bulks is placed under a PM. In this way, the experiment system is fit for different measurements, especially the measurement of maglev properties of superconductor samples with a PMG. The inertial force effects (tension) of the moving parts are decreased by four force balance chain wheels and corresponding counterweights. The x-y

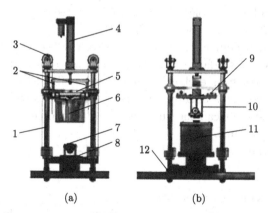

(a)   (b)

Fig. 6.19   The design scheme of the SCML-02 HTS maglev measurement system, there, (a) is upper liquid nitrogen vessel and (b) is bottom liquid nitrogen vessel. 1–Vertical support post; 2–Vertical force sensor; 3–Force balance chain wheel; 4–Servo motor; 5–Slip set; 6–Upper liquid nitrogen vessel; 7–Bottom clamp of cylinder PM or PMG; 8–x-y electro-motion seat; 9–Horizontal force sensor; 10–Upper clamp of cylinder PM or PMG; 11–Bottom liquid nitrogen vessel; 12–Optical bedplate.

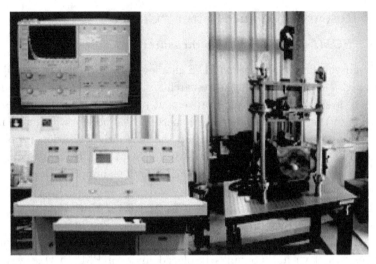

Fig. 6.20   Photograph of the SCML-02 HTS maglev measurement system: Levitation force measurement interface (top left); console control desk (bottom left); experiment system (right).

electromotion seat is fixed to the optical bedplate. Horizontal directions along the x and y axes can be moved together or separately. Thus, the measured HTS bulk samples can be moved in three dimensions.

Figure 6.20 shows a photograph of the SCML-02 HTS maglev measurement system. Four force sensors measure the vertical levitation force and the horizontal guidance force. The vertical sensor and the horizontal sensor are isolated from each other, and it is possible to avoid interference between the two kinds of force sensors. Therefore, the SCML-02 maglev measurement system can measure the levitation force and the guidance force at the same time.

The special functions of the SCML-02 maglev measurement system are:

(1) High precision measurement;
(2) Measurement of both a single or a number of HTS bulk samples;
(3) Measure the HTS bulk samples and the PM while they move along the three dimensions at the same time;
(4) The location of the HTS bulk and PM can be interchanged;
(5) Measurement of maglev properties between the measured HTS bulk specimens and a cylindrical PM or a PMG;

(6) Simultaneous measurement of the levitation force and guidance force for which the sample can be displaced vertically and horizontally at the same time;

(7) Synchronous measurement of the cross stiffness of the levitation force or the guidance force in a real-time independent measurement along the vertical or horizontal;

(8) Interaction of dynamic rigidity of levitation force and guidance force synchronous measurement;

(9) Relaxation time of levitation force and guidance force measurement along the vertical or horizontal directions.

The main technical specifications of the SCML-02 maglev measurement system are:

(1) Vertical maximal displacement of 150 mm;

(2) Horizontal maximal displacement of 100 mm;

(3) Position precision of 0.05 mm;

(4) Vertical maximal support force of 1000 N;

(5) Vertical force precision of 2‰;

(6) Horizontal maximal support force of 500 N;

(7) Horizontal force precision of 1‰.

The maglev properties of YBCO bulk are measured after the calibration of the SCML-02 HTS maglev measurement system. The levitation forces of a single YBCO bulk and an array of seven YBCO bulks above a PMG were measured by the SCML-02. The array of seven YBCO bulks are concentrically arranged, where three bulks share one axis along the axis line of the PMG. The diameter of the single YBCO bulk was 48 mm, and the diameter of each of the array of seven YBCO bulks was 30 mm.

Figure 6.21 shows the measurement results of the levitation forces of two YBCO bulks with diameter of 48 mm with a field cooling height (FCH) of 35 mm and vertical measurement ranges from 60 to 100 mm and backward. Figure 6.22 shows the measurement results of the guidance forces along the lateral direction of the PMG. The YBCO samples had a diameter of 48 mm, FCH of 35 mm, measurement height (MH) of 15 mm, and lateral displacement measurement range from -20 to 20 mm.

Figure 6.23 shows the measurement results of the levitation forces of three different arrangements of seven YBCO samples above the PMG. Each YBCO bulk has a diameter of 30 mm and the FCH is 35 mm. Figure 6.24

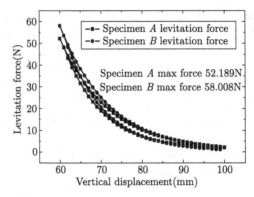

Fig. 6.21   Levitation forces of two YBCO bulk samples with diameter of 48 mm (FCH of 35 mm, measurement range of 60–100 mm).

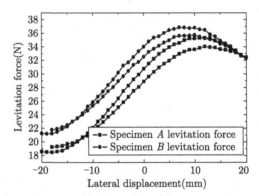

Fig. 6.22   Guidance forces of two YBCO bulk sample with diameter of 48 mm (FCH of 35 mm, measurement height of 15 mm).

shows the measurement results of the guidance forces of three arrangements of seven YBCO bulks above a PMG. The MH for this experiment was of 10 mm, and the lateral measurement range from −10 mm to 10 mm.

Figure 6.25 shows the measurement results of the array of seven YBCO samples above a PMG at different MHs. The FCH was set to 20 mm, the measurement heights ranged from 10 mm to 30 mm, and the lateral measurement ranged from −10 mm to 10 mm.

The levitation forces and guidance forces of YBCO bulks were measured by the SCML-02 HTS maglev measurement system. The measurement

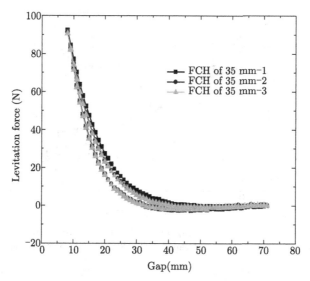

Fig. 6.23 Levitation forces of an array of seven YBCO bulk samples with diameter of 30 mm (FCH of 35 mm).

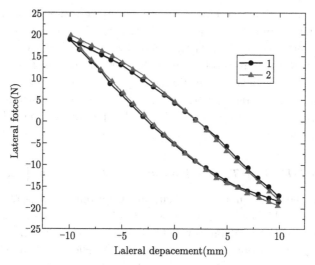

Fig. 6.24 Guidance forces of an array of seven YBCO bulks, and each has a diameter of 30 mm (FCH of 35 mm, measurement height 10 mm).

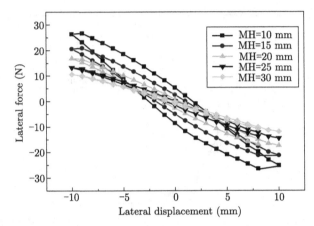

Fig. 6.25   The guidance forces of 7 YBCO bulks with diameters of 30 mm are measured by SCML-02 (FCH of 20 mm, MH range of 10–30 mm, lateral measurement range of −10 to 10 mm).

results of HTS maglev properties validated the accuracy and reliability of the SCML-02.

The SCML-02 measurement system can make real-time measurements and the data acquisition process is controlled by a computer. The main functions and specifications of the system are unique in the world. Many valuable experimental results were obtained from this system, including the maglev vehicle's levitation force, guidance force and their stiffness, and cross stiffness, and so on. All these experimental parameters are very helpful to evaluate the load capability of the HTS maglev vehicle.

6.9.2   SCML-03 HTS maglev dynamic test system [12, 45]

Although the HTS maglev measurement system SCML-02 has more functions and higher precision than SCML-01, it cannot measure the running performance of bulk YBCO samples above a PMG. For further engineering application of the HTS maglev vehicle, the dynamic properties of the maglev must be understood clearly. From this viewpoint, an SCML-03 HTS maglev dynamic test system was designed and successfully developed.

When the HTS maglev vehicle runs along the PMG, it is difficult to measure its dynamic properties. In SCML-03, the rotational motion of a circular PMG instead of the physical motion of the YBCO bulk was taken to be the equivalent measure of the dynamic interaction between the superconductor and the PMG. That is, the circular PMG can rotate to different speeds while the onboard HTS maglev equipment is fixed above the PMG and this will simulate the fact that the superconductor is traveling above a PMG. SCML-03 is composed of a vertical load, horizontal load, three-dimensional measurement systems, liquid nitrogen vessel, circular PMG, drive device, data acquisition and processing, autocontrol, and so on.

The main measurement functions included the dynamic stability of the HTS maglev equipment (liquid nitrogen vessel with HTS samples), the levitation force and guidance force of both single and multiple HTS bulk samples, the levitation force and guidance force rigidity of both single and multiple HTS bulk samples, the levitation force and guidance force change with the levitation gap, and so on.

The design scheme of the main part of SCML-03 is shown in Fig. 6.26. The airframe of the HTS maglev dynamic test system SCML-03 (not including the measurement control desk) is shown in Fig. 6.27. The total technical parameters of the principal part of the SCML-03 are 3.3 m long, 2.4 m wide

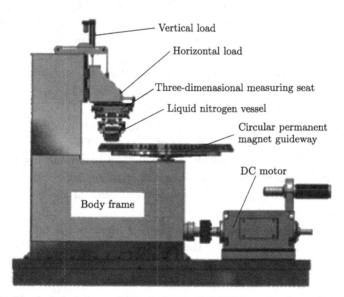

Fig. 6.26   The design scheme and principal parts of the HTS maglev dynamic test system SCML-03 (not including power supply and measurement control desk).

Fig. 6.27   The photo of the airframe of the HTS maglev dynamic test system SCML-03 (3.3 m long, 2.4 m wide and 3.15 m high with a total weight of 13.95 t).

and 3.15 m high. The total weight is 13.95 t, and includes the circular PMG disk which weighs 0.6 t. Figure 6.28 shows the measurement scene of the HTS maglev dynamic test system — the control desk is on the bottom right.

A DC motor is used to drive the circular PMG to rotate, and to control its rotational speed. The rotational direction of the DC motor is translated into the horizontal rotation direction of the circular PMG by a gear redirection case. The circular PMG is fixed along the circumferential direction of a big circular disk with a diameter of 1500 mm. The rotating unbalance of the big circular disk is less than 20 gm. The maximum linear velocity of the PMG is about 300 km/h when the circular disk rotates round the central axis at 1280 rpm. The rotation speed error of the circular disk is less than 3%.

The three-dimensional measuring seat is fixed on the horizontal load. The seat can be moved along the horizontal direction, which is perpendicular to the tangent direction of the circular PMG. Therefore, the HTS maglev dynamic guidance force can be measured at the same time. Six force sensors are used to measure the vertical, transverse, and longitudinal directional forces of the liquid nitrogen vessel. In order to measure the

Fig. 6.28   The test scene of the SCML-03 HTS maglev dynamic test system. The control desk is on the right hand side.

three-dimensional dynamic response, the liquid nitrogen vessel is connected to the three-dimensional measuring seat by elastomers.

The movement of the servo motor and the speed of the circular PMG are controlled by both the software program and related control card.

In order to calibrate measurement precision of the dynamic measurement system, the 11 channels Noise & Vibration Measurement System made by B&K was used. The vertical dynamic levitation force is measured when the circular PMG is rotated at different speeds.

The SCML-03 can measure the dynamic properties in opposite motion between the vehicle and the PMG. Main technical specifications include:

(1) Diameter of the circular PMG of 1500 mm;
(2) Maximal linear speed of the circular PMG of 300 km/h;
(3) Rotation speed precision of the circular PMG of ±3%;
(4) Vertical maximal displacement of 200 mm;
(5) Horizontal maximal displacement of ±50 mm;
(6) Position precision of ±0.05 mm;
(7) Vertical maximal support force of 3350 N;
(8) Horizontal maximal support force of 500 N;
(9) Force sensor precision of ±1%.

The most important property of the dynamic test system is its self-stability. The rotating unbalance of the PMG circular disk was measured to be less than 20 gm. The rotating unbalance value satisfies the system measurement's needs. The other important parameter of the dynamic test system is the self-stability of the liquid nitrogen vessel and the body frame. In order to confirm the self-stability, a 11 channel Noise & Vibration Measurement System was used. A 4507-004B accelerometer was fixed to the body frame, and four 4507-004B accelerometers were fixed at the clamp device of the liquid nitrogen vessel. A picture of the measurement scene of the SCML-03 is shown in Fig. 6.27.

The vibration spectra of the body frame and the clamp device of the liquid nitrogen vessel were measured at the rotation speeds of 50 rpm, 100 rpm, 200 rpm, 300 rpm, and 400 rpm respectively. The vibration spectra of the clamp device of the liquid nitrogen vessel include both the perpendicular direction and the horizontal direction. The vibration spectrum of the body frame at 400 rpm is shown in Fig. 6.29. The vibration spectra of the perpendicular direction and the horizontal direction of the liquid nitrogen vessel at 400 rpm are shown in Figs. 6.30 and 6.31.

The main distinction of the SCML-03 and the previous HTS maglev measurement system is the ability to measure the maglev performance at

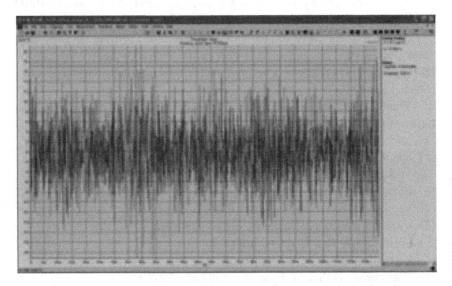

Fig. 6.29   Vibration spectrum of the body frame at 400 rpm.

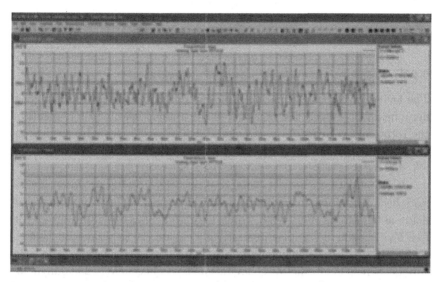

Fig. 6.30 Vibration spectrum along the perpendicular direction on the clamp device of the liquid nitrogen vessel at 400 rpm.

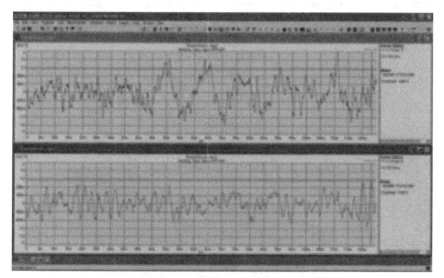

Fig. 6.31 Vibration spectrum along the horizontal direction on the clamp device of the liquid nitrogen vessel at 400 rpm.

the running case. By using SCML-03, both levitation forces and guidance forces are measured experimentally at different velocities.

6.9.3   Research results of dynamic maglev

The levitation force attenuation is closely related to the exposure to magnetic field history and the variation of the applied field, which depends on the rotational velocity of the circular PMG. Several cases were carried out for the investigation of the force transition of the HTS samples (Fig. 6.32) [46, 47].

We found that the zero-field cooling (ZFC) levitation force oscillation back and forth is mainly affected by the PMG vertical magnetic field, as shown in Fig. 6.33. Force oscillation causes vibrations of the free levitation body, but at present we focus on the general tendencies of the levitation force behavior and study its force attenuation. The measurement results of the ZFC levitation force between 0–238 km/h are drawn in Fig. 6.34. The change in levitation force was almost instantaneous in accordance to the varying of the vertical component of the applied field, and generally attenuated with running time. The sharp rising and declining of the levitation force in Fig. 6.34 were most likely caused by interference, and the corrected average curve is shown in Fig. 6.35.

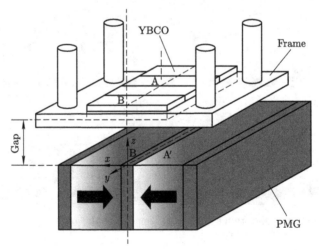

Fig. 6.32   Schematic diagram of the YBCO bulk array above the PMG in the SCML-03 measurement system.

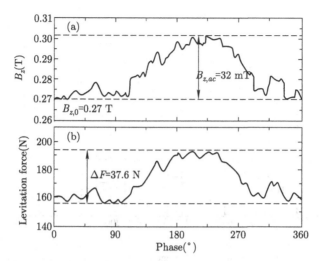

Fig. 6.33   The ZFC levitation force profile (b) with the PMG vertical circumferential magnetic field (a). The force-varying amplitude was 37.6 N, 22% of the force average 171.9 N in one rotational cycle caused by a 32 mT field perturbation amplitude. Note that the change in levitation force was almost instantaneous in accordance to the varying of the vertical component of the applied field.

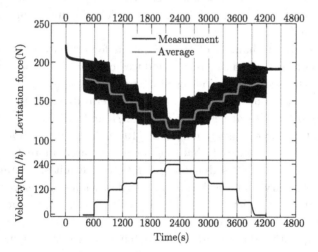

Fig. 6.34   Levitation force measurement and its average result. The grey line near the middle is the average line indicating the general tendency of the levitation force. Note that the sharp rising and declining of the levitation force were most likely caused by interference, and the corrected average curve is in Fig. 6.35.

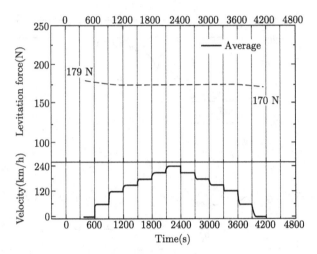

Fig. 6.35   Corrected levitation force average in steady velocities. There is no significant force attenuating over 60 km/h, and the joint transition among all velocity parts of the curve is smooth.

The average curve tendency indicated that the general levitation force attenuation saturated over 60 km/h, and the saturation was hardly affected by the changing of speeds, while the energy loss dominates in force decay under 60 km/h.

The decay mechanism of the levitation force was attributed to the declined performance parameter (J_i, induced current) of the bulk which was caused by the heat generation resulting from the energy loss. Figure 6.36 shows the levitation force relaxation after several dynamic measurements between 0 and 2700 s at velocities of 30 km/h and 119 km/h, respectively. The levitation force relaxing curves of 119 km/h locating above the ones of 30 km/h also indicated that the force decreased more for 30 km/h in between 0 and 2700 s dynamic operation (the initial levitation force of those four curves at the beginning of the dynamic measurement were approximately equivalent). It also indicates that force decays more at low velocities, as discussed above.

Consequently, the levitation force tends to reach a dynamic equilibrium in dynamic operation. This observed phenomenon is of great importance in the design and application of the HTS maglev system and more detailed research on the HTS dynamic performance is undergoing.

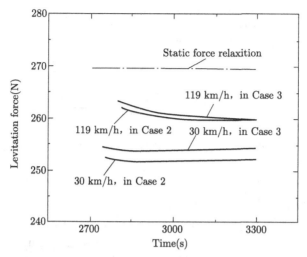

Fig. 6.36 Levitation force relaxation after several dynamic measurements in between 0 and 2700 s at velocities of 30 km/h and 119 km/h, respectively. Note that the initial levitation forces of those four curves at the beginning of the dynamic measurement were approximately equivalent, so the levitation force relaxation after operation at 30 km/h decreased more than that of operation at 119 km/h.

6.9.4 Pre-load process

The pre-load method was found to be effective in suppressing the levitation height or force decay and improving stability of the HTS maglev vehicle system [48–50]. The steps of the pre-load method are as follows: (1) the on-board HTS bulks and the related levitation part of the vehicle are descended to a height lower than the working height (WH) after field-cooling (FC) by adding load or other ways; (2) fixing this lower levitation position for some time; (3) releasing the levitation part of the vehicle; (4) finishing, and then better performance will be obtained.

Table 6.6 shows the advantage of the pre-load method through the experiment using a HTS maglev vehicle model. Experiments show that the pre-load method can help to suppress the levitation height decay. Moreover, the HTS maglev vehicle gets more stability after the pre-load: with higher stiffness, damping coefficient as well as better anti-vibration performances.

On the other hand, the pre-load method really reduces the 14% levitation force decay to 2.5% in the cases of the same repeated lateral movement

Table 6.6   Dynamic stiffness k and damping coefficient c of HTS maglev vehicle model by pre-load expriment.

Pre-load	k (n/m)	c (ns/m)
no	53740	144
yes	87050	214

[49]. Moreover, for the high-speed running, it is suggested to pre-load at low-speed for some time.

6.9.5   *Eddy-current damper* [51, 47]

As a good heat conductor, copper is widely used in the onboard cryostats of the HTS maglev vehicle system. However, because of its good current conductivity, copper is a good damper candidate to enhance the dynamic performance of the HTS maglev system. From the point of view of technological application, it is easy to incorporate copper as an eddy current damper because copper is the main material of the cryostat in the onboard HTS maglev equipment. There are two ways to fix the copper eddy current damper to the bulk superconductor. One way is to add copper under the HTS bulks, and the other way is to surround the HTS bulks with copper.

For comparison, vibration parameters of the HTS-PMG system without any eddy current damper are shown in Table 6.7.

It is found that an eddy current damper does not affect the quasi-static performance of the HTS maglev system, e.g., levitation force, load capacity and so on. It meets the requirement of stability during the installation and start-up process in the future HTS maglev application. And, it implies that the incorporation of an eddy current damper is feasible to improve the anti-disturbance capability of the HTS maglev system (as shown in Figs. 6.37 and 6.38). Figure 6.37 depicts results for the vertical incorporation method.

As is well-known, the magnetic field of the PMG grows exponentially with WH. The eddy current damper is more sensitive to the strong magnetic field at the lower position over PMG. Thus, in both the vertical and lateral vibration experiments, damping coefficient decreases with the increase of WH as well as FCH, as shown in Figs. 6.37 and 6.38.

Table 6.7   Vibration parameters of the HTS-PMG system without an eddy current damper.

Parameters of the freely levitated state		FCH of 40 mm WH of 23 mm	FCH of 30 mm WH of 19 mm	FCH of 20 mm WH of 15 mm
Vertical direction	Resonance frequency f (Hz)	6.5	7	9
	Dynamic stiffness k (N/mm)	5.54	6.42	10.62
	Damping coefficient c (Ns/m)	11.63	13.58	15.36
	Damping ratio	4.29%	4.65%	4.09%
Lateral direction	Resonance frequency f (Hz)	5.5	7	10
	Dynamic stiffness k (N/mm)	3.96	6.42	13.11
	Damping coefficient c (Ns/m)	9.67	9.37	15.27
	Damping ratio	4.22%	3.21%	3.66%

*FCH is field-cooling height. WH is the levitated working height after the field-cooling process. E.g., FCH of 40 mm, WH of 23 mm means that the onboard HTSCs are cooled at 40 mm height over PMG and then released to an equilibrium position of 23 mm height over PMG.

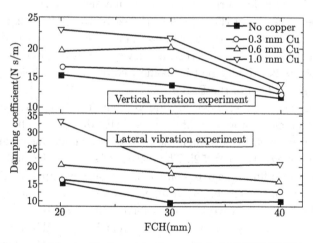

Fig. 6.37   Copper thickness effect on damping coefficient of HTS-PMG system in the vertical incorporation method.

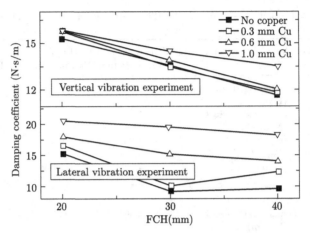

Fig. 6.38   Copper thickness effect on damping coefficient of HTS-PMG system in the side incorporation method.

6.9.6   *Pre-magnetized HTS bulks*

For present HTS maglev test vehicles or prototypes, directly field-cooling magnetization (DFCM) above the PMG is the most popular style. Due to the limited magnetic energy of permanent magnet material, the high flux-pinning performance of HTS bulks is limited by the PMG's magnetic field. To break this limitation, the feasibility of introducing pre-magnetized HTS bulk magnets into the present HTS maglev system is investigated. In such a method, the bulk HTS magnet is magnetized by magnetic fields in addition to the magnetic field of the PMG.

It has been experimentally found that the pre-magnetized bulk HTS magnet can stably levitate above PMG due to the flux-reform or re-magnetization effect, although the HTS bulk magnet was magnetized by a non-PMG magnetic field [52]. The feature of such a HTS bulk magnet is that the trapped flux is easy to control or increase by adjusting the magnetization field source. The higher trapped flux will lead to a bigger guidance force and smaller repulsive levitation force [53]. Therefore, the HTS bulk magnet is a feasible alternative to the present HTS maglev vehicle system because of the same low-temperature environment [54]. Besides the above conclusions, more investigations are very delightful for the present HTS maglev vehicle system [55, 56].

6.9.7   3D finite-element modeling of the HTS using T-method [13]

The prevailing 2D or cylindrical electromagnetic model for HTS commonly presumes that the circulating domain of the induced current is constrained within the ab-plan of the HTS [57–62]. However, the actual case is that the current in the HTS can also flow along the c-axis though its value of this direction is rather lower than that in the ab-plan [63], which necessitates the development of the numerical model for HTS in the 3D case when higher accuracy of the calculation is required. In this regard, aiming to the maglev transit using bulk HTS above a magnetic rail, a 3D model considering the anisotropic behavior of the critical current of HTS is reported [64]. In the context of this model, in order to describe the anisotropic behavior, the HTS is split into two different ingredients: one is a homogeneous conductor that is identical to an anisotropic superconductor, and the other is a virtual conductor whose conductivity survives only along the c-axis. Aside from this methodology, we describe the anisotropic behavior of the critical current by making recourse to a tensor resistivity (conductivity), by which the anisotropy of the HTS can be encapsulated into the electromagnetic governing equation reasonably. With the additional help of the T-method (where T is the current vector potential) [65], the 3D electromagnetic governing equations for the bulk HTS above a magnetic rail were deduced on the basis of Maxwell's equations and Helmholtz's theorem.

The 3D model was derived from a combination of Maxwell's equations with Helmholtz's theorem, and the anisotropic behavior of the critical current was represented by a tensor resistivity in deducing the governing equations [66]. The governing equations of the model in terms of the x-, y-, and z-components are described as follows:

$$\frac{1}{\sigma_{ab}}\left(-\frac{\partial^2 T_x}{\partial x^2} - \alpha\frac{\partial^2 T_x}{\partial y^2} - \frac{\partial^2 T_x}{\partial z^2} + (\alpha-1)\frac{\partial^2 T_y}{\partial x \partial y}\right) + \mu_0 C(P)\frac{\partial T_x}{\partial t}$$

$$+ \frac{\mu_0}{4\pi}\int_S \frac{\partial(\boldsymbol{n'} \cdot \boldsymbol{T'})}{\partial t}\frac{\partial}{\partial x'}\left(\frac{1}{R(P,P')}\right)dS' + \frac{\partial B_{ex}}{\partial t} = 0, \qquad (6.1)$$

$$\frac{1}{\sigma_{ab}}\left(-\alpha\frac{\partial^2 T_y}{\partial x^2} - \frac{\partial^2 T_y}{\partial y^2} - \frac{\partial^2 T_y}{\partial z^2} + (\alpha-1)\frac{\partial^2 T_x}{\partial x \partial y}\right) + \mu_0 C(P)\frac{\partial T_y}{\partial t}$$

$$+ \frac{\mu_0}{4\pi} \int_S \frac{\partial(\mathbf{n}' \cdot \mathbf{T}')}{\partial t} \frac{\partial}{\partial y'} \left(\frac{1}{R(P,P')} \right) dS' + \frac{\partial B_{ey}}{\partial t} = 0, \qquad (6.2)$$

$$\frac{1}{\sigma_{ab}} \left(-\frac{\partial^2 T_y}{\partial x^2} - \frac{\partial^2 T_y}{\partial y^2} - \frac{\partial^2 T_y}{\partial z^2} \right) + \mu_0 C(P) \frac{\partial T_z}{\partial t}$$

$$+ \frac{\mu_0}{4\pi} \int_S \frac{\partial(\mathbf{n}' \cdot \mathbf{T}')}{\partial t} \frac{\partial}{\partial z'} \left(\frac{1}{R(P,P')} \right) dS' + \frac{\partial B_{ez}}{\partial t} = 0, \qquad (6.3)$$

where σ_{ab} is the conductivity in the ab-plane of the HTS; α is the ratio of the critical current between in the ab-plane J_c^{ab} and along the c-axis J_c^c, determined by the anisotropic behavior of the HTS; $R(P, P')$ is the distance between the source point P' and the field point P, where the prime refers to the quantity at source point, $\mathbf{n}'$ is a unit vector out of the surface S'. The value of $C(P)$ is 1 when point P is in the HTS, $1/2$ when on the surface S' of the HTS, and zero elsewhere.

The angular dependence of the critical current due to anisotropic behavior was formulated by an elliptical model, which is one of the prevailing models used to describe the ferromagnetic anisotropy [67]. Based on this assumption, the critical current J_c, with respect to the angle φ between the orientation of the local applied field and c-axis, can be described as

$$J_c(\varphi) = \sqrt{J_{cx}^2 + J_{cz}^2} = J_c^{ab} \sqrt{\cos^2 \varphi + (\sin \varphi/\alpha)^2}, \qquad (6.4)$$

where J_{cx} and J_{cz} are the induced current densities in the ab-plane and along the c-axis, respectively.

The nonlinear E-J relation in the HTS can be simulated in this 3D model by either the power-law model, or the flux flow and creep model. The detailed formula of the two models can be referred to elsewhere [66].

The finite-element technique was adopted to numerically tackle the governing Eqs. (6.1)–(6.3) via Galerkin's method. A nodal tetrahedral element with four points was chosen to mesh the space domain of the HTS, while the domain related with air space was not meshed here due to the profound merit of the T-method: that the state variable outside the conductor is totally zero [65]. The final algebraic matrices associated with Eqs. (6.1)–(6.3) were compactly represented as follows:

$$[K_i(\sigma_{ab})]\{T_i\} + [K_{12}(\sigma_{ab})]\{T_{3-i}\} + [Q_{0i}] \left\{ \frac{\partial T_i}{\partial t} \right\}$$

$$+ [Q_{1i}] \left\{ \frac{\partial(\mathbf{n}' \cdot \mathbf{T}')}{\partial t} \right\} = \{L_i\} \quad (i = 1, 2, 3), \qquad (6.5)$$

where

$$[K_i(\sigma_{ab})] = \sum_e K_e(\sigma_{ab}^e)]_e = \sum_e \frac{1}{\sigma_{ab}^e} \int_{V_e} \left(\alpha_i \left[\frac{\partial N}{\partial x}\right]_e^T \left[\frac{\partial N}{\partial x}\right]_e \right.$$

$$\left. + \alpha_{3-i} \left[\frac{\partial N}{\partial y}\right]_e^T \left[\frac{\partial N}{\partial y}\right]_e + \left[\frac{\partial N}{\partial z}\right]_e^T \left[\frac{\partial N}{\partial z}\right]_e \right) dV,$$

$$[K_{12}(\sigma_{ab})] = \sum_e [K_{12}(\sigma_{ab}^e)]_e$$

$$= \frac{1}{2}(1-\alpha) \sum_e \frac{1}{\sigma_{ab}^e} \int_{V_e} \left(\frac{\partial [N]_e^T}{\partial y} \frac{\partial [N]_e}{\partial x} + \frac{\partial [N]_e^T}{\partial x} \frac{\partial [N]_e}{\partial y} \right) dV,$$

$$[Q_{0i}] = \sum_e [Q_{0i}]_e = \mu_0 \sum_e \int_{V_e} [N]_e^T [C(P)N]_e dV,$$

$$[Q_{1i}] = \sum_e [Q_{1i}]_e = \frac{\mu_0}{4\pi} \sum_e \int_{V_e} [N]_e^T \left(\int_S [N]_e \frac{\partial}{\partial u_i'} \frac{1}{R(P,P')} dS' \right) dV,$$

$$\{L_i\} = \sum_e \{L_i\}_e = -\sum_e \int_{V_e} [N]_e^T \frac{\partial B_{ei}}{\partial t} dV.$$

N is the shape function. The subscript e represents its detailed formula in each element. The conductivity for each element is represented with σ_{ab}^e. All the other parameters with a subscript including i is determined by Table 6.8.

In the process of designing the code, the algebraic matrices of three components in Eq. (6.5) were compressed into a single one with a shape represented by,

$$[K(\sigma_{ab})]\{T\} + [Q_0]\left\{ \frac{\partial T}{\partial t} \right\} + [Q_1]\left\{ \frac{\partial(n' \cdot T)}{\partial t} \right\} = \{L\}, \qquad (6.6)$$

Table 6.8

i	K_i	T_i	T_{3-i}	Q_{0i}	Q_{1i}	L_i	α_i	α_{3-i}	u_i'	B_{ei}
1	K_x	T_x	T_y	Q_{0x}	Q_{1x}	L_x	1	α	x'	B_{ex}
2	K_y	T_y	T_x	Q_{0y}	Q_{1y}	L_y	α	1	y'	B_{ey}
3	K_z	T_z	0	Q_{0z}	Q_{1z}	L_z	1	1	z'	B_{ez}

where

$$[K(\sigma_{ab})] = \sum_{i=1}^{2}[[K_i(\sigma_{ab})] + [K_{12}(\sigma_{ab})]] + [K_3(\sigma_{ab})],$$

$$[Q_0] = \sum_{i=1}^{3}[Q_{0i}], \quad [Q_1] = \sum_{i=1}^{3}[Q_{1i}], \quad \{L\} = \sum_{i=1}^{3}\{L_i\}.$$

The temporal domain of Eq. (6.6) was discretized by the finite-difference technique via the Crank-Nicolson-θ method with $\theta = 2/3$, and a detailed form for the nth time step can be written as

$$\left(\frac{[Q_0]}{\Delta t} + \theta[K(\sigma_{ab})]\right)\{T^n\} = \theta\{L^n\} + (1 - \theta)\{L^{n-1}\}$$

$$+ \left(\frac{[Q_0]}{\Delta t} - (1 - \theta)[K(\sigma_{ab})]\right)\{T^{n-1}\} - [Q_1]\{T^{n-1} - T^{n-2}\}.$$

$$(6.7)$$

For the sake of getting a sparse matrix in the left hand side of Eq. (6.7), we presume that the dense matrix $[Q_1]$ can be transferred to the right side without basically losing the accuracy when the time step Δt is sufficiently low.

The aforementioned 3D method has been experimentally checked for a HTS maglev system with a bulk HTS levitated above a PMG [68], and then was employed as a tool to optimize the geometry and configuration of the PMG, while combined with commercial software, to simulate the magnetic field of a PMG [13].

Figure 6.39 shows a comparison of the levitation force between experimental and computational results in the case of a bulk vertically moving in a loop from 60 mm to 8 mm above a magnetic track. Figure 6.40 shows a comparison of both the vertical and transverse forces between experimental and computational results in the case of a bulk laterally moving in a loop from the original position to 5 mm, to −5 mm and then back to 5 mm at a height of 18 mm above a magnetic track.

6.10   HTS maglev launch [69]

The HTS maglev launch system must consider a series of specialized technologies on the basis of research on the levitation and the guidance between YBaCuO and the PMG. It includes the stability and the reliability at

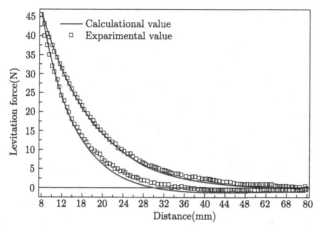

Fig. 6.39   A comparison of the levitation force between experimental and computational results in the case of a bulk vertically moving in a loop from 60 mm to 8 mm above a magnetic track.

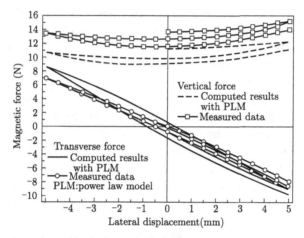

Fig. 6.40   A comparison of both the vertical and transverse forces between experimental and computational results in the case of a bulk laterally moving in a loop from the original position to 5 mm, to −5 mm and then back to 5 mm at a height of 18 mm above a magnetic track.

high acceleration, the acclivitous temperature device, integration technology between the vehicle and the low-temperature device, tests of the HTS maglev's performance when climbing up a large slope, techniques of the HTS maglev in the ETT, and so on. Therefore, the most important thing is to study the maglev system's efficiency of high acceleration and new drive technology using high T_c superconductors and PMG.

The laboratory-scale high-temperature superconducting maglev launch test system was developed with its predecessor's development experience and information from experiments on basic design considerations of the HTS maglev launch system include the PMG of about 10 m, the levitation force of 500–600 N, the guidance force of 60 N, and acceleration of 1–4 g.

This HTS maglev launch system has no intrinsic difference when compared to HTS maglev vehicle [5]. The launch system also includes the vehicle, the NdFeB PMG, and the linear motor. The NdFeB PM guideway is 10.05 m long and 0.68 m wide. There is a linear motor drive set between the two parallel PMGs. The frame of the PMG is 11.2 m long and 1.0 m wide. The PM guideway in the ETT has a longitudinal slope of 0 ~ 30 degree from the horizontal, and a transverse slope of 15 degrees.

The HTS maglev launch test vehicle is 1.0 m long and 1.0 m wide. There are four onboard levitation devices with HTS YBaCuO bulks in the launch vehicle. The gap between the linear motor and its induction board is 10 mm. Each linear motor is 695 mm long, 155 mm wide, and 80 mm thick. The total length of 15 single motors is about 10.4 m. The acceleration of the linear motor is 1–4 g.

Figure 6.41 shows the photograph of the PMG of the HTS maglev launch system. The linear motor is between the guideways. The linear motor control set is located in the top right corner of the photo.

In order to increase mechanical impact strength when decreasing the size and weight, both the launch vehicle and low-temperature device should be a single device. The low-temperature adiabatic device is a part of the entire launch vehicle. In order to increase dynamic strength, no liquid

Fig. 6.41   A photograph of the PMG of the HTS maglev launch system.

nitrogen cooling is used. There is no liquid nitrogen vessel, but only a tube in the vehicle. The bent shape of the tube is determined by the requirement of its application. The liquid nitrogen is allowed to flow into the adiabatic device through a tube before the experiment, and the liquid nitrogen flow is stopped when the temperature of the YBaCuO bulk reaches 77 K.

The launch vehicle includes four HTS maglev devices. Each HTS maglev device is composed of 16 YBaCuO bulks of diameter 30 mm, insulation materials, cooling tube. In order to increase the mechanical impact and the dynamic strength, some special and ingenious designs are used.

The total levitation forces of four HTS maglev devices on the No. 3 PMG (similar to the launch PMG) are listed in Table 6.9. The levitation gap is the distance between the top surface of PMG and the bottom surface of YBaCuO bulk. In the HTS maglev launch test vehicle, the net levitation gap includes the insulation thickness of about 3–5 mm. The total levitation forces at the levitation gap of 15 mm are 422 N at FCH of 30 mm and 743 N at ZFC, respectively. The total levitation forces at the levitation gap of 10 mm are 743 N at FCH of 30 mm and 1097 N at ZFC, respectively.

The total guidance forces at the field-cooling height (FCH) of 35 mm are 66 N at the levitation gap of 15 mm and 86 N at the levitation gap of 10 mm, respectively. The above total levitation forces and guidance forces of the HTS maglev launch test vehicle satisfy the design specification of the levitation force of 500–600 N and the guidance force of 60 N, which indicate the design validity of the HTS maglev launch system. Research results of quasi-static HTS on PMG in the past several years have provided strong technical support for the development of the HTS maglev launch test system.

The launch vehicle of liquid-nitrogen-free or solid state nitrogen is a good method. The merit of this method is to eliminate fluctuation of liquid nitrogen when the launch vehicle moves. This fluctuation will badly affect the stability of the launch vehicle.

Table 6.9   Total levitation forces of four HTS maglev devices on No.3 PM guideway.

Levitation gap (mm)	FCH of 20 mm (N)	FCH of 30 mm (N)	FCH of 40 mm (N)	ZFC (N)
20	73	213	3.8	506
15	230	422	534	743
10	506	743	866	1097
5	939	1191	1322	1540

6.11   Moderate/low-speed HTS maglev train

International research interest in the man-loading HTS maglev vehicle was
ignited after the first man-loading HTS maglev test vehicle in the world,
"Century", [5] was developed. The man-loading HTS maglev vehicles were
developed in 2004 in Germany [15] and Russia [16], respectively. The full-
sized HTS maglev train, as a substitute of a light track in the city, is under
development in Brazil [17, 18]. Moreover, Japan [19] and Italy [20], among
other countries, had also developed HTS maglev car models.

The trapped magnetic field of HTS bulk had achieved 17.24 T at 29 K
[70]. It implies not only that the HTS maglev vehicle has bright prospects,
but also that its practical application process is ahead of people's prospec-
tive schedule.

Compared with the urban light rail vehicle (LRV) for the same number
of passengers, the HTS maglev train has a lighter vehicle body, lighter supporting structure weight, smaller drive power, cheaper infrastructure, and
so on. The HTS maglev train does not need the elevated reinforced concrete
structure, but the LVR does. Therefore, the total weight of the HTS maglev
train line is lighter than that of the LVR, and the total construction and
operation costs are lower than the equivalent LRV system.

The Brazil group presented a study comparing the construction costs of
an HTS maglev line of 1.0 km inside the Campus of the Federal University of
Rio de Janeiro (UFRJ), with an LRV. Preliminary calculations have already
shown that this particular maglev line can be cheaper than a LRV. HTS
maglev cars are lighter than LVR cars for the same number of passengers;
the global efforts are approximately 75% lower. The weight of the supporting structure for the HTS maglev is 50% of the LVRs. Figure 6.42 shows
a real-scale HTS maglev vehicle prototype in Brazil [72, 73]. Figure 6.43
shows the onboard assembled HTS maglev cryostat [74–76] which was made
in ATZ, Germany. Table 6.10 listed specifications of the HTS maglev cryostat.

The German IFW group developed a Supra Trans II large-scale HTS
maglev vehicle [77] (Fig. 6.44). On July 1, 2009, the project of Supra Trans
II HTS maglev vehicle has been approved by the German government after
the Supra Trans I HTS maglev vehicle [15] in IFW, Germany in 2002–2004.
The specifications of Supra Trans II HTS maglev vehicle are as follows:

(1) Oval guideway length of 80.84 m;
(2) NdFeB magnets of 4.85 t;

Fig. 6.42 A real-scale HTS maglev vehicle prototype in Brazil. (The onboard HTS maglev cryostat is located at the lower right corner). (From Ref. [73], with permission).

(3) Superconductor of YBCO bulk;
(4) Levitation force of 8.9 kN at 8–10 mm;
(5) Vehicle mass of about 400 kg;
(6) Maximum speed of 20 km/h;
(7) Levitation height of 10 mm;
(8) Propulsion: synchronous short stator linear motor.

After the breakthrough of the first man-loading HTS maglev test vehicle in December 2000, the authors' group (ASCLab at SWJTU, China) began to conduct the test line design for the next step in the research. In January 2004, authors have presented the HTS maglev train project, and has been approved by the expert review meeting. In May 2008, ASCLab has completed a HTS maglev test line system design. The design calculation validates the total construction and operation costs of the HTS maglev

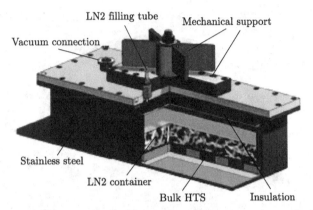

Fig. 6.43 Onboard HTS maglev cryostat in ATZ, Germany. (From Ref. [74], with permission).

Table 6.10   Specifications of the maglev cryostat in ATZ, Germany. (From [74], with permission.)

Geometry	440 mm × 180 mm × 120 mm
YBCO bulks	2 × 12 pcs, size (64 × 32 × 13)
HTS are	492 cm^2
Cooling	LN$_2$, storage capacity 2.5.1
Operat time	24~30 h, static
Cryostat weight	~17 kg
Magnetic distance	2 mm, distance bottom–YBCO
Levitation	~2500 N at 5 mm; fc30
	~4000 N at 5 mm Halbach PM
Force density	5 N/cm^2, 8 N/cm^2 (Halbach)

line system to be lower than the equivalent LRV system. The weight of the support structure of this HTS maglev guideway is about 50% of the LRV's. The design specifications of the HTS maglev vehicle are listed in Table 6.11. The test line length is 2 km, and the maximum test speed is 160 km/h.

In March 2013, a 45 m HTS maglev ring test line was successfully developed by ASCLab under the financial support of State Key Laboratory of Traction Power, SWJTU. Figure 6.45 presents the pictures of the whole scene of the ring line and a close-up of the second-generation HTS maglev vehicle [78].

The new maglev vehicle (2.2 m in length, 1.1 m in width) is designed for one passenger with a levitation height of 15~20 mm; the PMG (45 m

Fig. 6.44   Supra Trans II large scale HTS maglev vehicle in IFW, Germany. [From Ref. [77], with permission.]

Table 6.11   Design specifications of the HTS maglev vehicle in the ASCLab.

Specifications	Units	Value
Passengers/m	p/m	10
Passengers' weight/m	kgf/m	750
Vehicle tare/m	Kgf/m	400
Total weight/m	kgf/m	1150
Vehicle length	m	12
Total vehicle weight	t	13.8
Maximal running speed	km/h	100
Acceleration	m/s^2	0.8

in length, 0.7 m of track gauge) is a racetrack shape with a curve radius of 6 m; and the driving is accomplished by an induction linear motor with a maximum running speed of 50 km/h. This second-generation HTS maglev vehicle system is highlighted by the cost-performance and multi-parameter onboard monitoring function. The maximum load capacity is over 1000 kg at a levitation gap of 10 mm, but the cross-sectional area of the PMG is 3000 mm^2; the total length of the linear motor is 3 m, composed of four motors installed at one straight section of the track. On the vehicle, parameters of the levitation weight, levitation height, running speed, acceleration, lateral offset, and total running distance of the vehicle are real-time monitored and displayed on the onboard tablet computer [78].

                    (a)   (b)

Fig. 6.45   A 45-m-long HTS maglev ring test line developed by ASCLab of SWJTU, China in March 2013. The whole scene of the ring line (a) and a close-up of the second-generation HTS maglev vehicle (b). (From Ref. [78], with permission.)

6.12   Super-high speed HTS maglev train

When the speed of the normal ground traffic transportation is higher than 350 km/h, not only is the running noise of the ground traffic vehicle higher, but also, the 90% driving power is dissipated in the aerodynamic resistance. However, people hope to achieve higher speed, for example, over 3000 km/h. The high speed may be realized whence a HTS maglev vehicle runs in the ETT.

The first man-loading HTS maglev test vehicle in the world, "Century" [5] was successfully operated, and since then more than 40000 passengers have ridden on it. The total mileage is about 400 kilometers. The long-term stability in both the vertical and the lateral directions of the HTS maglev vehicle is a guarantee of the super-high speed train in the ETT without control system. D. Oster was convinced that [79] "automated, silent ETT works by removing resistance. Ultra lightweight, pressurized cabins travel in tubes on thin wheels, or maglev. No air is in the tube to cause resistance. Energy is recovered when slowing. Propulsion fuel is not carried onboard. High capacity at low cost is achieved using frequent vehicles instead of huge vehicles." M. Okano et al. [80] described theoretically a HTS maglev vehicle in a vacuum passage. The combination of the ETT and the HTS maglev will bring both advantages into full play, and perhaps it may be developed into a new mode of high speed or super-high speed ground transportation.

On January 12, 2004, the authors presented a super-high speed ETT-HTS maglev train project with a maximum speed of ground traffic of more

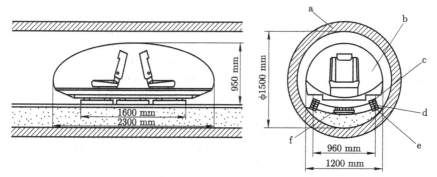

Fig. 6.46   The schematic structure of the super-high speed HTS maglev in the low-pressure tube (LPT) or ETT: a–Tube. b–Vehicle body. c–Connector. d–Low-temperature part (with superconductor inside). e–Permanent magnet guideway. f–Linear motor.

than 600 km/h. Unfortunately, this super-high speed [30, 81, 82] HTS maglev train project has not been approved by the experts of the review meeting.

D. Oster and his wife happily took the HTS maglev vehicle in 2003, and D. Oster, with authors, cooperated twice with the authors to develop the ETT-HTS maglev vehicle in 2003 and 2004, respectively after they visited the ASCLab in December 2002.

There are important differences between the HTS maglev train and other maglev trains. The levitation and guidance of the HTS maglev train is inherently stable, i.e. it does not need any control device. The control technology of the HTS maglev train system is used only in system operations. Therefore, the HTS maglev train is especially suitable for ETT transportations. The ETT–HTS maglev train (Fig. 6.46) is a new form of transportation, and it has clear characteristics, namely small vehicle size, small guideway, and high frequency dispatch.

Based on the experimental results of the HTS maglev and the theoretical calculation results of the levitation forces of a single onboard HTS maglev device of high quality HTS bulk on new guideway, the design consideration of the super-high speed HTS maglev is achieved [30].

The low-temperature rectangle-shaped liquid nitrogen vessel onboard has high mechanical strength at the temperature of 77 K. The continuous operating period is 12 h. The exterior size of the low temperature vessel is 150 mm × 516 mm; the interior size is 102 mm × 470 mm, and the height is 170 mm. The bottom thickness of the vessel is 5 mm in order to increase mechanical strength.

The total load of the maglev vehicle is 500 kg including two seats and the weight of two passengers, 150 kg. The vehicle body is a quasi-ellipsoidal shape, and the exterior size of the vehicle with shell is about 2.3 m in length, 0.95 m in height, and 1.2 m in width. The final design results of the maglev vehicle body are 5000 N of levitation force and 1000 N of guidance force at 15 mm of net levitation gap. The guideway length is 1000 m. The maximum speed of the vehicle is over 600 km/h when there are no men in the vehicle, and the manned speed is 100 km/h.

The vehicle is driven by a linear motor and is controlled by a control system on the ground. Drive acceleration is 5 g (about 49 m/s^2). The vehicle running states include the acceleration in 300 m, the uniform speed motion in 400 m, and brake section in 300 m.

Figure 6.47 shows a diagram of the design for the quasi-all-superconducting HTS maglev train with a super-high speed (EET not drawn). This project includes the uses of the HTS maglev vehicle [5], HTS maglev bearing [83], HTS maglev FESS [84], and HTS linear motors [85] for maglev vehicle propulsion. The quasi-all-superconductive HTS maglev system [86] is fit not only for the super-high speed maglev train but also for the HTS maglev launch system. The speed of the quasi-all-superconductive HTS maglev launch system is several kilometers per second. Future HTS maglev trains will be combined with the HTS linear motor [87] and the HTS cable [88, 89], to compose a full HTS system.

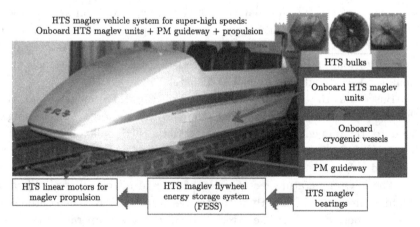

Fig. 6.47   Design diagram of the super-high speed HTS maglev train.

6.13   Postscript

The year of 2011 was an extraordinary year for these superconducting maglev experts!

In 1911, Heike Kamerlingh Onnes and his group in Leiden discovered superconductivity for which he received the Nobel Prize in 1913. In 2011, the Superconductivity Centennial Conference 2011 was held in Hague, the Netherlands.

In 1912, Emile Bachelet obtained an U.S. patent for levitating transmitting apparatus. In 2011, the 21st International Conference on Magnetically Levitated Systems and Linear Drives (MAGLEV 2011) was held in Daejeon, Korea.

Both superconductivity and the maglev have been with us for 100 years. This article is based on the invitation keynote speech, "Past, Present and Future R&D of HTS maglev in China", of MAGLEV 2011 and ASCLab's research. I sincerely dedicate this article to the commemoration of the 100th anniversary of both superconductivity and the maglev.

It took forty years for the LTS maglev train to become reality in Japan. How many years will it take the HTS maglev train?

Maybe a super-high speed HTC maglev train will take you to the next Centennial Conference at 3000 km/h.

Acknowledgments

The authors would like to thank the staff of the Applied Superconductivity Laboratory (ASCLab), Southwest Jiaotong University, China. Thereinto, Jun Zheng, Guangtong Ma, Zigang Deng, Wei Liu, Lu Liu and Wei Wang et al. offered the summary of their research work about HTS maglev vehicles. The authors would like to thank Zhongyou Ren, He Jiang, Min Zhu, Changyan Deng, Youwen Zeng, Qixue Tang, Hiyu Huang, Xiaorong Wang, Xuming Shen et al. for their important and substantive contributions to the development of "Century", and Yiyun Lu, Jing Li, Honghai Song, Xingzhi Wang, Ya Xhang, Rongqing Zhao, Rong Zeng, Minxian Liu, Yujie Qin, Qingyong He, Hua Jing, Siting Pan, Qunxu Lin, Donghui Jiang, and Changqing Ye, Jina Cen, Yonggang Huang, Jian Zhang, Lulin Wang, and Ming Jiang et al. for their important contributions to the development of the HTS maglev after "Century". The authors are extremely grateful to Hongtao Ren and his group for providing high quality HTS bulk materials.

The HTS maglev vehicle project was supported by the 863 Program and by the National Natural Science Foundation in China from 1990 to 2010.

References

[1] F. Hellman, E. M. Gyorgy, D. W. Johnson, Jr., H. M. O'Bryan and R. C. Sherwood, *J. Appl. Phys.* **63**, 447 (1988).

[2] P. N. Peters, R. C. Sisk, E. W. Urban, C. Y. Huang and M. K. Wu, *Appl. Phys. Lett.* **52**, 2066 (1988).

[3] E. H. Brandt, *Science* **243**, 49 (1989).

[4] H. J. Bomemann, A. Tonoli, T. Ritter, C. Urban, O. Zaitsev, K. Weber and H. Rietschel, *IEEE Trans. Appl. Supercond.* **5**, 618 (1995).

[5] J. S. Wang, S. Y. Wang, Y. W. Zeng, H. Y. Huang, F. Luo, Z. P. Xu, Q. X. Tang, G. B. Lin, C. F. Zhang, Z. Y. Ren, G. M. Zhao, D. G. Zhu, S. H. Wang, H. Jiang, M. Zhu, C. Y. Deng, P. F. Hu, C. Y. Li, F. Liu, J. S. Lian, X. R. Wang, L. H. Wang, X. M. Shen and X. G. Dong, *Physica C* **378–381**, 809 (2002).

[6] F. C. Moon, *Superconducting Levitation*, New York: Wiley (1994).

[7] J. R. Hull, *Sci. Technol.* **13**, R1 (2000).

[8] K. B. Ma, Y. V. Postrekhin and W. K. Chu, *Rev. Sci. Instrum.* **74**, 4989 (2003).

[9] M. Murakami, Materials Developments and Applications of Bulk Re-Ba-Cu-O Superconductors, In: Anant Narlikar Ed., *Frontiers in Superconducting Materials*, Springer Verlag, Germany, 869–884 (2005).

[10] J. S. Wang and S. Y. Wang, Synthesis of Bulk Superconductors and Their Properties on Permanent Magnet Guideway. In: Anant Narlikar Ed., *Frontiers in Superconducting Materials*, Springer Verlag, Germany, 885–912 (2005).

[11] H. H. Song, J. S. Wang, S. Y. Wang, Z. Y. Ren, X. R. Wang, O. de Haas, G. Fuchs and L. Schultz, Studies of YBCO Electromagnetic Properties for High-Temperature Superconductor Maglev Technology. In: Barry P. Martins Ed., *New Topics in Superconductivity Research*, Nova Science Publishers, Inc., 107–156 (2006).

[12] J. S. Wang and S. Y. Wang, High Temperature Superconducting Maglev Measurement System. In: Milind Kr Sharma Ed., *Advances in Measurement Systems*, InTech Publisher, 51–80 (2010).

[13] G. T. Ma, J. S. Wang and S. Y. Wang, 3-D finite-element modelling of a maglev system using bulk high-Tc superconductor and its application. In Adir Moysés Luiz (Ed.), *Applications of High-Tc Superconductivity*, InTech Publisher, 119–146 (2011).

[14] Y. Zhang and S. G. Xu, Proceeding of Fifteenth International Conference on Magnet Technology, In: L. Z. Lin, G. L. Shen, L. G. Yan Ed., Science Press, Beijing, 763–766 (1998).

[15] L. Schultz, O de Haas, P. verges, C. Beyer, S. Rohlig, H. Olsen, *et al.*, *IEEE Trans Appl. Supercond.* **15**, 2301 (2005).

[16] K. L. Kovalev, S. M.-A. Koneev, V. N. Poltavec, *et al.*, Magnetically levitated High-speed carriages on the basis of bulk HTS elements," in Pro. 8th Intern. Symp. Magn. Susp. Technol. (ISMST'8), Dresden, Germany, **51** (2005).

[17] R. M. Stephan, A. C. Ferreira, R. de Andrade, Jr. M. A. Neves, M. A. Cruzmoreira, M. A. P. Rosario, *et al.*, "A superconducting levitated small scale vehicle with linear synchronous motor," in 2003 IEEE Intern. Sym. Indust. Electro. **1**, 206–209 (2004).

[18] G. G. Sotelo, D. H. Dias, O. J. Machado, E. D. David, R. de Andrade Jr., R. M. Stephan and G. C. Costa. Experiments in a real scale maglev vehicle prototype. *Journal of Physics: Conference Series* **234**, 032054 (2010).

[19] M. Okano, T. Iwamoto, M. Furuse, S. Fuchino and I. Ishii. Running performance of a pinning-type superconducting magnetic levitation guide, *J. Phys. Conf. Ser.* **43**, 999–1002 (2006).

[20] G. D' Ovidio, F. Crisia, G. Lanzara. A 'V' shaped superconducting levitation module for lift and guidance of a magnetic transportation system, *Physica C, Jul.* **468**, 1036–1040 (2008).

[21] J. S. Wang, S. Y. Wang, Y. W. Zeng, C. Y. Deng, Z. Y. Ren, X. R. Wang, H. H. Song, X. Z. Wang, J. Zheng and Y. Zhao. The present status of HTS Maglev vehicle in China. *Supercond. Sci. Technol.* **18**, S215–S218 (2005).

[22] J. S. Wang, S. Y. Wang and J. Zheng, *IEEE Trans. Appl. Supercond.* **19**, 2142 (2009).

[23] J. S. Wang, S. Y. Wang, J. Zheng, Z. G. Deng, Y. Y. Lu, G. T. Ma, Y. Zhang and F. Yen, *Chinese J of Low Temp. Phys.* **3**, 142 (2009).

[24] http://www.wyzxsx.com/Article/Class4/201108/252441.html

[25] 赵建柱, http://wenku.baidu.com/view/d3e98309581b6bd97f19ea01.html? from=rec&pos= 2&weight=44&lastweight=42&count=5

[26] K. Nagashima, T. Okata, K. Mizuno, Y. Arai, H. Hasegawa and T. Sasakawa, The 21st International Conference on Magnetically Levitated Systems and Linear Drives (MAGLEV 2011), Daejeon, Korea, October 2011, Paper No. VLG-04.

[27] J. S. Wang, S. Y. Wang, Z. Y. Ren, H. Jiang, M. Zhu, X. R. Wang, X. M. Shen and H. H. Song, *Phys. C* **386**, 431 (2003).

[28] J. S. Wang, S. Y. Wang, Z. Y. Ren, M. Zhu, H. Jiang and Q. X. Tang, *IEEE Trans. Appl. Supercond.* **11**, 1808 (2001).

[29] S. Y. Wang, J. S. Wang, Z. Y. Ren, M. Zhu, H. Jiang, X. R. Wang, X. M. Shen and H. H. Song, *Phys. C* **386**, 531 (2003).

[30] J. S. Wang, S. Y. Wang, C. Y. Deng, Y. W. Zeng, H. H. Song, J. Zheng, X. Z. Wang, H. Y. Huang and F. Li, *IEEE Transac. on Appl. Supercond.* **15**, 2273 (2005).

[31] S. Y. Wang, Q. X. Tang, Z. Y. Ren, M. Zhu, H. Jing and J. S. Wang, *Cryogenics and Supercond.* **29**, 14 (2001).

[32] S. Y. Wang, J. S. Wang, X. R. Wang, Z. Y. Ren, Y. W. Zeng, C. Y. Deng, H. Jiang, M. Z., G. B. Lin, Z. P. Xu, D. G. Zhu and H. H. Song, *IEEE Transactions on Appl. Supercond.* **13**, 2134 (2003).

[33] M. Zhu, Z. Y. Ren, S. Y. Wang, J. S. Wang and H. Jiang, *Chinese J. Low Temp. Phys.* **24**, 213 (2002) (In Chinese).

[34] J. S. Wang and S. Y. Wang, *Cryogenics and Supercond.* **27**, 1 (1999) (In Chinese).

[35] J. S. Wang, S. Y. Wang, G. B. Lin, H. Y. Huang, C. F. Zhang, Y. W. Zeng, S. H. Wang, C. Y. Deng, Z. P. Xu, Q. X. Tang, Z. Y. Ren, H. Jiang and M. Zhu, *High Technol. Lett.* **10**, 56 (2000) (In Chinese).

[36] Wang Jiasu, Suyu Wang, Zhongyou Ren, Min Zhu, He Jiang and Qixue Tang, Levitation force of a YBaCuO bulk high temperature superconductor over a NdFeB guideway. *IEEE Trans. on Appl. Supercond.* **11**, 1801 (2001).

[37] Z. Y. Ren, J. S. Wang, S. Y. Wang, H. Jiang, M. Zhu, X. R. Wang and H. H. Song, *Physica C* **384**, 159 (2003).

[38] H. Jiang, J. S. Wang, S. Y. Wang, Z. Y. Ren, M. Zhu, X. R. Wang and X. M. Shen, *Physica C* **378–381**, 869 (2002).

[39] X. R. Wang, H. H. Song, Z. Y. Ren, M. Zhu, J. S. Wang, S. Y. Wang and X. Z. Wang, *Physica C* **386**, 536 (2003).

[40] H. H. Song, O. De Haas, Z. Y. Ren, X. R. Wang, *et al.*, *Physica C* **407**, 82 (2004).

[41] J. S. Wang, S. Y. Wang, Z. Y. Ren, X. R. Wang, M. Zhu, H. Jiang, H. H. Song, X. Z. Wang and J. Zheng, *IEEE Transact. on Appl. Supercond.* **13**, 2154 (2003).

[42] X. R. Wang, Z. Y. Ren, H. H. Song, X. Z. Wang, J. Zheng *et al.*, *Sci. Technol.* **18**, S99 (2005).

[43] W. Liu, S. Y. Wang, Q. X. L, P. Yang, H. F. Liu, G. T. Ma, J. Zheng and J. S. Wang, *Chinese J. Low Temp. Phys.* **3**, 85 (2009) (In Chinese).

[44] S. Y. Wang, J. S. Wang, C. Y. Deng, Y. Y. Lu, Y. W. Zeng, H. H. Song, H. Y. Huang, H. Jing, Y. G. Huang, J. Zheng, X. Z. Wang and Y. Zhang, *IEEE Trans. Appl. Supercond.* **17**, 2067 (2007).

[45] J. S. Wang, S. Y. Wang, C. Y. Deng, Y. W. Zeng, L. C. Zhang, Z. G. Deng, J. Zheng, L. Liu, Y. Y. Lu, M. X. Liu, Y. H. Lu, Y. G. Huang and Y. Zhang, *IEEE Trans. Appl. Supercond.* **18**, 791 (2008).

[46] L. Liu, J. S. Wang, Z. G. Deng, *et al.*, *IEEE Trans. Appl. Supercond.* **20**, 920 (2010).

[47] L. Liu, J. S. Wang, S. Y. Wang, *et al.*, *IEEE Trans. Appl. Supercond.* **21**, 920 (2011).

[48] Z. Deng, J. Zheng, J. Zhang, J. Wang, S. Wang, Y. Zhang, *et al.*, *Physica C* **463–465**, 1293 (2007).

[49] G. Ma, Q. Lin, J. Wang, S. Wang, Z. Deng, Y. Lu, *et al.*, *Supercond. Sci. Technol.* **21**, 065020 (2008).

[50] W. Wang, J. S. Wang, W. Liu, J. Zheng, Q. X. Lin, S. T. Pan, Z. G. Deng, G. T. Ma and S. Y. Wang, *Physica C* **469**, 188 (2009).

[51] J. Zheng, Z. G. Deng, Y. Zhang, W. Wang, S. Y. Wang and J. S. Wang, *IEEE Trans. Appl. Supercond.* **19**, 2148 (2009).

[52] J. Zheng, Z. Deng, S. Wang, J. Wang and Y. Zhang, *Phys. C* **463–465**, 1356 (2007).

[53] J. Zheng, J. Li, Z. G. Deng, H. H. Song, S. Y. Wang and J. S. Wang, *Mater. Sci. Forum* 2079–2084 (2007).

[54] Z. Deng, J. Zheng, J. Li, Y. Zhang, S. Wang and J. Wang, *Rare Metal. Mater. Eng* **37** (Suppl. 4), 350 (2008).

[55] W. Liu, J. S. Wang, X. L. Liao, S. J. Zheng, G. T. Ma, J. Zheng and S. Y. Wang, *Phys. C* **471**, 156 (2011).

[56] W. Liu, J. S. Wang, X. L. Liao, C. Q. Ye, G. T. Ma, J. Zheng and S. Y. Wang, *J. Supercond. and Novel Magne.* **24**, 1563 (2011).

[57] M. Uesaka, Y. Yoshida, N. Takeda and K. Miya, *Int. J. Appl. Electromagn. Mater.* **4**, 13 (1993).

[58] Y. Yoshida, M. Uesaka and K. Miya, *IEEE Trans. Magn.* **30**, 3503 (1994).

[59] Y. Luo, T. Takagi and K. Miya, *Cryogenics* **39**, 331 (1999).

[60] M. J. Qin, G. Li, H. K. Liu, S. X. Dou and E. H. Brandt, *Phys. Rev. B* **66**, 024516 (2002).

[61] A. Sanchez, D. X. Chen, N. Del Valle, E. Pardo and C. Navau, *J. appl. Phys.* **99**, 113904 (2006).

[62] X. F. Gou, X. J. Zheng and Y. H. Zhou, *IEEE Trans. Appl. Supercond.* **17**, 3795 (2007).

[63] M. Murakami *et al.*, *IEEE Trans. Magn.* **27**, 1479 (1991).

[64] Y. Y. Lu, J. S. Wang, S. Y. Wang and J. Zheng, *J. Supercond. and Novel Magn.* **21**, 467 (2008).

[65] K. Miya and H. Hashizume, *IEEE Trans. Magn.* **24**, 134 (1998).

[66] G. T. Ma, J. S. Wang and S. Y. Wang, *IEEE Trans. Appl. Supercond.* **20**, 2219 (2010).

[67] A. D. Napoli and R. Paggi, *IEEE Trans. Magn.* **19**, 1557 (1983).

[68] G. T. Ma, J. S. Wang and S. Y. Wang, *IEEE Trans. Appl. Supercond.* **20**, 2228 (2010).

[69] J. S. Wang, S. Y. Wang, C. Y. Deng, J. Zheng, H. H. Song, Q. Y. He, Y. W. Zeng, Z. G. Deng, *et al.*, *IEEE Trans. Appl. Supercond.* **17**, 2091 (2007).

[70] M. Tomita and M. Murakami, *Nature* **421**, 517 (2003).

[71] E. G. David, R. M. Stephan, G. C. Costa, Jr R. Andrade and R. Nicolsky, Maglev 2006, Dresden, Germany.

[72] G. G. Sotelo, D. H. Dias, O. J. Machado, E. D. David, R. de Andrade Jr. and R. M. Stephan, *J. Phys.: Conference Series* **234**, 032054 (2010).

[73] R. M. Stephan, R. de Andrade Jr., A. C. Ferreira, O. J. Machado, M. D. A. dos Santos, G. G. Sotelo and D. H. Dias, The 21st International Conference on Magnetically Levitated Systems and Linear Drives (MAGLEV 2011), Daejeon, Korea, October 2011, Paper No. DPO-8.

[74] F. N. Werfel, U. Floegel-Delor, R. Rothfeld, T. Riedel, B. Goebel, D. Wippich and P. Schirrmeister, *IEEE Trans. Appl. Supercond.* **21**, 1473 (2011).

[75] F. N. Werfel, U. Floegel-Delor, T. Riedel, D. Wippich, B. Goebel, R. Rothfeld and P. Scirrmeister, The Superconductivity Centennial Conference EUCAS/ICEC/ICMC2011 (SCC2011), No. 4-LB-010, Hague, the Netherlands, Sep. 2011, 18–23.

[76] R. Stephan, E. David, R. Jr Andrade, O. Machado, D. Dias, G. G. Sotelo, O. de Haas and F. Werfel, Proc. 20th Int. Conf. on Magnetically Levitated Systems and Linear Drives (MAGLEV) (San Diego, Dec.) 2008, P27.

[77] L. Schultz, L. Kuehn, O. de Hass, D. Berger and B. Holzapfel. Supra Trans II-research facility for large scale HTS applications. The Superconductivity Centennial Conference, EUCAS/ICEC/ICMC2011 (SCC2011), Paper No. 4-LB-12, Hague, the Netherlands, Sep. 2011, 18–23.

[78] Zi-Gang Deng et al., Private communication, 2013.

[79] Daryl Oster, Private correspondence and www.et3.com.

[80] M. Okano, T. Iwamoto, et al., Phys. C 386, 531 (2003).

[81] J. S. Wang, S. Y. Wang, C. Y. Deng, Y. W. Zeng, H. H. Song and H. Y. Huang, Int. J. Mod. Phys. B 1–3, 399 (2005).

[82] J. S. Wang, Past, Present and Future R&D of HTS Maglev in China. The 21st International Conference on Magnetically Levitated Systems and Linear Drives (MAGLEV 2011), Invitation Keynote Speech No. 2, Daejeon, Korea, October 2011 (Invitation keynote speech).

[83] Z. G. Deng, J. S. Wang, S. Y. Wang, J. Zheng, Q. X. Lin and Y. Zhang, Trans. China Electrot. Soci. 24, 1 (2009) (In Chinese).

[84] Z. G. Deng, J. S. Wang, S. Y. Wang, J. Zheng, G. T. Ma, Q. X. Lin and Y. Zhang, Trans. China Electrot. Soci. 23, 1 (2008) (In Chinese).

[85] F. Yen, J. Li, S. J. Zheng, L. Liu, G. T. Ma, J. S. Wang, S. Y. Wang and W. Liu, Supercond. Scie. Techn. 23, 105015 (2010).

[86] J. S. Wang. The written application of the National Natural Science Foundation in China, in March, 2008 (In Chinese).

[87] C. Y. Lee, J. H. Lee, J. M. Jo, C. B. Park, W. H. Ryu, Y. Do Chung, Y. J. Hwang, T. K. Ko, Se-Y. Oh and J. Lee. IEEE Trans. Appl. Supercond. 24, 3600304 (2014).

[88] H. Ohsaki, Z. Lv, N. Matsushita, M. Sekino, T. Koseki and M. Tomita, IEEE Trans. Appl. Supercond. 23, 3600705 (2013).

[89] M. Tomita, M. Muralidhar, Y. Fukumoto, A. Ishihara, K. Suzuki, Y. Kobayashi and T. Akasaka, IEEE Trans. Appl. Supercond. 23, 3601504 (2013).

Printed in the United States
By Bookmasters